ELECTRONIC
MATERIALS

ELECTRONIC MATERIALS

EDITED BY NICHOLAS BRAITHWAITE
AND GRAHAM WEAVER

MATERIALS DEPARTMENT
OPEN UNIVERSITY, MILTON KEYNES, ENGLAND

Butterworths
LONDON BOSTON SINGAPORE
SYDNEY TORONTO WELLINGTON

First published 1990

British Library Cataloguing in Publication Data
Electronic Materials.
  1. Materials
  I. Braithwaite, Nicholas   II. Weaver, Graham   III. Series
620.1′1

ISBN 0-408-02840 8

Library of Congress Cataloging-in-Publication Data
Electronic Materials/edited by Nicholas Braithwaite and Graham Weaver.
    p. cm.–(Materials in action series)
  This text forms part of an Open University Materials Department course.
  Bibliography: p.
  Includes index.
  ISBN 0-408-02840 8
  1. Materials.   I. Braithwaite, Nicholas   II. Weaver, Graham
(Graham H.)   III. Open University. Materials Dept.
IV. Series.

Butterworth Scientific Ltd

Part of Reed International P.L.C.

Designed by the Graphic Design Group of the Open University

Typeset and printed by Alden Press (London & Northampton) Ltd, London, England

This text forms part of an Open University course. Further information on Open
University courses may be obtained from the Admissions Office, The Open University,
PO Box 48, Walton Hall, Milton Keynes MK7 6AB.

ISBN 0-408-02840 8

Series Preface

The four volumes in this series are part of a set of courses presented by the Materials Department of the Open University. Although each book is self-contained, the first volume, *Materials Principles and Practice* is an introduction to the ideas, models and theories which are then developed further in the separate areas of the other three. It assumes that you are just starting to study materials, and that you are already competent in pre-university mathematics and physical science.

Unlike many introductory texts on the subject, this series covers materials science in the technological context of making and using materials. This approach is founded on a belief that the behaviour of materials should be studied in a comparative way, and a conviction that intelligent use of materials requires a sound appreciation of the strong links between product design, manufacturing processes and materials properties.

The interconnected nature of the subject is embodied in these books by the use of two sorts of text. The main theme (or story line) of each chapter is in larger, black type. Linked to this are other aspects, such as theoretical derivations, practical techniques, applications and so on, which are printed in green. The links are flagged in the main text by a reference such as ▼Assessing hardness▼, and the linked text, under this heading, appears nearby. Both sorts of text are important, but this format should enable you to decide your own study route through them.

The books encourage you to 'learn by doing' by providing exercises and self-assessment questions (SAQs). Answers are given at the end of each chapter, together with a set of objectives. The objectives are statements of what you should be able to do after studying the chapter. They are matched to the self-assessment questions.

This series, and the Open University courses it is part of, are the result of many people's labours. Their names are listed after the prefaces. I should particularly like to thank Professor Michael Ashby of Cambridge University for reading and commenting on drafts of all the books, and the group of student 'guinea pigs' who worked through early drafts. Finally, thanks to my colleagues on the course team and our consultants. Without them this project would not have been possible.

Further information on Open University courses may be had from the address on the back of the title page.

Charles Newey
Open University
February 1990

Preface

An electronic engineer assembling a circuit does not need to know what the circuit components are made of; on the other hand, a physicist finding models for the behaviour of materials does not need to know what can be made from the materials. However, those engineers and designers who are conversant with both ends of the spectrum may make gadgets which use materials in novel or clever ways, and so get the patents or advance their companies' opportunities.

Our aim is to provide engineers with some of the important aspects of science which underpin the hardware of our information age. In the space available, and at the intended level, we can reach neither erudition in the purer science of materials, nor detail in the more applied science of devices. The consequent danger of saying nothing about anything has been avoided by considerable discrimination in our topic selection. Our main theme is the development and use of materials to meet the needs of transduction, memory and display. To gather and manipulate electronic information, many materials other than semiconducting materials are used; we devote only one chapter to semiconductors. Various specific cases are drawn from among ceramic oxide transducers, magnetic and semiconductor memories and liquid crystal displays. The book ends with a brief look at the potential of one of materials science's latest novelties — warm superconductors. We have assumed throughout that the reader is familiar with the ideas in the companion volume *Materials Principles and Practice* (ed. Charles Newey and Graham Weaver).

The structure of the book has changed only a little from the Course Team's original plan. We are very grateful to Peter Lilley for his assistance during this phase and in subsequent stages for substantial contributions to early drafts of Chapters 5 and 6. In addition we have been ably assisted with the text by Frances Saunders and Vas Deshmukh of RSRE (Malvern) and by contacts in the Plessey companies, particularly David Cardwell at the Caswell Research Centre and Jon Burnie at Roborough. John Briggs, Andrew Rivers and Charles Newey have also made valuable contributions to the preparation of this book.

In spite of the high calibre of our advisors and thorough, painstaking editing by Allan Jones, there may yet be misunderstandings or errors and these are ours alone. Finally, we acknowledge the extremely high standard of professional support received from many others at the Open University.

Nicholas Braithwaite
Graham Weaver

Open University Materials in Action course team

Contents

Chapter 1 Goals and gambits

1.1 First thoughts

Materials for electronics are, rather obviously, those which have useful electrical (and magnetic) properties. The problem is that all materials have electrical properties but only some are 'useful' in technological and, ultimately, commercial contexts. Which they are is not often self-evident. This book aims to explore what properties there are, to explain how each material comes to exhibit its individual suite of properties and to show how the connections between these can be manipulated. On Figure 1.1 this is represented by the 'property–principle' line. Particular properties imply the possibilities of particular products. So for example semiconduction goes with transistors and the product–principle–property face of Figure 1.1 becomes a focus. Following that example you will remember that to make transistors, silicon is needed as high purity single crystal — processing is vital.

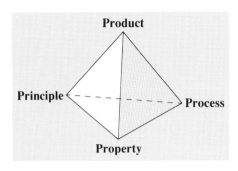

Figure 1.1

Throughout this book there are examples in which this four-cornered perspective is applied implicitly in assessing whether or where or when an alternative product design should be used. What, for example, is required of a computer memory? Different parts of a computer need different qualities of memory and the 'tetrahedron' approach can be used again and again for each distinct set of products, processes, properties and principles. The important thing is that the product, whatever it is, should achieve its function adequately.

We could argue that people do not really want to buy electrical equipment at all. They want what I will call **soft functions**. By this I mean their aspirations to speak to someone at the other end of the country, to store food in a cold place, or to have home entertainment on the TV. These are 'goals' for engineers. The 'gambits' or tricks used to achieve them are **hard functions**. For example in a windmill, the action of wind on a sail which turns a shaft and the shaft turning a grindstone are the hard functions providing the soft function of corn ground to flour. Today we often stack together electrical hard functions to reach our goals. So electrical solutions must have some special recommendations — are they inevitable? Let's look into that question.

1.1.1 Why electricity?

It is useful to distinguish between electrical and electronic engineering. In a nutshell, electrical engineering covers energy conversion (or power) systems — motors, dynamos, heating, lighting — while electronic engineering involves information processing either for signalling, control, storage or calculation. Looking broadly like this it is easy to see pre-electrical devices providing equivalent soft functions to modern electrical and electronic equipment.

Power machines.   Animals, water wheels, windmills, steam engines, internal combustion engines and catapults all convert energy into a useful form.

Information stores.   All kinds of writing can be done on many different materials (clay tablets, stone slabs, paper) with several instruments (chisels, pens, burnt sticks), as can pictures, diagrams, maps etc. Note the need for information to be **coded** (alphabets, words, map symbols) so that information can be put into and retrieved from the store. Pianola rolls and musical box cylinders are data stores designed for mechanical information retrieval.

EXERCISE 1.1   Write lists of nonelectric methods for (a) lighting and (b) telecommunications (sending information further than you can shout).

Many of these functional demands are ancient and providing for them without electricity has involved the ingenious use of a range of materials. By comparison however, the electrical solutions generally offer significant improvements of performance: 'better things are electric' as one advertising jingle had it.

For example, electric motors are quiet and clean, emit no exhaust, can be easily controlled, may be built for power outputs ranging from a watt to a megawatt and are generally compact compared with other engines. These merits apply to electric heaters and lighting too.

For transmitting information, the electric systems (radio, telephone) offer faster data transmission with high reliability and low error rate. Electrical signals are relatively insensitive to the environment (unlike, for example, smoke signals on a windy day) and are very long range.

It is not necessary for me to list advantages of every provision of the electrical world. Nor indeed should every example be regarded as 'a better thing'. But you will find, if you reflect upon some systems, that their advantages are covered by these generalities:

- Versatility — power, control and information can all be handled in the same electrical medium. Information about virtually any physical condition (such as stress, temperature, light intensity, chemical composition) can be reproducibly and sensitively converted into an electrical analogue signal. (This is basic atomics. All stimuli to matter affect electron interactions between atoms so an electrical response of some sort is bound to exist.)

- Speed — electric currents can be varied or switched on and off very quickly so an electrical system can respond to control decisions and it can handle high frequency data. (Music goes up to 20 kHz but this is very slow in electrical contexts.) Binary logic (yes/no or on/off statements) can therefore be processed rapidly. The same binary code permits electrical calculating machines to be made.

- Complexity — speedy processing becomes useful if large quantities of data are being derived from several sources. High signal-to-noise ratios permit information to be processed with low associated power. This in turn allows compact circuitry of great complexity (silicon chips especially) to meet these goals.

- Reliability — compared with mechanical, hydraulic and pneumatic control, electronic systems do not wear: there are no moving parts. They can give very reliable service.

- Electromagnetic waves — there is the unique connection between alternating electric currents contained within circuits and electromagnetic radiation broadcast away. Electromagnetic radiative connection between equipment is the basis of several functions, like broadcast radio, which could scarcely be contemplated in any other medium.

The development of a worldwide dependence on electricity was quite unforeseen when electricity was first discovered. We now have holes in the walls of our houses and factories which allow us to use electric and electronic solutions whenever we want.

Perceptions of what hard functions were possible, and how they could be put together to meet soft functions, grew slowly as many specific electrical/magnetic gadgets proved their worth. With this came a growth in the knowledge of materials properties, principles and processes for making electrical products. Though we are now 150 years into the 'Electric Age' neat, hard-function innovation is still a hallmark of the industry. Perhaps that is why we consumers expect it to be able to provide for every soft-functional aspiration.

EXERCISE 1.2   Compare pen and paper with an electronic word processor as convenient media for creative writing. Is either 'better' than the other?

1.1.2 Research, development and industry

The pace at which fundamental discovery is translated into commercial product has quickened somewhat over the two centuries of industrialization. In 1800 Volta invented his battery, which allowed the study of electric current rather than just static electricity. ▼Currents and chemistry▲ revises principles here. Surprisingly it was 1820 before Oersted noticed that a magnetized needle could be deflected by the passing of an electric current through a nearby wire. ▼Electricity and magnetism▲ reminds you of some important ideas in this area. The scene was then set for the invention of an 'electric telegraph'. The need for long-distance communication between railway signalmen provided the impetus and the finance for this first electric gambit. ▼Cooke and Wheatstone▲ describes the working of the elegant instrument patented in 1837 and pictured in Figure 1.2. The instrument had five indicator needles, two of which were activated to define each letter by pointing along the diagonal grid lines. This arrangement dispensed with six letters of the alphabet, which must have given some interesting spelling!

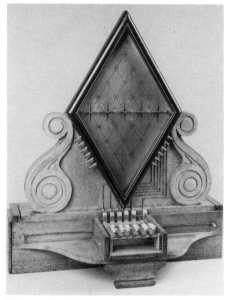

Figure 1.2 The Cooke and Wheatstone telegraph (courtesy of the Trustees of the Science Museum)

▼Currents and chemistry▲

The connection between electricity and chemistry, as you know well from previous study, is the electron. Chemistry has all its rich variety from interactions between the outer surfaces of the electron clouds of atoms. Among the possibilities are

(a) In metals some electrons do not 'belong' to specific atoms; they are free to move. An electric current is an organized flow of charge so metals can support this.

(b) Ionic chemistry involves electrons changing allegiance from one atom to another making both into charged ions — the one that has lost electrons becomes positive and the one that has gained electrons becomes negative. If the ions can move there is another opportunity for charge flow or current. Conductors in which charge is transported by ions are called 'electrolytes'.

(c) Covalent chemistry locks up all the electrons in molecules. They move only in a restricted space defined by the molecular 'orbitals', like atomic quantum states but involving the positive nuclei of more than one atom. Then the only possibility of current comes from bonds being broken

(thermal agitation might do this) leaving an electron free to move. Depending on how strong are the covalent bonds we get insulating or semiconducting properties from these compounds.

Batteries such as the one Volta invented are devices put together to encourage chemical reactions between metals and electrolytes. SAQ 1.1 will remind you of their action and let you see the thermodynamic origin of their electromotive force (e.m.f.).

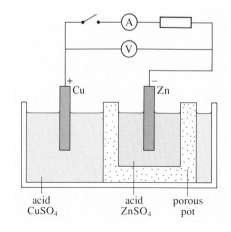

Figure 1.3

SAQ 1.1 (Objective 1.2)
(a) Figure 1.3 shows a battery driving a current in an outside circuit. Describe what happens at the electrodes to produce the battery voltage and to drive the current.

(b) When the switch in Figure 1.3 is open, no current flows and the chemical reactions at the electrodes come to their individual equilibria. Explain in free energy terms how contact potentials between the electrodes and their surrounding electrolyte arise and why they depend on the concentrations of metal ions supplied in the electrolytes.

(c) Now close the switch. The two electrode reactions which were separate thermodynamic systems have become joined to make a single thermodynamic system. How does this view explain the flow of electric current? What is the thermodynamic meaning of a 'flat battery'?

Much of the manipulation of materials to get desirable electrical properties is chemical, doping silicon with boron or phosphorus to get p-type or n-type extrinsic semiconduction respectively is an example you'll be familiar with. Many more chemical tricks will be introduced through this book.

▼Electricity and magnetism▲

Figure 1.4 Iron filings showing the magnetic field surrounding a bar magnet

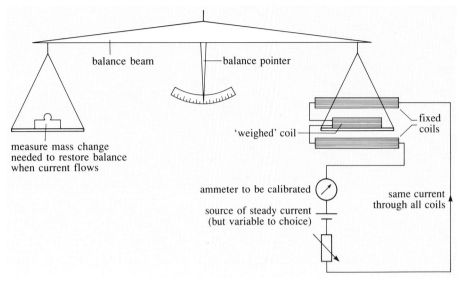

Figure 1.6 Current balance with ammeter to be calibrated in series

balance beam — balance pointer

'weighed' coil — fixed coils

measure mass change needed to restore balance when current flows

ammeter to be calibrated

source of steady current (but variable to choice)

same current through all coils

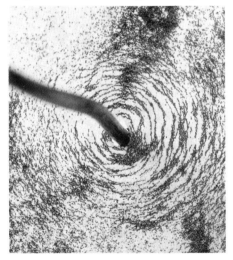

Figure 1.5 Iron filings showing the magnetic field generated by a current

Magnets exert forces on magnets. The model is of a magnetic 'field of influence' surrounding a magnet (Figure 1.4). Oersted's discovery that an electric current in a wire moves a nearby magnet suggested immediately that the current also generates a magnetic field. Figure 1.5 shows the iron filings pattern due to a current.

Given this model, it isn't surprising that one current exerts a force on another. This was Ampère's discovery and is the basis of the definition of the unit of electric current bearing his name.

'The ampere is that constant current which, when maintained in two parallel, infinitely long, straight conductors of negligible cross section, placed one metre apart in vacuum, produces between these conductors a force of 2×10^{-7} N per metre length.'

By also defining the ampere as a charge flow rate of one coulomb per second, the unit of charge, the coulomb is decided.

The 'current balance' illustrated in Figure 1.6 uses more compact geometry than the definition specifies to get better forces, and is a practical method of calibrating ammeters. More importantly, these interactions are the basis of all electromagnetic machines: motors, in which conductors carrying currents through magnetic fields are made to move, and generators, where conductors are moved in magnetic fields causing currents to flow. Figure 1.7 shows how the magnetic field due to a current-carrying coil is concentrated and steered by a piece of iron. Making the most of magnetic materials in these and other situations is to be studied in Chapter 3.

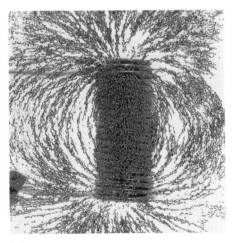

(a)

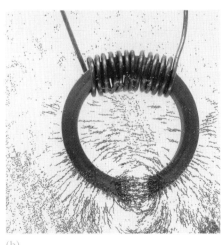

(b)

Figure 1.7 Iron filings showing the field due to a current-carrying coil (a) without core (b) with a core of soft magnetic material in the shape of a gapped ring

▼Cooke and Wheatstone▲

1 indicating needle
2 deflected needle
3 axle to which both needles are fixed
4 deflector coils on wooden former
5 stop for deflected needle
6 front panel carrying letters
7 rear support rail
8 front support
9 axle bearing adjustment screw
10 pointed end of axle in cupped screw

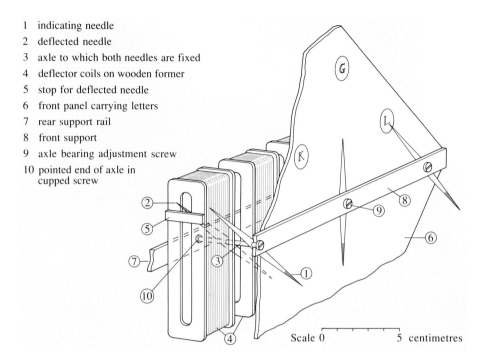

Figure 1.8 The heart of the Cooke and Wheatstone telegraph

William Fothergill Cooke was the entrepreneur and Professor Charles Wheatstone the scientist who 'did by petition humbly represent unto his late most Excellent Majesty King William the Fourth, that after considerable application and expense [they had] invented Improvements in Giving Signals and Sounding Alarms in Distant Places by Means of Electric Currents transmitted through Metallic Circuits'. They were awarded a 14 year patent on their apparatus.

Each of the five indicator needles of the telegraph is the visible part of a pair of magnetized needles (1 and 2 in Figure 1.8) fixed to a common axle (3). The axle passes through a narrow gap between the two coils (4). Each coil carries 800 turns of fine wire and is shaped to allow the inner needle to swing inside until stopped by the bar (5). When the coils are energized, the magnetic field deflects the inner needle which carries the outer needle with it because both are rigidly fixed to the axle. The direction of deflection is determined by the direction of the current flow in the coils and the outer needle points to the required letter on the screen (6). The indicator needle of each pair is mounted so that its north pole points in the opposite direction to that of the inner needle. This makes the pair insensitive both to the fields of adjacent needle pairs and to the Earth's magnetic field. The end of the outer needle is weighted to make it hang vertically when there is no current.

It is interesting to note that the patent covers not only the design of the coils and pointers, but also describes the switching arrangement which allows current to flow both ways through the coils, and it has a major section covering the use of insulated outdoor long-distance cabling. Nowadays these are matters of complete familiarity and unconcern.

SAQ 1.2 (Objective 1.1)
Identify each of the following statements about the Cooke and Wheatstone telegraph under the categories 'principle', 'property', 'soft function' and 'hard function'.

(a) Pressing the outer keys transmits letter A.
(b) The needles are deflected by the magnetic field surrounding a current-carrying wire.
(c) Copper is a good enough conductor for the job.
(d) A voltaic pile of 30 cells allows needles in Slough to be deflected by a signaller at Paddington.
(e) Zinc/copper cells generate 1.1 V e.m.f.
(f) 'Joe, the up-train has just passed me two minutes early.'

This earliest example of a successful electrical solution to a functional problem illustrates the essential flow from fundamental research through development and on to deployment which has always been characteristic of the electrical/electronic industry. It is not that this progression has no place in other industries. But perhaps, for example in mechanical engineering, nature provided examples for technology to emulate: wind and water power, levers, springy sticks and rolling stones. Useful electrical and magnetic phenomena have, by contrast, only been discovered by searching. And their material manifestations, for example metal wires and batteries, are not mimics of nature. So the way forward has often been called for the deployment of new materials specifically chosen for their electrical or magnetic properties. New materials

(a) come essentially from materials research,
(b) may require revision of an industrial supply base,
(c) call for innovation in the business context.

The first two points become clear by example. Copper had been found to be a good conductor and was fairly readily available (as sheet perhaps for cladding ships). Did the advent of the electric telegraph so increase demand that mining output was raised? Or at least, was a wire-drawing factory needed? For tungsten lamp filaments a new process had to be developed to enable this 'best' material to be used.

The third point, concerning innovation, requires discussion.

Figure 1.9 shows routes to profit for companies in a competitive

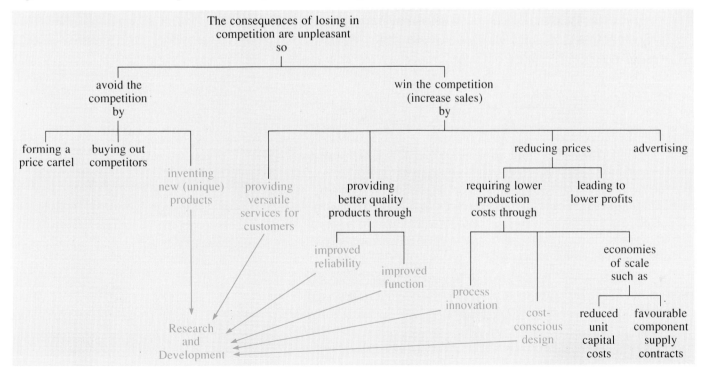

Figure 1.9

economy. There may often appear to be safer or surer options than hoping for innovation to emerge from expensive research. So it is perhaps surprising that such speculations should be the basis of so successful an industry. Again there are three observations to be made:

- Small companies may be profitable because they can break new ground by narrowly concentrating research and development before their capital resources run out. This is the entrepreneurial spirit.

- Large companies, to survive, have to invest in research and development (or in buying in new expertise) since in the long run all technologies become obsolete.

- Sponsorship for uncertain activity helps. Military budgets are often seen in this role in the context of electronics. Advances in radar, for example, are easier to justify on a military rather than a civil budget.

So electronics can use exotic materials, refined in composition and microstructure, grown by extravagant processes and deployed with delicate precision, to perform particular magic. These materials and the devices which incorporate them are the quintessence of laboratory based engineering.

SAQ 1.3 (Objective 1.2)
Turbogenerator sets for power stations are very large, complex and expensive machines. They are built in small numbers and are expected to last a long time. In response to demand from the electricity supply industry, their individual power capacity has grown over the years from a few tens of megawatts to a thousand megawatts. Which factors do you think promote, and which inhibit, innovation in the manufacture of such machinery?

1.2 The electronic world

1.2.1 Principles

The electronic world is totally artificial. It is constructed of materials deliberately selected to perform in sundry specific ways. To insulate, to conduct, to be permanently magnetized and to produce voltage were the specific behaviours demanded of materials in the Cooke and Wheatstone telegraph. We have seen how these properties produced a 'hard' function (the deflection of needles at a distance) which in turn was used to satisfy a 'soft' functional need (telegraphy).

Figure 1.10 represents this world and defines its boundaries. Inside those bounds is a place of 'circuits'. Currents are driven by voltages through 'conductors' and are prevented from mixing by 'insulators'. Power ($V \times I$) is transferred and information (coded voltage changes for example) is processed. Hard functions such as signal and power amplification, logically determinant switching, heating and memory are achieved by appropriate designs of circuits. The many devices which are

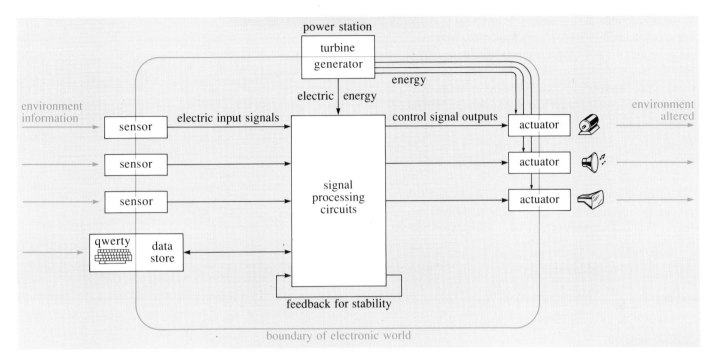

Figure 1.10 The electronic world

networked to make these circuits work — resistors, conductors, capacitors, diodes, transistors and so on — deploy many different materials, as you will see as this book goes along. The following questions will remind you of the sizes of quantities which might be involved.

SAQ 1.4   (Objective 1.4)
A current of 10 A flows through a load of 10 Ω. What is the potential drop across the load? What power is dissipated in the load?

SAQ 1.5   (Objective 1.4)
A current of 1 μA flows through 1 m of round wire conductor of 1 mm diameter. The insulation is 100 μm of plastic with a resistivity of 10^{16} Ω m. What fraction of the current is leaked if the conductor is at 100 V potential above the outer environment?

SAQ 1.6   (Objective 1.4)
The d.c. set-up of a simplified transistor amplifier is shown in Figure 1.11. What current is drawn and how much power is supplied by the battery? Ignore any effects of the signals entering the base of the transistor and leaving the output.

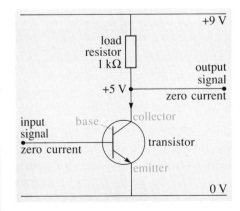

Figure 1.11 A simplified transistor amplifier

If the electronic world were self contained it wouldn't achieve much. It needs to be coupled to its surroundings. In essence we usually require the electronics to take in information from the environment and use it to control something that will modify the environment. Transducers convert signals or energy from one form to another. Chapters 3, 4 and 6 will have much to say about the specialized materials by whose mediation these translations are accomplished.

Transducers at the input side (Figure 1.10) of electronics which convert information from any form (temperature, pressure, smokiness, humidity) into electrical signals are 'sensors'. At the output end transducers are 'actuators' — they act to cause some change to occur. Often, but not necessarily, actuators will be electric power converters, exemplified in Figure 1.10 by an electric motor, a loudspeaker and a cathode ray tube.

SAQ 1.7 (Objective 1.8)

Identify the parts of the following systems which correspond to sensors, actuators, and so on in the electronic world as shown in Figure 1.10:

(a) A radio receiver.
(b) A bimetal switch thermostat in an electric oven.
(c) A washing machine controller: an electric motor drives a camshaft which opens and closes switches in a preset sequence. Is this sufficient to achieve the soft function?
(d) A spacecraft sending pictures back from a distant planet.

Do the descriptions fit the offered general model of electronic systems equally well?

In a later section of this chapter we will look in some detail at three systems which meet very different soft-function goals to see what sort of electrical properties of materials are needed. But already it is clear that the ability of some materials to conduct electricity and or others to prevent its flow is a paramount classification. In your earlier studies you will have learnt several facts about this phenomenon and met some explanatory models. A review of mechanisms of conductivity, or lack of it, is essential.

1.2.2 A review of conduction mechanisms

Electric current is a flow of charged particles through a material. The particles are driven down the potential gradient ($- \, \mathrm{d}V/\mathrm{d}x = E$, electric field) and the geometry of the conductor is split from its material property (conductivity σ_e) by defining the resistance of the conductor (Figure 1.12) as

$$R = \frac{l}{\sigma_e A}$$

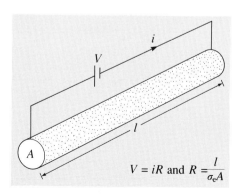

$$V = iR \text{ and } R = \frac{l}{\sigma_e A}$$

Figure 1.12

Alternatively, the resistivity of the material, $\rho_e = 1/\sigma_e$ can be used, giving

$$R = \frac{\rho_e l}{A}$$

Recasting this you can see the units of resistivity are $(\Omega \times m^2)/m$, or Ω m. Conductivity is therefore expressed in $\Omega^{-1}m^{-1}$.

Figure 1.13 shows the vast range of conductivities at our disposal. Excluding superconductors, there is a range of some 25 orders of magnitude running from 10^8 $\Omega^{-1}m^{-1}$ for silver to 10^{-17} $\Omega^{-1}m^{-1}$ for polystyrene. As our subject develops in this book, we shall see materials deployed for several other electrical properties in addition to their conductivity. Sometimes the suite of properties, including conductivity, demanded for an application will be easily achieved. In other instances conductivity will have to be controlled to a suitable level in order to be able to exploit another property. To control conductivity we have to know what goes on inside materials to give their various conductivities.

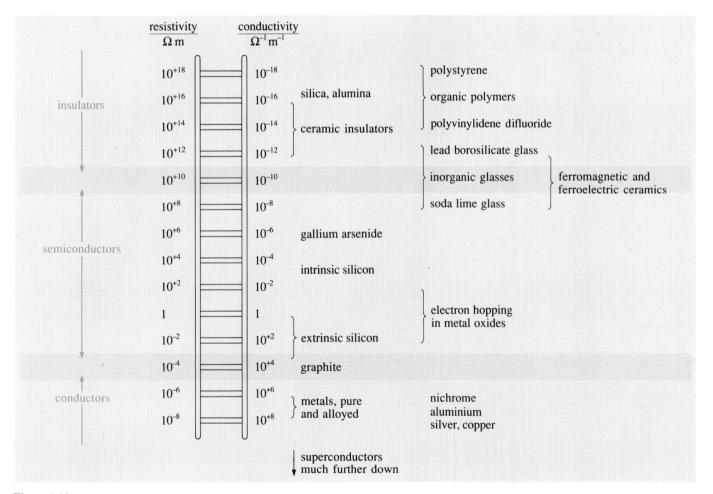

Figure 1.13

SAQ 1.8 (Objective 1.5)

Recall the equation which relates σ_e to the charge transport mechanisms within the material. What do the terms in the equation mean?

In general, the conductivity of a material may be the result of motion of several species of charge carrier and can be expressed as the sum (Σ) of the contributions of each:

$$\sigma_e = n_1 q_1 \mu_1 + n_2 q_2 \mu_2 + \ldots = \Sigma n_i q_i \mu_i$$

where the subscript i identifies each species and, appearing on n, q and μ, emphasises that different populations, charges and mobilities may apply to each species.

One point worth clarifying immediately is how the summation works when there are carriers of opposite signed electric charge. In an electric field, positive and negative charges will drift in opposite directions. Their mobilities, defined as drift velocities per unit of potential gradient, are vector quantities and therefore also carry a sign. If, arbitrarily, we say that positive carriers have positive mobility then negative charges must have negative mobility. The contributions of the two sorts of charge to the current then add together rather than cancel:

$$\sigma_e = [n(+q)(+\mu)]_{\text{positive charge}} + [n(-q)(-\mu)]_{\text{negative charge}}$$

Once you have appreciated this you can then forget it. The equation is simply used by inserting the magnitudes of the charges and mobilities without having to worry about their signs.

This equation is the best way of discussing whether a material is an insulator or a conductor. It separates what might drift nq from the mechanisms of their motion μ and doesn't beguile us into imagining that electrons are the only things that ever move. Whether or not this equation has to be applied with full rigour depends on the material. In some materials, just one species dominates conduction. In other materials the separate contributions are comparable. ▼Familiar cases▲ reminds you of some of these.

Though electrons are certainly not the only carriers of charge, they are the dominant charge carriers in many situations. Let us therefore remind ourselves of the ideas of electron conduction. The band theory of electrons in solids models the numbers of electrons available for conduction in different classes of material, while separate concepts indicate the origins of the 'drag' which limits mobility.

▼Familiar cases▲

Example 1 — ordinary metal

Electrical conduction is by 'free' electrons, those delocalized from their atoms as the metallic chemical bond is formed. Ostensibly then $\sigma_e = ne\mu$ with n free electrons, charge e and mobility μ. But what of the metal ions which remain when the electrons become free? For a monovalent metal (one freed electron per atom) they have the same n and opposite charge $(+e)$. Having opposite charge sign they will tend to move in the opposite direction under an applied field so will add to the current. So strictly

$$\sigma_e = (nq\mu)_{\text{electrons}} + (nq\mu)_{\text{ions}}$$

Here the second term is negligible and usually ignored because the ions, locked in their crystal lattice, have much lower mobility than the electrons. Their effect as contributors to conduction is invisible. In other systems where there are no free electrons no such dismissal of ion conduction can be made.

Example 2 — superconducting metal

$$\sigma_e = (nq\mu)_{\text{electron pairs}} + (nq\mu)_{\text{single electrons}} + (nq\mu)_{\text{ions}}$$

The famous theory of superconduction in metals ascribes the apparently infinite conductivity to a rather small population of electron pairs which cooperate in such a way as to avoid being scattered. They have $\mu = \infty$ so in spite of their small numbers they completely dominate conduction. Not only conduction by ion motion but even the ordinary electronic conduction is negligible by comparison.

Example 3 — intrinsic silicon semiconductor

Silicon is a covalently bonded solid with all the outer electrons of every atom involved in 'electron sharing' bonds. Thermal 'rattle' in the crystal disrupts a small proportion of the bonds, generating free electrons and 'holes' (damaged bonds which are missing electrons) in equal numbers:

$$\sigma_e = (nq\mu)_{\text{electrons}} + (nq\mu)_{\text{holes}}$$

Is either term negligible? The mobilities at 300 K quoted in data books are 0.14 m² $\text{V}^{-1}\,\text{s}^{-1}$ for electrons and 0.05 m² $\text{V}^{-1}\,\text{s}^{-1}$ for holes. The electron and hole charges are respectively $-e$ and $+e$. So, with equal populations, the electrons are contributing some 75% of the current.

EXERCISE 1.3 Set out the complete equation for an n-type extrinsic silicon semiconductor displaying terms for free electrons, free holes and ionized dopant atoms. Which terms are small? Why?

SAQ 1.9 (Objective 1.6)
Outline the principal features of the band theory of electrons in solids, explaining how the theory distinguishes between good conductors, extrinsic semiconductors, intrinsic semiconductors and insulators. A few sketches of the 'bands' and how they are filled may serve to answer this.

The band theory only tells us the number of electrons available as mobile charge carriers. It does not indicate what determines their mobility. Quantum concepts of electrons treat them as waves, giving a better description of their 'collisions' with a lattice of thermally vibrating atoms. The collisions are modelled in terms of interactions between electron waves and the elastic waves (phonons) which represent the atom vibrations. Experiments confirm that conductivity by a fixed population of free electrons falls off with rising temperature.

The other way the progress of an electron can be impeded is for it to fall into a 'trap'. For example it can fall back into a level in the valence band, repairing a thermally broken chemical bond. Bonding in transition metals and alloys is seen as part covalent (localized states) and part metallic, so electrons switching roles within this bonding regime are either 'free' or 'trapped'. As temperature rises we can expect traps to be less successful in catching and holding electrons. Balancing the opposing temperature sensitivities of trapping and phonon scattering by careful composition choice can provide transition metal alloys with virtually zero temperature coefficients of conductivity. Manganin (84% Cu–16% Mn) is used to confer temperature stability to resistors.

Applied to insulators, the band theory can give some false impressions. Polymers and ceramics, which are the customary insulators, are covalently and/or ionically bonded. So their electrons are all bound to particular atoms and not 'free' in the sense that metals have free electrons and their valence bands are full. Perhaps, though, there may still be enough electrons in the conduction band to confer some residual conductivity — it depends on how large the energy gap to the conduction band is, relative to the average energy of thermal agitation. ▼Insulator band gaps▲ looks at this proposition.

A full valence band and a large gap are necessary but not sufficient conditions for a material to be a good insulator. In nonmetals, conduction by electrons is swamped by conduction from other charge carriers because there are many more of them. Thermal diffusion of ions biased by applied electric fields is the common source of electrical conductivity. In crystalline oxides, oxygen ion diffusion using lattice vacancies is often the dominant mechanism of conduction. This is because these large ions, which form the 'cages' housing the cations, can move with fairly small activation energy into an adjacent vacancy. They do not have to move either much closer or much further from the cations to make the jump. In oxide glasses it is usually the cations which drift. The open and irregular structure of glasses holds ions in large cages with easy routes from one to another. In polymers too, conduction is apparently by traces of ionic impurities left from their manufacture.

The variable valencies of transition elements can produce another conductive mechanism in their ionic compounds. If an element occurs as ions with different charges it becomes possible for electrons to hop from one ion to another. Thus the magnetic iron oxide magnetite (Fe_3O_4) which contains two sorts of iron ion, conducts electricity by electrons jumping from Fe^{2+} to Fe^{3+}. These conduction mechanisms make the d.c. resistivities of insulators dependent on their process histories, so they are quoted by range rather than as precise values of σ_e. Chapters 2 and 4 will revisit these topics.

1.3 Materials for function

As the previous sections have shown, we shall need to employ materials for a much wider range of properties than just conductivity. Especially in transducer roles it is vital to find materials which respond well to particular stimuli while remaining unaffected by others. A strain gauge for example would best use a material whose resistance was strongly dependent on strain but invariant with temperature. The extent to which you can have the one without the other has a lot to do with how properties can be independently manipulated. But problems may also be avoided by the design of the device. Here we shall look at three systems which call for very different kinds of materials properties in order to do their jobs: electronic control for car engines; memories for computers; and TV screens.

1.3.1 Electronic control in a petrol engine

Petrol engines for cars convert energy in the fuel into mechanical energy. This is done by burning some petrol in a closed chamber, one end of which has a movable piston connected to a crank (Figure 1.14). An inlet valve opens to admit petrol and air to the chamber and an electric spark initiates combustion. When the petrol burns, the temperature of the gas rises sharply. So the pressure increases and pushes the piston down, turning the crank. As the piston rises again an exhaust valve opens to allow the burnt mixture to escape so the chamber is ready to perform again. The timing of the valve and spark operation is an important part of the engine's design. During the history of cars, effective ▼Mechanical control of petrol engines▲ has evolved.

Over the past few years mechanical control has often been replaced by electronic systems. The strongest influence for change was antipollution legislation in the USA. If the mixture is too rich in fuel there is not enough oxygen to burn the fuel to carbon dioxide and water. Ideally:

$$2\,C_8H_{18} + 25\,O_2 \rightarrow 16\,CO_2 + 18\,H_2O$$

But with inadequate oxygen, some fuel is unburnt and carbon monoxide (CO) rather than CO_2 is formed. With excess oxygen, some of the nitrogen of the air is also burnt. For example:

$$N_2 + 2\,O_2 \rightarrow 2\,NO_2$$

But nitrogen forms several oxides, usually written NO_x and referred to colloquially as 'nox'. The Americans legislated for very low levels of these pollutants when it was discovered that a platinum catalyst in the hot exhaust gas could get rid of them, provided that (a) the original combustion was kept close to exact stoichiometry, meaning neither excess nor deficiency of oxygen (Figure 1.15) and (b) the spark timing is exactly right to give the fuel time to burn. (Note that most of the world's platinum is produced from a particular geological structure in South Africa, and that the largest similar structure is in Antarctica. The demand for platinum might therefore have political implications for policies on both South Africa and mining in Antarctica.)

It turned out that the mechanical system could not deliver the precision needed under all engine conditions and engines were liable to go 'out of tune' as parts wore. To provide the required short-term versatility and long-term stability, electronic control was looked to. What is involved?

The inlet and exhaust valves remain coupled mechanically to the crankshaft. The driver's action on the accelerator still adjusts the air flow via a flap valve. Electronics starts by replacing the mechanically operated switch in the distributor by an electronic (transistor) one. Also, to give the electronics a way of controlling the mixture, the carburettor is abandoned in favour of a fuel injection method. Essentially this is a pumped constant-pressure pipeline with a nozzle to spray petrol into the air intake. Control is then achieved by opening a tap in the line for an appropriate time. To control the spark-timing and mixture-composition, four items of data are needed.

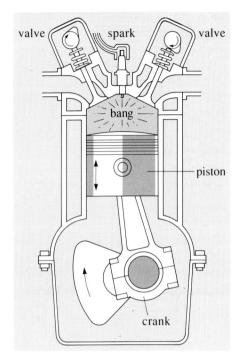

Figure 1.14 The petrol engine

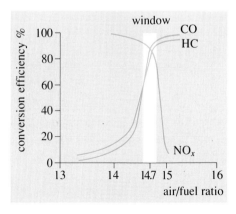

Figure 1.15 Conversion efficiency of the catalyst in the exhaust pipe is only adequate over a narrow band of mixture

▼Mechanical control of petrol engines▲

The timing of the opening and closing of the inlet and exhaust valves, and the timing of the electric spark which initiates combustion, must all be controlled to keep them in correct relation to the motion of the piston. In a mechanically controlled engine, this is done by direct drives from the crankshaft to cams operating the valves and the spark mechanism.

The power output of the engine depends on the amount of air and fuel mixture which is admitted. Air and fuel are drawn in when the inlet valve is open and the piston is moving down the cylinder. Because only a very short time is available to refill the cylinder and because the air has to pass through some pipework, the pressure of the mixture is much less than atmospheric. By depressing the accelerator pedal, the driver opens a flap in the air inlet pipe so that more air can be drawn in as the cylinders refill.

Petrol is mixed with the air in the carburettor. As the air enters it passes through a constriction and this reduces its pressure (the Venturi effect). Suction then pulls petrol through a fine jet as shown in Figure 1.16. As more air is taken in, the Venturi draws in more petrol, giving a bigger bang in a bigger mass of air and hence more power. The proportion of fuel to air is determined by the characteristic of the suction system and the jet orifice. Auxiliary jets allow the appropriate mixture to be set for varying air-flow rates.

The high voltage for the spark is created across an inductive component (the 'coil') when current through it is abruptly cut off

by the opening of the contact breaker 'points'. This switch is operated by a camshaft in the distributor driven directly from the crankshaft.

The final subtlety concerns the spark timing. Calling for more power may either cause the engine to speed up (when you want to go faster) or it may just match an increased load requirement (when you come to a hill and climb it at unchanged speed). Both circumstances need the spark timing to be adjusted to come earlier in the cycle. At higher engine speeds, combustion has to start earlier to give the fuel time to burn completely before the piston is poised

to descend again. Similarly, if there is more fuel to burn, the reaction will take longer, so must start earlier.

Within the distributor are two devices to 'advance' the spark timing by moving the cam follower forward relative to the cam. A centrifugal device (the bob-weight) responds to the rotation speed of the distributor shaft and operates gearing to move the contact breaker points relative to the shaft. The extra suction in the Venturi when the engine is on load deflects a diaphragm ('vacuum advance') to do the same thing in response to increased fuel flow.

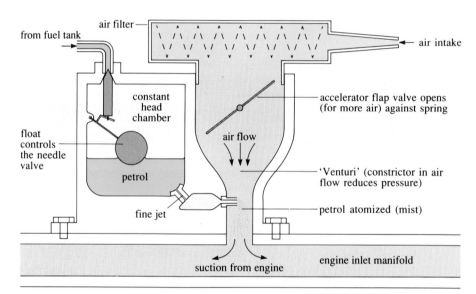

Figure 1.16 Carburettor operation

SAQ 1.11   (Objective 1.9)
Identify the information needed to specify the engine condition sufficiently to set the spark timing and mixture. (Refer to 'Mechanical control of petrol engines'.) Fill in the blanks in Table 1.1

Some of the gadgets used to collect these data are described in ▼Engine sensors▲. Having collected the data, what is to be done with them? There is a problem. Both the spark timing and the mixture have to be controlled by two variables. The electronics has to handle conditional instructions:

IF engine speed is S and IF inlet pressure is p, THEN advance the spark to z milliseconds before the piston is at top dead centre.

EXERCISE 1.4   Write an equivalent sentence for mixture control.

Table 1.1

What information?	Why?
(a)	
(b)	
(c)	
(d)	

▼Engine sensors▲

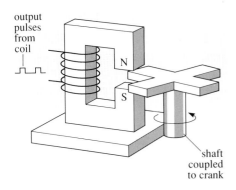

output pulses from coil

N

S

shaft coupled to crank

Figure 1.17 Magnetic reluctance sensor

Where is the piston?

The spark circuit needs an electrical trigger related to piston position. A magnetic reluctance sensor can provide this. A shaft coupled to the crank rotates in fixed relation to the piston and carries a steel tab between the poles of a U-shaped magnet (Figure 1.17), with a coil wound on it. When the tab closes the magnet's air gap, the magnetic flux changes and a voltage pulse is induced in the coil. To get a strong signal, an appropriate magnetic material is needed (this will be covered in Chapter 3).

How fast is the engine running?

Successive spark trigger pulses can start and stop a 'clock' to decide how fast the engine is running. An electronic clock is a generator of regular and rapid voltage pulses and a counting circuit. The pulse generator is most probably governed by a slab of 'piezoelectric' ceramic which vibrates mechanically and electrically at a characteristic frequency (Chapter 4).

What is the inlet pressure?

In some materials, small mechanical strains cause changes in electrical resistivity. The inlet manifold pressure and atmospheric pressure are presented to opposite faces of a silicon membrane carefully machined by chemical etching (Figure 1.18).

Deflection of the membrane is detected by resistors which have been 'ion implanted' into the silicon. (This technique will be described in more detail in Chapter 5.) Silicon is a favoured material for transducers because circuitry can be integrated onto the transducer. This permits digital or amplified analogue outputs from the sensor.

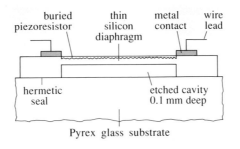

buried piezoresistor   thin silicon diaphragm   metal contact   wire lead

hermetic seal   etched cavity 0.1 mm deep

Pyrex glass substrate

Figure 1.18 Silicon diaphragm pressure sensor

How much air is coming in?

The resistance of a wire depends on its temperature. So if the wire is hot because a constant electric current is driven through it, the voltage across it will depend on its heat losses. Air flow rate is detected by this principle. The instrument is shown in Figure 1.19. There are various criteria for choosing the wire material apart from the

temperature-dependence of its resistivity. It has to stay clean so that the cooling rate is not reduced by surface dirt. The best way to do this is to run it hot and burn off the dirt. And the wire material also has to be corrosion resistant at the operating temperature. Platinum is best. Nichrome might do and would be very much cheaper.

Each sensor must be sufficiently

- accurate
- sensitive
- stable
- reliable
- cheap.

There is plenty of scope for competitive design.

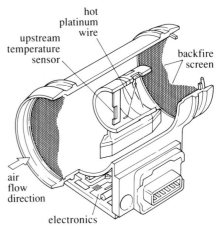

hot platinum wire

upstream temperature sensor

backfire screen

air flow direction

electronics

Figure 1.19 Hot wire sensor for manifold air flow

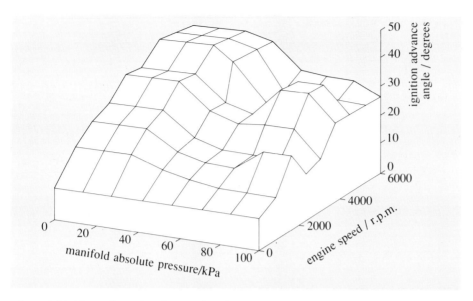

Figure 1.20 Three-dimensional control surface for ignition advance

The easiest way of achieving this sort of control is to refer the signals from the sensors to a 'look-up' table held in the electronic memory. Figure 1.20 shows the map to be stored for the spark advance of one engine. It dictates the amount by which the spark time is shifted relative to the signal from the piston position sensor. The engine management electronics has now grown by another dimension. It needs a memory.

Elaborate though the system is, it is still 'open loop control'! The objective was to set the stoichiometry of the inlet gas and to ensure complete combustion. The strategy so far is to set controls which should give the correct behaviour. To close the control loop, we need to give the system information about its success or failure, upon which it can act. For this a further measure is needed. The exhaust-gas oxygen sensor provides it. If the mixture is too rich in fuel every vestige of oxygen will be used up. If the mixture is too lean the exhaust gas will still contain one or two percent oxygen. The sensor converts this difference into an electrical signal to feed into the control.

You should now be able to see how the necessary system fits the Figure 1.10 model of the electronic world. Several transducers inform about engine conditions. Signal processing circuits combine the information to produce appropriate control instructions. Actuators driven by these carry out the desired changes. A wide range of materials is used to make a system which is sensitive, rapidly responsive, stable over a long time and reliable even in the heat, dirt and vibration of an engine compartment.

1.3.2 Memories for computers

Sitting down to write this on my personal computer, I realize what a closed piece of electronic world it is. I switched it on and it asked me for a program. I inserted a disc carrying instructions for 'word processing' and then ordered it to open Chapter 1 and go to page 18. Then I typed these words and now I shall command it to save them. The keys are not transducers. They merely close switches so that signals from somewhere inside the machine can get to another part of it.

Of course this magic all depends on memory. In the work just described, three types of memory have been used. Two are silicon and the other is iron oxide — as usual a variety of materials.

First, when I switched it on, the computer knew what to do with the disc. That means that there must be a memory which

- was loaded by the manufacturers
- does not disappear when the power is off
- cannot be accessed by the user.

This is a Read Only Memory (ROM). It is built in silicon 'hard wired' so that a set sequence of events occurs when the memory receives power. (This is what the 'look-up-table' for the petrol engine controller is.)

There is another silicon memory, large and fast, which is the 'working memory'. It holds the program for the current application (word processing), the documents I am using just now and the bits and pieces I may temporarily cut out to stick elsewhere. It has to be 'write and read' since I need to be able to alter the document by writing some more. I may want to call up a different document for reference, or quit word processing and replace it with a spreadsheet program. It is a random access memory (RAM) meaning that the computer can get at all stored pieces of information with equal facility. It is also a 'volatile' memory — its contents are lost if power is switched off. Chapter 5 will take you into the design of such a memory.

The third type of memory is the 'floppy disc' by which I presented the application program and my documents to the machine. Its contents were copied to the process memory when required. You will know that this is a magnetic memory so its materials are quite different from the others. Materials for discs and reading heads are studied in Chapter 3.

SAQ 1.12   (Objective 1.9)
The floppy disc memory characteristics are right for its function.
Which of these features apply?

(a) Read only
(b) Read and write
(c) RAM
(d) Large capacity
(e) Fixed to machine
(f) Portable
(g) Volatile
(h) Permanent
(i) Transferable between different machines
(j) Compact data store

The memory types are chosen, we see, with their functions in mind. The speed and high capacity of the process memory are vital. Its volatility is a snag we choose to live with. The comparative slowness of magnetic disc memory is counterbalanced by its combination of permanence, portability and read/write ability.

The principle of memory is simple. Any two-state device will do. Blocks of bistable devices can perform the hard function of storing a binary code, though the machine has to know the code — 11000 is the letter A in a telegraph code or the number 24 in binary arithmetic!

Silicon memories are electronic circuits in which a particular point flips to voltage V or flops to zero voltage depending on a signal input. On the disk, discrete areas are magnetized one way or another. The memories chosen for personal computers are now well adapted to their function, but earlier in computer history much less convenient designs have been tolerated. Memory capacity was for a long time a limiting factor in the usefulness of small computers. They were bulkier, less easily addressed and used a good deal of power. An example from 1950s/1960s technology is the ▼Ferrite core memory▲.

1.3.3 Television pictures

Television receivers decode input signals and reconstruct them into pictures of acceptable quality. The question of what quality is acceptable is decided by viewers' opinions of competing technologies — and these may change with time, in terms of both quality and price. The soft function is to allow a family group to watch from around their living room at a typical range of say 1.5 m to 3 m. Aspects of quality are that the picture should

(a) be large enough
(b) be sufficiently detailed
(c) be bright enough
(d) discriminate darker and lighter areas (contrast)
(e) have full colour
(f) allow continuous motion to be seen
(g) be visible at angles up to 45°.

For many years now it has been said that thin, flat TV receivers which would hang on the wall like a picture were 'just around the corner'. It seems the corner is a long time coming up. The cathode ray tube (CRT), reaching some 40 cm from front to back, still leads the competition. ▼Human factors dictate picture quality▲ shows how tightly constrained is the specification. The problem for any other system of display is to be as good as that old CRT at a competitive cost and with similar reliability.

▼Ferrite core memory▲

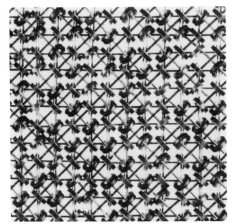

Figure 1.21 Ferrite core memory

Ferrite core memories were rings of a magnetic ceramic mounted on sheets, each ring having three wires through its hole (Figure 1.21). The ceramic was specially formulated to have a square magnetic hysteresis loop (explained in Chapter 3). Binary data was recorded by the magnetization being clockwise or anticlockwise around the ring. Rings receiving a pulse on only one wire could not switch, because the field did not reach the value demanded by the square loop (Figure 1.22a). But if current pulses on two wires arrived simultaneously at a ring, with their field opposing the direction of magnetization as shown on Figure 1.22(b) their field could switch the magnetization of the ring, inducing a voltage pulse in the 'read' wire. (The memory then had to be reset by another pair of pulses.) Rings recording the opposite data symbol did not switch because their magnetization is already in the same direction as the field due to the currents, Figure 1.22(c). (This idea of matrix addressing is used again in Chapter 6 in a visual display context.)

Quite large memories were constructed of these rings. But since every switching dissipated energy, the whole memory took a lot of power and had to be well ventilated. As the years went by smaller and smaller rings were used to raise the data density and lower the power demand. You can imagine what a fiddly job assembly must have been, hence the development of alternatives.

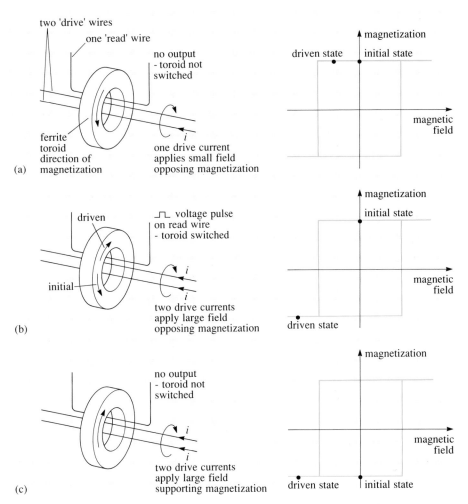

Figure 1.22 Working of the ferrite core memory

▼Human factors dictate picture quality▲

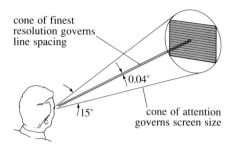

cone of finest resolution governs line spacing

0.04°

15°

cone of attention governs screen size

Figure 1.23

The soft function statements can be converted into a hard function demand once a few facts about eyesight are explained.

How big should the picture be?

The picture should just fill the central area of vision to which our brains give most attention. Then we'll not be distracted by things outside the edge of the picture nor shall we have to hunt around to see all the action. At the suggested maximum range that is about 75 cm wide, representing an angle of 15° for the 'cone of attention' (Figure 1.23). This is one reason why TV screens, having flirted with bigger sizes as a marketing gimmick, have settled to around this size.

What resolution?

The difficulty of building the picture will depend on how fine its detail is. There is no point in producing detail the viewers cannot see. The minimum angle of resolution is fixed by the aperture of the eye and by the wavelength of light, according to wave diffraction theory. It turns out that the cellular structure of the eye's retina (our input transducer) is reasonably matched to these predictions. A reasonable figure for the angular resolution is 0.04° — this means that we can distinguish two lines 0.1 mm apart at 150 mm, our closest focusing range, or 1 mm apart at 1.5 m. So 1 mm spacing of detail (for example the lines on the screen) won't be seen at our 'usual' TV range. With a 25 inch screen and the European standard 625 lines per frame the spacing is about 0.6 mm and the lines become visible at around 1 m range — try it.

These aspects of our eyes' performance go some way towards specifying the picture. On standard 4×3 rectangular shape, an adequate picture could be built of 625 lines each with $\frac{4}{3} \times 625$ dots, a total of 527 000 picture cells or 'pixels' as jargon must have it. The suggestion for deployment of materials is that the screen should have half a million devices on it each able to operate independently — already a tall order.

Colour perception

The sensation of colour of all hues can be induced by mixing different proportions of three 'primary' colours (red, green and blue) on a single retina cell. So with each pixel of the screen further subdivided into three areas, each of which can emit one primary colour, the minimum resolvable area can be made any colour if the fractions of light of each primary can be controlled. Not only has this just multiplied the number of devices on the screen by three (perhaps requiring three different materials to be organized in the array) but has introduced a need for controlling their relative outputs. The contrast demand is a further control factor.

Contrast

A scene is made of light and shade and this is a vital ingredient in our perception of shape. It turns out that an acceptable illusion of continuous variation of shading can be given if there are 64 different levels of brightness. So in addition to the relative brightness of the three subpixels to induce the colour sensation, the absolute brightness of each whole pixel must be controlled on this 64-point 'grey scale' to imply light and shade in the picture.

What picture rate?

Like other transducers, eyes have a response time. When a retina cell has sent a signal to the optic nerve it takes about 50 ms to reset. That is very useful for the TV screen designer. It means that there are 40 ms (a bit less than the reset time) to build a picture. If the picture changes every 40 ms the changes will be seen as continuous motion by our eyes. Things changing faster than this, bees wings for example, appear blurred. If the pictures took longer to form and be replaced they would appear to flicker. Smooth motion then becomes jerky, like early film.

By the same token if a picture persists too long even if 'overwritten' by a new image there will seem to be 'ghosts' of the previous picture.

Brightness

Finally, now we have defined the size of pixels and the rate at which they have to change, the question of their optical energy output can be asked. If black is zero energy then at the next point on the grey scale the pixel must emit enough light to be visibly distinct. Since we don't want to watch in a darkened room, this minimum energy must be defined for a daylight-adapted eye (with the pupil partly contracted).

There is no shortage of ways of transducing from electric to visible signals. Eventually a control signal is presented at a screen where light is either modulated or generated. Here are some options, each with an example of an application.

Light-emitting methods

The most obvious way of converting electricity to light is to heat the material electrically until it is so hot that it radiates light. Other methods involve excitation of electrons to higher energy levels — light is emitted as they fall back to their lowest energy. Table 1.2 lists a number of examples of light-emitting methods for information display.

Table 1.2   Light-emitting methods for information display

Energy input	Material	Example	Effects
electrical current in solid	tungsten wire	filament lamp	heating to incandescence
electric current in gas	sodium vapour	street lamp	collisional excitation of atoms gives characteristic spectrum as the electrons fall back to lower energy levels
	neon vapour	advertising signs flat 'plasma display panel' for portable computer	
electric current through specially designed semiconductor diode	GaAs–GaP	coloured on/off indicators on video recorders used in early digital watches	minority carriers (holes or electrons) which are injected by forward bias recombine and radiate
electric current through semiconducting phosphor (electroluminescence)	ZnS–Mn	yellow numerical displays, for example on instrument panels	electrons are knocked into higher energy levels and emit light when they fall back
electron beam impact on phosphor (cathodoluminescence)	ZnO–Zn	screen of cathode ray tube vacuum fluorescent displays such as blue-green numbers on video recorders	electrons are knocked into higher energy levels and emit light when they fall back

Light-modulating methods

Modulating light means using a control signal to modify the intensity of a light beam. For example, changes in reflectivity or light transmission can readily be achieved by mechanical shutters. Using materials technology we can shrink the 'moving parts' to the level of molecules or ions within a crystal structure, to produce faster, lower power light modulating (electro-optic) effects. One way of doing this is to use a material which rotates the plane of polarization in polarized light.

A sheet of polarizing filter transmits light of one polarization only (Figure 1.24(a)). The light will only pass through a second sheet if the light's plane of polarization is aligned with that of the sheet (Figure 1.24(b)). A material which rotates the light's plane of polarization in response to an electrical signal, placed between the two polarizers, can modulate the intensity of the light passing through (Figure 1.24(c)).

For a television display, the light source would be behind the screen and the picture signal would change the amount of light being transmitted through each pixel. Table 1.3 lists some other examples of light-modulating methods for information display.

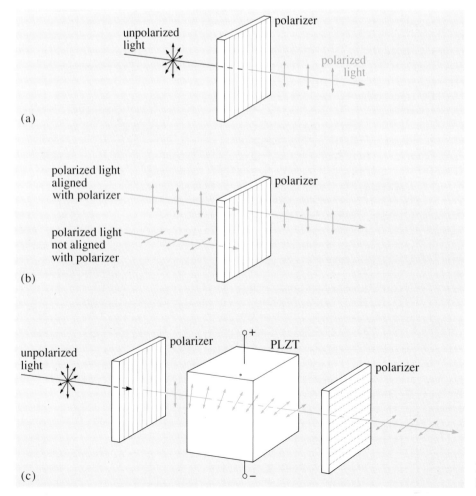

Figure 1.24 Light modulation by polarization

Table 1.3   Light-modulating methods for information display

Input signal	Material	Example	Effect
electric current or voltage	painted surface	electromechanical announcement boards	reflected colours
electric field	lead lanthanum zirconium titanate (PLZT)	flash goggles (Chapter 4)	electric polarization of transparent ferroelectric ceramic
electric field	liquid crystals, for example cyanobiphenyls	grey numbers on digital watches and other flat displays	reorientation of molecules
magnetic field	amorphous rare earth–transition metal alloy, for example Gd–Fe	read/write compact disc (Chapter 3)	polarization of thin film (Kerr effect)

Quality and choices

If there are so many possible ways of producing optical images in response to electrical signals, why is it that only one of them, the cathode ray tube, still provides nearly all the colour TV pictures in our living rooms? The problem is in finding methods and materials which will do these tricks on a screen to give adequate quality.

Given the tight constraints on picture quality, perhaps the amazing thing about TV reception is that it is possible at all, let alone is satisfied so well by our present cathode ray tubes. But for more about this you must wait till Chapter 6.

SAQ 1.13   (Objective 1.9)
Make a general comment about what makes a transducer suitable for its job. Why might 'suitability' be more easily achieved for the transducers in a car engine than for a television picture?

Goals and gambits for electronic materials

The 'goals' set for electronic engineers by customers' aspirations now far exceed the performance capabilities of any earlier technology in the field of information handling. Even the few examples cited here prove this, and demonstrate the great versatility of electronics.

A 'gambit' is defined in one dictionary as 'an initial move, especially one with an element of trickery'. The trickery is to steer materials to the appropriate property suites to permit devices with useful characteristics — conductors and insulators; resistors and capacitors; transistors and transducers — the whole bag of bits which are built into electronic systems by electronic engineers. In chess the tighter definition of 'gambit' is 'the offer of a sacrifice for the sake of advantage'. You will see such compromises again and again through this book as you study and come to understand how materials are manipulated for electronic purposes.

Objectives

After reading this chapter (and revising earlier study as necessary) you should be able to:

1.1 Distinguish the ideas of 'hard' and 'soft' function and suggest reasons for present day reliance on electronics. (SAQ 1.2)

1.2 Discuss the relevance of research and development in electronic innovation. (SAQ 1.3)

1.3 Describe, in terms of the chemical reactions at the electrodes, the generation of e.m.f. in a battery. Explain the equilibrium conditions of these reactions as a competition between increasing enthalpy and entropy, using the free energy equation to balance these factors. (SAQ 1.1)

1.4 Do calculations based on Ohm's Law and the power formulae. (SAQ 1.4, SAQ 1.5, SAQ 1.6)

1.5 Explain several mechanisms of conduction in terms of $\sigma = ne\mu$, distinguishing between the roles of carrier population n and mobility μ. (SAQ 1.8)

1.6 Describe the mobile electron populations of metals and semiconductors in terms of the band theory correctly using the terms 'valence band', 'conduction band' and 'energy gap'. (SAQ 1.9)

1.7 Perform a calculation of the electron population of a conduction band given appropriate data. (SAQ 1.10)

1.8 Discuss the nature of the 'electronic world' including input and output transducers. (SAQ 1.7)

1.9 With reference to the three systems studied, illustrate

(a) the use of electronics to improve control precision and to cope with several sensors simultaneously (SAQ 1.11)

(b) the virtues of different types of memory (SAQ 1.12)

(c) the nature of the problem of choice between several methods of meeting a single goal. (SAQ 1.13)

1.10 Define or explain the following terms and concepts
actuator
coded information
hard function
sensor
soft function
transducer.

Answers to exercises

EXERCISE 1.1

(a) One basic nonelectric method of lighting — flames — covers most of the examples you might have chosen. Candles, oil lamps, gas mantles and so on are just devices for different intensities and control flexibility.

(b) Nonelectric telecommunications might fall into three categories. Audible methods include the alpenhorn and drums. Visible methods include beacons, smoke signals, semaphore and mirrors reflecting the sun. Messages can also be carried from place to place on a recording medium, for example a written message sent by carrier pigeon, or a spoken message taken by a human messenger.

EXERCISE 1.2 Pen and paper is cheap, portable and convenient. Manufacturers of word processors are trying hard to emulate these features with laptop machines for use while travelling. But so far the need for batteries, keyboard and a readable screen mean that it is still a lot easier (and cheaper) to carry a notebook and pencil to capture those moments of literary inspiration.

Word processors offer tidy editing which helps to keep text organized. Dictionary, thesaurus and spelling check facilities can be included. Printed output is usually more legible than handwriting — most publishers will not accept (handwritten) manuscripts so if you want to publish, your work will have to be typed at some stage. Some publishers now take text on floppy disc for electronic publishing.

Nevertheless there may be some circumstances, for example in a personal letter, where the reader would prefer handwritten text.

If you type well, typing will be faster than handwriting. Some people argue that the speed of creative thought is much slower than either, so it doesn't matter. But however fast the ideas flow, with a word processor, rewriting and polishing is usually neater and easier even for a slow typist.

Magnetic disks are notoriously easy to damage into illegibility.

For myself, I'd not want to be without either facility.

EXERCISE 1.3

$$\sigma_e = (nq\mu)_{\text{electrons}} \\ + (nq\mu)_{\text{holes}} \\ + (nq\mu)_{\text{ions}}$$

The first term dominates the second because n-type silicon is doped to make $n_e \gg n_h$.

Although their mobilities are comparable, the enhanced population of electrons makes these 'majority' carriers the most important. The population of donor ions is almost equal to that of the free electrons (that's where the electrons have come from). But the mobility of these ions is tiny. They are bound into the crystal lattice and can only move by atomic diffusion. So both the second and third terms are 'small', one because of small n and the other because of small μ.

EXERCISE 1.4 IF air mass flow is M, and IF inlet pressure is p THEN open the fuel injection tap for t milliseconds.

Answers to self-assessment questions

SAQ 1.1 You should have got some of the following.

(a) The battery voltage is the difference of contact potentials between electrode and electrolyte generated by electrochemical reactions at the two interfaces. Current flow in the metal part of the circuit is by electrons and in the acid solution (electrolyte) is by ions. It is the chemical reactions at the electrodes which convert ion flow to electron flow. At the zinc electrode zinc atoms go into solution in the acid as Zn^{2+} ions leaving surplus electrons in the metal. These move through the metal circuit to the copper plate where they discharge Cu^{2+} ions. The current is transported across the porous pot (and much of the electrolyte) by hydrogen ions.

(b) There are two increases in energy to get ions into solution. Firstly, there is an increase in chemical potential energy as the metal ions detach from the sea of electrons which constitute the strong metallic bond. The magnitude of this energy change depends on the metal involved (electrochemical series) and the solvent used. Water, having polar molecules, reduces the energy increase because they can bond weakly to the metal ions.

Secondly, to remove the positive ions from the metal, leaving an excess of electrons, calls for an increase in electrical potential energy. It is the build up of this electrical potential as more and more ions go into solution that becomes the 'contact potential' between metal and electrolyte.

The increasing entropy of the system counters these energy changes, as metal atoms disperse into solution rather than sit in their regular lattice sites. This means that $\Delta G = \Delta H - T \Delta S$ is negative because the negative second term outweighs the positive ΔH and so the free energy falls. G reaches a minimum when the energy increase $+ \Delta H$ for one more ion just balances the decrease of $- T \Delta S$ for that change. Because of the statistical nature of entropy ($S = k \ln W$) the entropy increase for one extra ionization gets progressively smaller as the concentration of ions in solution increases. Meanwhile $+ \Delta H$ grows progressively as the contact potential grows. So for two solutions of different initial concentrations, comparatively few ions can move into the stronger solution before the balance is

struck. Contact potentials between a given metal and a given electrolyte therefore depend upon the concentration of the reacting ions in the electrolyte.

(c) When the switch is closed the two electrode reactions can find a mutual equilibrium which gives the combined system an even lower free energy. The flow of electrons through the wire represents one reagent, electrons, getting from one reaction site to the other. The new condition of thermodynamic equilibrium exists when there is no more chemical need for further reactions so no electrons flow. That is when we say the battery is 'flat'.

SAQ 1.2

(a) Hard function
(b) Principle
(c) Property
(d) Hard function
(e) Property
(f) Soft function

SAQ 1.3 Factors which might inhibit innovation are:

• It costs a lot to design a complex machine like a turbogenerator. Since the production rate is low, the design cost per machine can only be kept down by using the same design for a long period.

• Feedback about successful or inadequate design accumulates slowly.

• Tooling costs can be spread if the production run is longer.

Factors which might promote innovation are:

• Customers (the electricity suppliers) have changing needs, for example for bigger machines.

• The customers are prepared to collaborate in financing research and development.

SAQ 1.4 Potential drop

$$V = I R = 10 \text{ A} \times 10 \text{ } \Omega = 100 \text{ V}$$

Power dissipated

$$W = I^2 R = (10 \text{ A})^2 \times 10 \text{ } \Omega$$
$$= 1 \text{ kW}$$

(or $W = VI = 100 \text{ V} \times 10 \text{ A} = 1 \text{ kW}$)

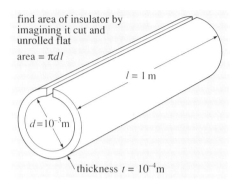

find area of insulator by imagining it cut and unrolled flat

area $= \pi d l$

$l = 1$ m

$d = 10^{-3}$ m

thickness $t = 10^{-4}$ m

Figure 1.25 Insulator resistance

SAQ 1.5 Refer to Figure 1.25.

The formula for the resistance of the insulation on one metre of cable becomes

$$R = \frac{\rho_e t}{\pi d l}$$

So the insulation provides resistance

$$R = \frac{10^{16} \Omega \text{m} \times 10^{-4} \text{m}}{\pi \times 10^{-3} \text{m} \times 1 \text{m}}$$
$$\approx 3 \times 10^{14} \text{ } \Omega$$

The current leaked through this resistance would be

$$I = \frac{V}{R} = \frac{100 \text{ V}}{3 \times 10^{14} \Omega} \approx 3 \times 10^{-13} \text{ A}$$

The current passing along the wire is 10^{-6} A so less than a millionth of it is leaked.

SAQ 1.6 The current is defined by the stated d.c. voltages of the battery and the collector with the chosen load resistor (1 kΩ).

$$I = \frac{V}{R} = \frac{(9 - 5) \text{ V}}{1000 \Omega} = 4 \text{ mA}$$

(This current sets the transistor in a condition where a signal presented at its base can be amplified without distortion. A different current and load resistor might suit a different transistor.)

The power dissipated the resistor and transistor combined is

$$W = 9 \text{ V} \times 4 \text{ mA} = 36 \text{ mW}.$$

SAQ 1.7

(a) In a radio receiver, the aerial is the input transducer sensing a modulated high frequency electromagnetic wave and converting it into equivalent varying voltages. The circuitry separates the audio frequency voltage modulation from the radio frequency carrier, and amplifies it until it has sufficient amplitude to modulate the drive current to the loudspeaker — the actuator.

(b) In the bimetal switch, the sensor and actuator are a single thermomechanical device. The bimetal strip gets hot and bends, breaking the switch in the electric power line. No electronic processing of the information 'too hot' is needed.

(c) The electrically driven camshaft is just a clock which works switches in the right sequence and at the right time intervals. The switches control current to the heater, the pump, the spinner etc. However not all the programme can be decided by elapsed time. Sensors for water pressure and temperature pass information to switch the cam-motor on/off when these conditions are correct.

(d) The camera senses the intensity and colour of the scene. The information, a vast amount, has to be coded by the electronics in a predetermined way. Logic circuits and memory are needed. Radio frequency power is supplied to the transmitting aerial and is modulated by the coded information.

It seems to me that the model is a bit glib. Item (b) has no signal processing. Item (c) uses a clock to decide some steps rather than an environmental signal. Item (d) reminds me that information can be carried in electronic memory as well as transduced from outside.

SAQ 1.8

You should remember from earlier study the equation

$$\sigma_e = nq\mu$$

which says that the conductivity is the result of the movement of n particles per cubic metre each carrying charge q and having mobility μ. 'Mobility' is defined as the 'drift velocity' per unit electric field. In other words, as the carriers accelerate down the voltage gradient and are repeatedly stopped by collisions, the mobility is their average speed divided by the driving field. By referring to an average speed the individual stopping events are replaced by an overall drag which opposes the accelerating force.

The units of mobility are $m^2\ V^{-1}\ s^{-1}$ since this is the same as $(m\ s^{-1})/(V\ m^{-1})$ or velocity per electric field strength as just defined. To get the same units from the formula calls for a little cunning.

$$\mu = \frac{\sigma_e}{qn}$$

so its units are $(\Omega^{-1}\ m^{-1})/(C\ m^{-3}) = m^2\ C^{-1}\ \Omega^{-1}$. Replace the coulomb by A s and the Ω by V A^{-1}, leaving the $m^2\ V^{-1}\ s^{-1}$.

SAQ 1.9

This theory identifies the energy levels available for electrons in solids. In isolated atoms the levels are sharply discrete and well separated in energy. But when atoms are packed close together, their mutual effect upon one another changes the energy spectrum. In a solid, the energy spectrum consists of 'bands' of very closely spaced levels, with 'gaps' in which there are no electron states. Electrons involved in the valence bonding of the material occupy levels in the 'valence band'. Electrons which have gained enough energy to release them from the constraint of a chemical bond are in the 'conduction band'. (See Figure 1.26 (a).)

The differing conductivities of the classes of material are discussed in terms of the size of the energy gap between these two bands. In metals the bands overlap, (Figure 1.26 (b)), so valence electrons and conduction electrons are not distinguished and they can all move easily to higher energies under the stimulus of an electric field. This essentially says that the electrons of the metallic chemical bond are delocalized.

Insulators and semiconductors have their valence bands fully occupied at zero temperature and an energy gap of width E_g separates this band from the conduction band (Figure 1.26 (c)). Thermal agitation breaks a small proportion of the N chemical bonds, which hold the crystal together. The electrons of the broken bonds are excited to levels in the higher band. The population n of the conduction band is temperature controlled; $n \approx N \exp(-E_g/2kT)$. It changes enormously as E_g varies, giving semiconductive or insulating properties. In silicon, E_g is about 1 eV, which would make pure silicon a pretty good insulator up to 500 K. But dissolving group III and V elements, so that some of the silicon atoms are replaced, leads to extrinsic semiconduction. Figures 1.26 (d) and (e) model this behaviour. Other covalent and ionic substances, which have electrons tied more tightly, have a large E_g and so ought to be insulators.

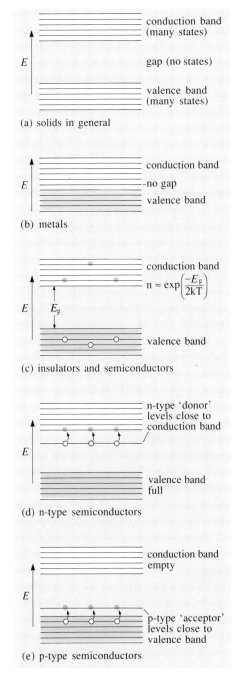

(a) solids in general

(b) metals

(c) insulators and semiconductors

(d) n-type semiconductors

(e) p-type semiconductors

Figure 1.26 Energy bands

SAQ 1.10

$$E = \frac{hc}{e\lambda}$$

$$= \frac{6.6 \times 10^{-34} \times 3 \times 10^8}{1.6 \times 10^{-19} \times 300 \times 10^{-9}} \, \text{eV}$$

$$= 4.1 \, \text{eV}$$

So

$$n \approx N \exp\left(\frac{-E_g}{2kT}\right)$$

$$\approx 10^{29} \exp\left(\frac{-4.1 \times 40}{2}\right) \text{m}^{-3}$$

$$\approx 2.4 \times 10^{-7} \, \text{m}^{-3}$$

and

$$\sigma_e = nq\mu$$

$$\approx 2.4 \times 10^{-7} \times 1.6 \times 10^{-19} \times 0.1$$

$$\approx 4 \times 10^{-27} \, \Omega^{-1} \, \text{m}^{-1}.$$

SAQ 1.11   See Table 1.4

SAQ 1.12   Floppy discs are (b), (c, providing we tolerate a slight delay as the disc and heads move) (d), (f), (h) and (j). Large capacity is a relative term. A $3\frac{1}{2}$ inch disc can hold 800 kilobytes of data, sufficient for a few hundred pages of text. Compared with a book it is compact. (Silicon memories have a similar storage density to that of discs.) 'Permanent' means that the disc does not need power to hold its data, but it can be deliberately erased for rewriting, or erased accidentally if a strong magnetic field gets at it. It is the portability of discs which makes them perfect for their purpose.

SAQ 1.13   This rather vague question should have made you think about the system in which a transducer is set. Transducers are the interfaces between the 'electronic world' and its environment. The primary criterion of suitability is that any transducer should perform the necessary translation of energy to or from the 'electronic world'. It has to achieve this function within constraints imposed from either side of the interface. The constraints may impose several conditions upon the quality of the transducers' actions:

• being sensitive enough to environmental change
• being insensitive to irrelevant stimuli
• being fast enough
• being stable in its performance

The problem with the TV picture transducer is that it must work within multiple constraints set by an unalterable part of the system — the performance of human eyesight. In contrast, some shortcomings of car engine transducers may be compensated for by improvements in the electronics.

Table 1.4

What information?	Why?
(a) instantaneous piston position	to instruct the electronic switch when to open
(b) engine speed	to advance spark as engine speeds up
(c) inlet manifold pressure	to advance spark as more fuel is presented and to infer suction on carburettor jet
(d) air mass flow rate	to infer how much petrol to allow in

Chapter 2 Materials for circuits

2.1 What are circuits?

The main plot of this chapter is how to choose and deploy insulators and conductors for modern circuits, and to draw out the manifold materials requirements of their production methods. We shall find clues which also give insight into designing resistors and capacitors.

The fantastic market surge in electronics in recent decades reflects a circuit building revolution. Expectations of function are ever increasing, demanding more complex, more reliable, more compact and cheaper circuits than were possible in the early days of electronics (▼Contrast▲).

In the earliest circuits, large components (valve holders, capacitors, etc.) were built into a chassis and connected by a nest of colour-coded wires. In Figure 2.1 the central circle is a loudspeaker about 20 cm in diameter. Putting this lot together required a small army of solderers. The invention of transistors around 1950 heralded the revolution. The size of active devices was cut a thousandfold, and they needed only one-tenth of the voltage to run them, allowing much smaller capacitors to be used. And in circuits with no functional need for power (such as signal amplifiers), introducing transistors reduced the power by a factor of a hundred; so resistors became smaller. All this miniaturization was well served by a new form of circuit construction in which the wires were replaced by a pattern of conductors fixed onto one side of a **substrate**. (This new word, which we shall use often, emphasises that the board is responsible for both the insulation between conductors and for their mechanical support.) The wire terminations of the devices poked through holes from the blank side of the board and were soldered to the conducting regions on the other side. This sort of circuit assembly is known as a **printed circuit board** (PCB) (see Figure 2.2). The formation of the pattern of conductors was to some extent a mechanized process, but at this stage each device still had to be placed and soldered by hand.

▼Contrast▲

Time was when a radio receiver was the ultimate consumer electronic gadget. It comprised a power supply, a few stages of amplification, and a local oscillator together with an aerial (input transducer) and one loudspeaker (output transducer). A 1930s model (Figure 2.1) used a handful of thermionic valves as active devices, dozens of resistors and capacitors, and no small amount of copper wire, and worked well enough. It was heavy, consumed about 40 watts and cost a couple of months' wages to buy. Today, for an equivalent amount, you can buy a lightweight, 'lap-top' computer containing millions of transistors and passive devices.

Figure 2.1 1930s radio (courtesy of Robert Hawes, British Vintage Wireless Society)

Figure 2.2 Printed circuit board in a car radio (topside and underside) from around 1960

To reach today's levels of circuit complexity, twin problems of productivity and reliability had to be solved. To make circuits of present-day complexity, production lines of people making solder joints one at a time would be prohibitively slow and expensive, and the consequences of dud joints rather nasty. Suppose as few as one in ten thousand joints is no good. If systems containing a thousand joints are in production then, on average, only 10% would not work and the fault may be easily found. But if each job contains ten thousand connections there's a poor chance that any will work and the fault finding is more difficult.

Modern circuitry exemplified in Figure 2.3 contains three distinct technologies. The main subassembly is still the printed circuit board. But many of the components mounted on it are much more than individual devices. You can see two 'sub-subassemblies', which are complex circuits in their own right. These are the **integrated circuits** (ICs) and the **hybrid circuits** labelled on Figure 2.3. The three types (printed, integrated and hybrid) are principally distinguished by the choice of substrate and by radical differences in construction methods flowing from this. Originally they also served different functions but the boundaries are now more blurred. Let us look briefly at their major design philosophies before studying in detail the interlocking materials requirements of each type.

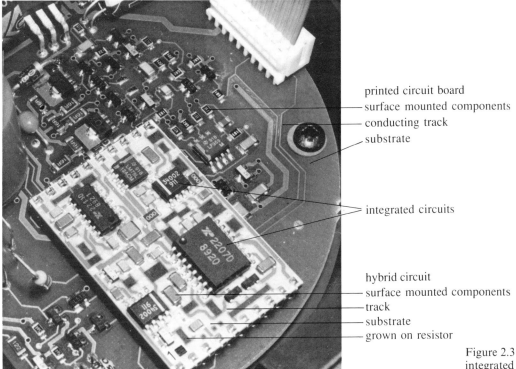

printed circuit board
surface mounted components
conducting track
substrate

integrated circuits

hybrid circuit
surface mounted components
track
substrate
grown on resistor

Figure 2.3 Modern circuit board with integrated and hybrid circuits. The cross-head screw (upper right) is about 4 mm in diameter

Probably there are more companies designing and building customized printed circuit boards than are engaged in any other activity in the electronics industry. This is the bread-and-butter of the business. Though the substrate is still a polymeric insulating board, the conducting tracks still copper, and soldering the main method of fixing devices, PCBs have come a long way since the 1950s. Efficient use of area is the key to good design, and to this end a company's first requirement is good computer-aided design (CAD) software. A great impact has been made by the possibility of 'multilayer' boards having several layers of conductors and insulators on top of one another. At the very least, provision of buried power and earth planes, which can be reached by connectors going into the plane of the board, avoids much awkward conductor meandering. Also, multilayering allows conductors to cross without connecting. (These observations apply also to ICs and hybrid circuits.) Many of the components are now very small or, as with ICs, have many connections, so track spacing is narrow. A spacing of 1 mm is considered generous. Except for fault correction, you can therefore forget about people manipulating these parts into place and soldering them with an iron. Most PCBs are now assembled by robot placement and soldered in a single pass through a soldering machine. The benefits of speed, accuracy, reliability and reduced labour more than compensate for the capital investment needed to set up production facilities.

Integrated circuits, that is 'silicon chips', represent the glamorous end of the industry, and very few manufacturers can make the necessary investment to join in the game. We shall postpone discussion of the product–process–property links in silicon IC technology until Chapter 5, by when you will be better equipped to understand them. Here we merely note the principles and applications of ICs.

The substrate of an IC is a wafer of single-crystal silicon onto which the devices and connections are gradually grown in a series of processing steps. At each step, several thousand components are simultaneously processed to the next stage of development. The whole intention of the technology is to make very large numbers of exceedingly small devices, packed very close together. Thus ICs offer by far the greatest density of components (see Figure 2.4), so these circuits are suited for logic processing and memory in computers, where enormous numbers of devices are needed. Very dense circuitry is needed for high-speed digital processing because even conductor lengths are significant. Electric signals are slower than light in a vacuum, which travels at a foot per nanosecond. A nanosecond is a long time in computing.

At such high circuit densities, manufacturers must ensure that each device dissipates only the minutest power. Fortunately logic and memory have no functional requirement of power (this is the debate taken up in Chapter 5). But it is also perfectly possible to build an audio-frequency power amplifier on a chip. Here the transistors capable of handling large currents, resistors able to dissipate power, and capacitors of low impedance at low frequencies will have to cover a large area of substrate. Between these extremes are a legion of standard circuit packages, operational amplifiers, oscillators and so on which designers now regard as devices to be built into bigger circuits.

Hybrid circuits mix the ideas of ICs and PCBs. Some of the circuit devices are simultaneously *grown* onto the substrate, while others are *placed* on and soldered (Figure 2.5). Hybrids use ceramic substrates; conducting tracks, resistors and insulating layers are made by depositing glassy enamels on the substrate. These enamels are placed by screen printing as 'inks' and fired to fix them and bring them to their functional condition (conducting or insulating). The substrate, being refractory, can withstand these firings. Part of the attraction of this method, compared with PCBs, is the simpler logistics of component handling. Hybrids also have a reputation for reliability, can dissipate significant power, and are remarkably robust in chemically, thermally and mechanically hostile environments. They can be advantageous for analogue circuits needing very precise resistor values (0.1%) and are also good at microwave frequencies (above 1 GHz).

PCBs and hybrids have in common the idea that metal conductors are laid onto an insulating substrate, either polymer or ceramic. Conductors and insulators each have their own implications for materials-selection issues. We reviewed conductivity in Chapter 1. Let us now look at the behaviour of insulators, in both steady and alternating electric fields.

0   50 μm

Figure 2.4 A 741-type integratred circuit. This widely used, general-purpose amplifier has 24 transistors, 11 resistors and a capacitor of about 30 pF. The large, irregular area is the capacitor. (Courtesy of D. K. Hamilton, Dept of Engineering Science, Oxford University)

Figure 2.5 Hybrid circuit (courtesy of Menvier Hybrids Ltd)

SAQ 2.1   (Objective 2.1)
In what ways will capacitors, resistors, inductors and transistors differ in the circuit technologies outlined above? Address each component in turn for each of the three technologies.

2.2 Insulators in steady and alternating fields

Making ever more compact circuitry demands good quality insulators, not only for substrates but also for intervening layers in multilayer circuits. Figure 1.13 denotes as insulators all materials having resistivity greater than 10^{10} Ω m, and shows polymers and ceramics well into this zone; hence the choice of these materials for PCB and hybrid substrates. But we can't just pick any of these material classes willy-nilly. When effective insulation is required from very thin sheets, compromises for the sake of processing, assembly or environmental factors in service may reduce an insulator's performance well below what the data books would have us believe. It is important to understand the materials science of electrical insulation in order to cope with these other factors.

It isn't just the resistivity of insulators which is important. That only tells us the 'resistance' to charge being transported through a material under modest electric field strength (or voltage gradient). Under high voltage-gradients, resulting from accidental or unavoidable overload, insulators may break down. This may happen, for example, in a motor as a commutator reverse current flow in the inductive winding; or when a component fails, and causes voltages elsewhere to go outside their specified values. We shall see that 'breakdown' mechanisms have very little to do with normal conduction.

Bear in mind that most electronic applications involve alternating currents. These may be at power-supply frequencies (50–60 Hz), audio frequencies (up to 20 kHz) and on up the spectrum through radio frequencies and digital data processing frequencies, up to the tens of gigahertz used in microwave circuits. Though very few of the atomic charged particles in insulators may actually be transported, all can *oscillate* in response to applied alternating electric fields. This gives rise to **dielectric** properties, which enable alternating current to flow and dissipate heat and so are relevant to the performance of insulators.

An archetypal a.c. device is a **capacitor**. It is two conductors held apart by a thin slice of insulator; the *amount* of capacitance C is determined by geometry and materials properties. (The fact that circuits essentially consist of conductors separated by insulators means we must expect to find capacitance everywhere.) With a potential difference V between the conductors, a capacitor holds charge Q given by $Q = CV$. If V varies (for example, when there is an alternating voltage), Q responds and current dQ/dt is passed.

In this section we look further at the internal mechanisms of insulation at atomic and microstructural scales to determine whether a material will make a good insulator. These ideas are developed later in this and subsequent chapters as we look for ways of controlling the conductivity of a range of nominally insulating materials.

2.2.1 DC response of insulators

In Chapter 1 we ascribed the slight conductivity of insulators to random thermally activated diffusion of ions under a biasing electric field. Thermal agitation continually causes ions to jump between sites and, in the absence of an electric field, these jumps will be randomly directed. But if there *is* an electric field, a bias will appear in the directions. There will be a greater probability that cations will jump towards the negative side of the applied voltage and that anions will jump the opposite way. This statistical preference emerges as a small electrical mobility. To model the conductivity by $\sigma_e = ne\mu$ we have to examine the populations of the things which might diffuse (n) and the factors regulating diffusion rates and hence the mobility (μ). There are two such factors. First a diffusing particle must gain sufficient energy to break free from its chemical bonds, and secondly it can only jump from its existing position if there is somewhere for it to go to.

The energy requirement is governed by Arrhenius's classic 'thermal activation' law. The fraction of ions which at any instant have enough energy E_a to jump is $\exp(-E_a/kT)$. Figure 2.6 is a sketch of this function. Since this fraction climbs with temperature we can expect conductivities to behave similarly. But, of course, ions may be held by bonds of various strengths in different materials so the fraction 'activated' will be dramatically different.

For the second factor we can imagine two extreme conditions and an intermediate one:

- lots of ions available to jump but few places for them to go;
- few ions present but no shortage of empty sites;
- plenty of ions and quite a lot of empty places.

Exercises 2.1 and 2.2 develop these thoughts and suggest what might be done to make the best insulators.

EXERCISE 2.1   Calculate $\exp(-E_a/kT)$ at room temperature and at 1000 K for activation energies of 0.1 eV and 1.0 eV. Boltzmann's constant $k = 86 \ \mu\text{eV K}^{-1}$. Comment.

EXERCISE 2.2   Classify the following insulators according to the three ionic conditions described just before Exercise 2.1.

(a) An amorphous polymer insulator.

(b) A crystalline oxide.

(c) Ordinary window glass.

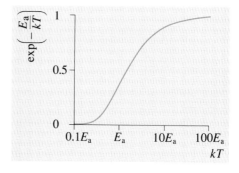

Figure 2.6 Graph of $\exp(-E_a/kT)$ against kT

For polymer insulators, since there is never any shortage of 'destinations' for ions the aim should be to reduce the number of 'travellers'. Many of the catalysts used to promote polymerization are ionic substances so the aim is to clean up the polymer thoroughly immediately after it has been made and before it is shaped. Alternatively, a non-ionic mode of polymerization could be used. Silicate glasses must have a high proportion of metal ions and a loose structure if they are to be workable at reasonable temperatures. How can this be arranged without provoking too much conductivity by diffusion of the positive metal ions (the cations)? ▼Making an insulating glass▲ tackles this question.

▼Making an insulating glass▲

The structure of amorphous silica is built of SiO_4 tetrahedra linked covalently at every corner through shared oxygen atoms, Figure 2.7(a). This glass has a very high softening temperature and practical glasses are made by introducing ionic metal oxides. The completely covalent bonding is modified with a high proportion of the tetrahedron corners becoming negative ions and with the cations lying in the irregular spaces of the network (Figure 2.7b). Clearly this structure is very 'defective' in the crystal sense and the gaps are large enough and so closely spaced that cations can diffuse rather more easily than in crystals. Common window glass, in which the added ions are Na^+ and Ca^{2+}, becomes a good conductor at slightly raised temperatures, say 300°C, because the alkali ions become quite mobile. To get high resistivity glass which still softens at reasonable working temperatures, three modifications make sense. First replace some of the network-forming silica by B_2O_3 or Al_2O_3. This reduces the number of covalent shared corners and so loosens the bonding. Such a glass would still be wholly covalently bonded so there should be no charge carriers. But there are no practical glasses in this composition without added ions. The second step is therefore to make those ions immobile. Double-charged ions, bonding more strongly than Na^+, need greater activation energy to break out of their sites so their diffusion is restricted. Thirdly, if we choose *big* double-charged ions, such as Ba^{2+} or Pb^{2+}, cation diffusion is further impeded because the surroundings have to distort more to let big ions pass. The room-temperature resistivity of lead borosilicate glass is a million times greater than that of soda-lime glass!

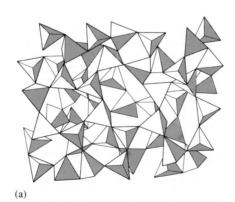

(a)

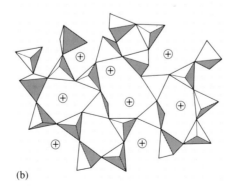

(b)

Figure 2.7 (a) Silica glass. (b) Effect of cations opening up glass structure

Crystalline oxides such as alumina are good insulators because their strong bonds make for high activation energies, both for ions to jump and for the formation of vacancies; see ▼Thermal defects in ionic crystals▲. Later, when we look at perovskites, we shall come upon less well bound oxides and run into compositional reasons for their much higher vacancy populations. Then we shall have to think again. In crystal oxides, in contrast to glasses, it is often the diffusion of the negative oxygen ions (anions) which gives conductivity. Incidentally, to apply $\sigma_e = ne\mu$ to oxide conduction it is customary to regard this conduction as due to the motion of a small number of vacancies with high mobility rather than to the motion of a large number of ions with small mobility. It makes no real difference because the ions can only move by filling vacancies, so the motion of an ion one way is exactly equivalent to the motion of a vacancy in the opposite direction. It is, however, then necessary to ascribe a charge and a sign to a vacancy: negative oxygen

▼Thermal defects in ionic crystals▲

At any temperature above absolute zero a crystal will contain vacant sites in its lattice. That's because the free energy G,

$$G = H - TS$$

can always be made lower by having some randomly distributed vacancies to provide increased entropy. The equilibrium is decided by the countering tendency of the enthalpy term H to rise as the presence of vacant sites weakens the bonding of the assembly. Simple analysis for a crystal made of N atoms of a single type (an element, in other words) gives the equilibrium population n of vacant sites as

$$n \approx N \exp\left(\frac{-E_a}{kT}\right)$$

with E_a the activation energy for vacancy formation. In an ionic oxide crystal, a further constraint is the demand that the crystal stays electrically neutral; that means that an O^{2-} vacancy will be accompanied by an electrically equivalent number of cation vacancies. Thus we may imagine two types of defect, depicted in Figure 2.8. The **Schottky defect** comprises stoichiometrically balanced empty anion and cation sites (there will be equal numbers in a substance of formula $M^{2+}O^{2-}$ shown in the diagram). **Frenkel defects** are those where an ion has shifted to an interstitial site. The populations of all these defects as temperature varies

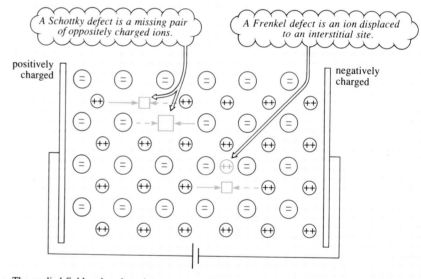

A Schottky defect is a missing pair of oppositely charged ions.

A Frenkel defect is an ion displaced to an interstitial site.

positively charged

negatively charged

⟶ The applied field makes these jumps more probable.

--→ The applied field makes these jumps less probable.

Figure 2.8 Schottky and Frenkel defects in ionic crystals

follow an Arrhenius law, just like simpler crystal vacancies.

For Schottky defects, the activation energy depends mainly on the strength of bonding in the crystal. So, for example, at a given temperature, there would be more Schottky defects in a weakly bonded ionic crystal of single-charged ions (such as NaCl) than in a stronger bonded assembly

of double-charged ions (such as MgO).

Activation of Frenkel defects also involves energy — to distort the lattice to accommodate ions on interstitial sites. Frenkel defects of displaced *cations* are therefore more common because these smaller ions can usually be accommodated with less distortion than big oxygen ions would cause.

ions diffusing through neutral vacancies are equivalent to positive vacancies moving among neutral oxygen ions. You may find this convention confusing, but much of the literature is written this way. It is wise to be forewarned. It does have the advantage of concentrating the mind on what has to be done to control conductivity, namely 'do something' about vacancy populations.

2.2.2 Dielectric breakdown

The **dielectric strength** of an insulating material is a statement of the greatest electric field intensity at which the forces restraining the charged particles can still 'hold on'. Typically it is expressed in kilovolts per millimetre (kV mm^{-1}). Above this threshold the material is a conductor, so dielectric strength sets practical limits to circuit voltages and dimension. A 500 kV transmission-line insulator may be 5 cm thick to get the voltage gradient safely within the dielectric strength: a capacitor with dielectric layers 5 µm thick charged to 50 volts is at similar risk.

The extreme model of dielectric breakdown is of the minute population of mobile carriers being so accelerated by the field that their collisions cause atoms to ionize. That creates more carriers, which accelerate, and . . . positive feedback does the rest! One practical insulator to approach this limit is silica grown as a thin film of high purity in integrated circuit manufacture. Its very high breakdown strength is a vital property for this technology. Such breakdown would affect the whole body of the material; but in bulk insulators, microstructural imperfections usually cause local breakdown at a small fraction of the intrinsic strength. It is still a positive-feedback effect — usually a thermal avalanche.

In ceramic insulators, porosity is the most common origin of breakdown. At sufficient field strength the gas in the pores ionizes and begins to carry current. This in itself is an avalanche effect because the ions, accelerating between collisions, may pick up enough kinetic energy to cause further ionization. Things are worse if the internal surfaces of pores are not smooth (Figure 2.9). Sharp features are sites of concentrated charge and intensified field which encourage early discharge. (This is 'corona discharge', the pointed-lightning-conductor effect — Figure 2.10.) Once the gas starts to conduct, current must enter and leave the pore and sharp points also localize the track of this current in the ceramic. Now, either by dielectric or resistive loss, the temperature will rise, perhaps quite steeply since the track is so tightly localized and thermal conduction is poor. At the raised temperature, conduction by ion diffusion locally increases, so more heat is generated and the resistance falls further. The conductive zone broadens as heat spreads away from the initial track and soon the insulation breaks down along a route joining several pores. Fortunately, the long time constants of the thermal parts of this process enable insulators to withstand voltage *pulses* much in excess of their continuous rating.

0   10 µm

Figure 2.9 SEM of rough interior of ceramic pore

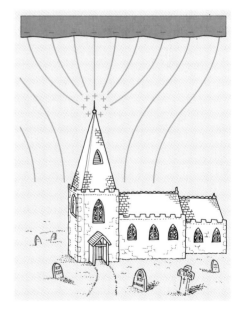

Figure 2.10 Sharp points concentrate field and charge. Thus a lightning conductor offers an easy discharge route for accumulated charge on a thunder cloud

With polymeric insulators the initial current track is often on the surface, perhaps because of salts from a sweaty fingerprint or flux residue. Again, local heating activates further conduction — resulting in further heating. Eventually some decomposition to carbon provides sufficient conductivity for the circuit to 'short'. Breakdown of high-voltage insulation of distributor caps in car engines, not the cleanest places, is not uncommon and is a particularly irritating example.

Such imponderables make dielectric breakdown difficult to predict, but at least they indicate the steps needed to avoid it: non-porous ceramics constituted of ions which do not mobilize very easily; and PCBs kept clean of ionics, protected from moisture and made of chemically stable polymers.

2.2.3 AC properties of insulators

Even when atomic-scale charged particles do not drift under the influence of an external field, they will still experience a force. Positive and negative constituents will experience opposite forces and will be displaced relative to one another. The material becomes **polarized**. During the instant of polarization, the moving charges constitute a transient current. So if the field switches its direction, so does the polarization and again a transient current flows.

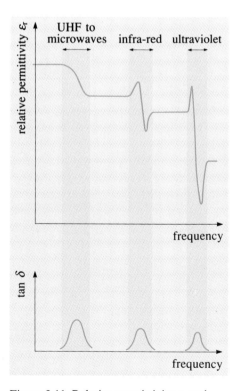

Polarization can be compared to the elastic response of a solid subject to mechanical stress. Just as a solid strains until the interatomic forces balance the applied load, so charges are displaced until the internal attractions balance the externally applied electrical force. Some materials give a big polarization for a small field; this is the electrical equivalent of a floppy material, which strains a lot under modest stress. There are also electrically 'stiffer' materials, in which charges are not displaced much when the field comes on. ▼Dielectric definitions▲ defines the relative permittivity or dielectric constant of a material, the technically useful description of these effects. It is not quite as simple as the stress–strain ratio which defines Young's modulus.

Table 2.1 lists a selection of these relative permittivities. What is the reason for this variety? Imagine some of the effects at the atomic scale. In every atom, the centre of the electron cloud will not be exactly at the nucleus. In ionic solids the positive ions will move slightly one way, the negative ions the other. Dipolar molecules will try to rotate into alignment with the field. The very high value for ferroelectrics in Table 2.1 needs further explanation; see ▼Perovskites▲, a class of mixed oxides which will occupy our attention later in this chapter and in Chapter 4. If the imposed field is alternating, the charged particles will attempt to oscillate in sympathy. How successfully they can do this will be frequency dependent. Electrons, with their low mass, can 'follow' much higher frequencies than heavier ions or nuclei. Polarizations arising

Figure 2.11 Relative permittivity ε_r and tan δ for a hypothetical material

Table 2.1 Typical dielectric properties of various categories of material at low frequency and room temperature

Material	Dominant charges displaced	ε_r	$\tan \delta$	$\dfrac{\sigma_e}{\Omega^{-1}\,m^{-1}}$
Vacuum	none	1	0	0
Gases (air)	not many	1	0	—
Hydrocarbons (polystyrene)	electrons	2	0.001	10^{-18}
Polar polymers (polyester)	polar groups on molecules	4	0.004	10^{-16}
Water (distilled)	small polar molecules rotate	80	large (≈ 0.1)	10^{-4}
Crystalline ceramics (alumina)	ions; small displacement	9	0.001	10^{-16}
Inorganic glasses	cations; displacements not so small	6 to 20	0.01	10^{-12}
Ferroelectric at Curie temperature ($BaTiO_3$)	ions; displacements large	10^4	0.02	10^{-12}
Metal (copper)	electrons; displacements very large	∞	∞	10^8

from molecular rotation are still slower in response. The relative permittivity will also therefore change with frequency and for a hypothetical material containing several displacement mechanisms may follow the upper curve on Figure 2.11.

It may be that the particles' oscillations are not perfectly elastic, and energy is lost to another form. This **dielectric loss** is a heating effect, to be distinguished from the joule heating of a direct-current flow. We can expect various kinds of oscillation loss. At low frequencies there are relaxation effects, where the orderly response of the charge to the field is opposed by thermal agitation. Polar molecules suffer such a loss as they rotate relative to their neighbours. At higher frequencies vibrating ions can couple energy out to the lattice. Then there are resonance effects. Electrons changing energy levels, for example, can also extract energy from the applied field. The loss curve on Figure 2.8 shows how relative permittivity and dielectric loss relate. The loss factor of a dielectric is universally expressed as the tangent of an angle! Such a weird practice needs explanation — see ▼ Tan δ ▲. It expresses the ratio

$$\frac{\text{energy lost per cycle}}{2\pi \times \text{maximum energy stored}}$$

The smaller the number the less the oscillating charges dissipate energy as heat. Table 2.1 includes this parameter too.

SAQ 2.2 (Objective 2.2)
Consider whether sea water should be classed as a dielectric material or a conductor.

SAQ 2.3 (Objective 2.3)
What is it about a material which gives it a high dielectric constant?

▼Dielectric definitions▲

Figure 2.12(a) represents a parallel plate capacitor connected to a battery which maintains the metal plates at potential difference V (volts). Suppose the space between the plates is vacuum. To establish the potential difference some electrons have moved from one metal plate to the positive end of the battery and some have moved from the negative battery terminal to the other plate. The attraction across the gap of the opposite charges exactly balances the repulsion which so many like charges experience by being crowded onto each plate. Evidently if plates of larger area were used, more charge could be pushed on by the same voltage. Similarly, if the plates were closer, the attraction would be stronger and again the device would hold more charge for a given potential difference. As you will know, the capacitance of this 'charge-holding' bucket is defined by the charge per unit potential difference, and can be further expressed as a combination of geometry (A and d on Figure 2.12a) with a 'material factor' for the vacuum between the plates:

$$C = \frac{Q}{V} = \frac{\varepsilon_0 A}{d}$$

Here ε_0 can most easily be thought of as a dimension-changing constant, making coulombs per volt (that is farads) on the left of the defining equation compatible with the dimensions (metres) of the fraction A/d. So ε_0 has units of farads per metre, and the experimentally determined value of 8.85×10^{-12} F m^{-1}. However the two equations for C turn sideways to give an expression for the electric field strength $\mathscr{E}$, which is the voltage gradient in the gap:

$$\mathscr{E} = \frac{V}{d} = \frac{Q}{\varepsilon_0 A}$$

or in other words the surface charge density, $\jmath = Q/A$, on the plates relates to the field between them by

$$\jmath = \varepsilon_0 \mathscr{E}$$

Now, we have already seen that one way of increasing the surface density of the charges is to move the plates closer together. Is another way to change the medium in the gap? If instead of vacuum the space contains matter, the applied electric field will polarize it by displacing positive and negative particles in opposite directions as we have seen. Perversely the electrical analogue of Young's modulus, the **dielectric susceptibility** χ, of the material is defined the other way up:

$$\chi = \frac{\text{electric strain}}{\text{electric stress}} = \frac{\mathscr{P}}{\varepsilon_0 \mathscr{E}}$$

where $\mathscr{P}$ is the **polarization** of the medium and $\mathscr{E}$ the applied electric field. How might $\mathscr{P}$ be defined? See Figure 2.12(b). If we think of electrons and nuclei being displaced oppositely by the field, atoms which were electrically symmetrical have become **dipoles** with the nuclear charge $+Ze$ and the electron charge $-Ze$ displaced some little distance x from each other (Z here stands for atomic number). The product Zex is called the **dipole moment** of the atom and the effect of the field has been to produce some dipole moment per unit volume in the material. This is how the polarization is defined. With dipole moment having units of coulomb metres, $\mathscr{P}$ is coulomb metres per cubic metre or, cancelling down, coulombs per square metre. So $\mathscr{P}$ is also defined by the surface density of charge which appears on the faces of the specimen placed in the field $\mathscr{E}$. Since $\varepsilon_0 \mathscr{E}$ has already defined the surface-charge density of the vacuum-filled plates, the dielectric susceptibility becomes the dimensionless ratio of two surface-charge densities.

The effect of introducing the dielectric material into the capacitor can now be seen. The polarization charges on the surface of the dielectric attract further charges from the battery (Figure 2.12c). The charge Q on the plates is larger for the same V, resulting in a bigger value of C. So the capacitance is:

$$\begin{aligned} C = \frac{Q}{V} &= \frac{A(\jmath + \mathscr{P})}{\mathscr{E} d} \\ &= \frac{\varepsilon_0 \mathscr{E} A(1 + \chi)}{\mathscr{E} d} \\ &= \frac{\varepsilon_0 \varepsilon_r A}{d} \end{aligned}$$

defining the **relative permittivity** (otherwise, loosely, called the **dielectric constant**) of the material as $\varepsilon_r = (1 + \chi)$ — again a dimensionless number. It is useful sometimes to use the **permittivity** ε, defined by $\varepsilon = \varepsilon_0 \varepsilon_r$ and thus having units of farads

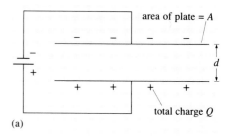

area of plate = A

d

total charge Q

(a)

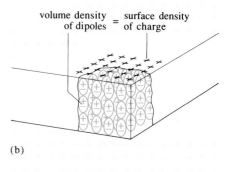

volume density of dipoles $=$ surface density of charge

(b)

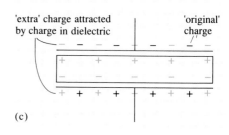

'extra' charge attracted by charge in dielectric

'original' charge

(c)

Figure 2.12 (a) Capacitor connected to battery. (b) Polarization in dielectric. (c) Polarized dielectric in capacitor, with dipoles giving surface charge and extra charges attracted on to plates

per metre, rather than the dimensionless relative permittivity. The constant ε_0 is called the permittivity of vacuum so ε_r is comparing other materials 'relative' to vacuum. More conveniently, the relative permittivity is thought of as the multiplier by which the capacitance of a given plate geometry in vacuum can be increased by putting a polarizable medium between the plates. Obviously to make compact capacitors, a prime requirement is to find some very easily polarizable materials which, in consequence, have large values of ε_r.

▼Tan δ▲

To understand the significance of tan δ as a description of dielectric loss it is necessary to conceive an equivalent circuit for a real capacitor. If the charge oscillations dissipate energy (that is to say in jargon that they are 'lossy') we can picture a resistor R_{ac} in parallel with a 'perfect' capacitor C and imagine a current through it to be the seat of the power loss (Figure 2.13). (There will also be some slight charge transport as well as charge oscillation, so a real capacitor should be imagined with another high-value parallel resistor, R_{dc}. Suppose for this analysis that the dielectric is a perfect insulator against d.c. so R_{dc} is infinite and can be ignored.) That leaves a parallel RC combination to be discussed.

SAQ 2.4   (Revision of a.c. theory)
The defining equation for capacitance relates the voltage across the dielectric to charge it carries. Instantaneously, the rate of change of charge (current) is:

$$\frac{dq}{dt} = C\frac{dv}{dt}$$

Show that with an alternating applied voltage $v = V \sin \omega t$ the instantaneous current can be written as

$$i_C = \frac{V}{1/\omega C} \sin\left(\omega t + \frac{\pi}{2}\right)$$

(The $\pi/2$ term is the radian measure for a quarter-cycle, or 90°.) Hence express the 'reactance' of a perfect capacitor (that is, its equivalent to a resistor's 'R' for relating current and voltage) and state the phase relationship between voltage and current in a capacitor.

The currents in the RC combination are displayed on a phasor diagram in Figure 2.14. The resultant current is not at 90 to

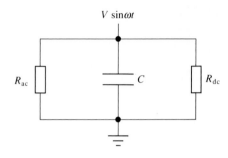

Figure 2.13 Perfect capacitor C in parallel with R_{ac} and R_{dc}

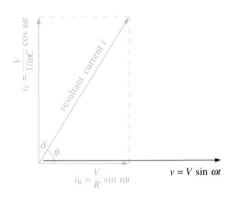

Figure 2.14 Phasor diagram for resistor R and capacitor C in parallel

the voltage but is displaced by a small angle δ given by the ratio of the magnitudes of the currents:

$$\tan\delta = \frac{V/R}{V\omega C} = \frac{1}{\omega CR}$$

Now I must show that this tan δ relates in a defined way to the losses in the real capacitor modelled by the a.c. thought to be bypassing C via R_{ac}.

The energy held by the perfect capacitor builds from zero, when $v = 0$ and at any

instant is

$$\text{Energy} = \int_0^\tau vi_C \, dt$$

$$= \int_0^\tau V^2\omega C \sin\omega t \cos\omega t \, dt$$

$$= \int_0^\tau \frac{V^2\omega C}{2} \sin 2\omega t \, dt$$

$$= -\frac{V^2\omega C}{4\omega}[\cos 2\omega t]_0^\tau$$

$$= -\tfrac{1}{4} CV^2[\cos 2\omega\tau - 1]$$

This reaches its maximum value of $\tfrac{1}{2} CV^2$ when the cosine function is -1. Turning attention to R_{ac}, the instantaneous power dissipation is vi_R so integrating gives:

Loss of energy per cycle

$$= \frac{V^2}{R_{ac}} \int_0^{2\pi/\omega} \sin\omega t \sin\omega t \, dt$$

$$= \frac{V^2}{\omega R_{ac}} \int_0^{2\pi} \tfrac{1}{2}(1 - \cos 2\omega t) \, d(\omega t)$$

$$= \frac{V^2}{\omega R_{ac}} [\tfrac{1}{2}(\omega t - \tfrac{1}{2}\sin 2\omega t)]_0^{2\pi}$$

$$= \frac{V^2}{\omega R_{ac}} [\tfrac{1}{2}(2\pi - 0 - 0 + 0)]$$

$$= \frac{V^2\pi}{\omega R_{ac}}$$

We can now set down the ratio

$$\frac{\text{energy lost per cycle}}{2\pi \times \text{maximum energy stored}}$$

and discover, amazingly, that it is

$$\frac{V^2\pi}{\omega R_{ac}} \div \frac{2\pi CV^2}{2} = \frac{1}{\omega CR_{ac}}$$

$$= \tan\delta$$

▼Perovskites▲

The extraordinary dielectric constant of barium titanate makes it interesting as a material for capacitor dielectrics. Its high value stems from its perovskite crystal structure, shown in Figure 2.15. You can see it comprises Ti^{4+} ions contained in octahedral 'cages' of O^{2-} ions joined corner-to-corner. That makes a large cuboctahedral hole (in the middle of the cube pictured) where the Ba^{2+} ion sits. How well do the cations fit their sites? That is the key to understanding the odd properties of $BaTiO_3$ and other perovskite mixed oxides we shall meet later in this book. The rule for building oxygen-ion cages is that the oxygen ions may not touch each other (like charges repel) so there is a minimum-size cation for a given cage shape.

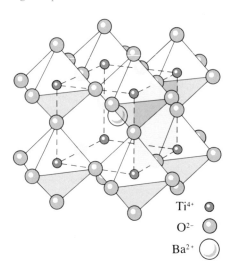

Ti⁴⁺ ◐
O²⁻ ◯
Ba²⁺ ◯

Figure 2.15 The perovskite structure is built of corner-shared octahedral cages

EXERCISE 2.3 Oxygen ions have radius $R_O = 0.14$ nm. What is the radius $R_{cat.}$ of the smallest-sized cation which can be held in an octahedral cage? Figure 2.16(a and b) will help.

Ti^{4+}, with radius 0.068 nm, will do nicely to form octahedral cages with the oxygen ions held apart. This will define an O–O edge length x of the cages in Figure 2.16(c). From this we can discover the width of the space available in the cuboctahedral cages using the geometry of Figure 2.17.

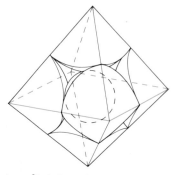

(a) An octahedral cage is built from 6 anions

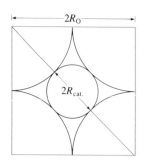

(b) Geometry for Exercise 2.3

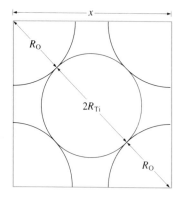

(c) Geometry for Exercise 2.4. Pythagoras yields $x = (R_O + R_{Ti})\sqrt{2}$

Figure 2.16

EXERCISE 2.4 Use Figure 2.16(c) to calculate the O–O edge length. The radius of the oxygen ion R_O is 0.14 nm; the radius of the titanium ion R_{Ti} is 0.068 nm. Use Figure 2.17 to evaluate the radius of an ion that is a tight fit in the cuboctahedron.

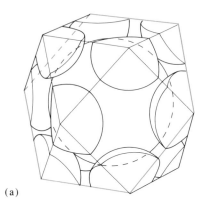

(a)

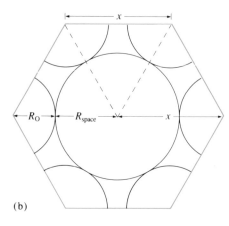

(b)

Figure 2.17 (a) Cuboctahedral cages are built from 12 oxygen ions at the centres of the cube edges. The green hexagon bisects the cage. (b) Because the hexagon can be made from equilateral triangles, its centre-to-point dimension is x. The radius of the available space is therefore $x - R_O$

Thus Ba^{2+} with a radius of 0.135 nm will be a sloppy fit. You can think of the barium ion 'pretending' to fill this cage at high temperature by rattling around in it and the symmetrical cubic structure is maintained. But when the temperature falls it won't be a surprise that the crystal becomes unstable. At a critical temperature of 125°C the octahedral cages distort and the positive ions move to off-centre positions. The crystal takes up a tetragonal form (cube stretched parallel to an edge). The positive and negative charges have different centres of symmetry so the whole crystal is polarized. (Because of this phenomenon the material is described as **ferroelectric**, that is it spontaneously becomes an electric dipole in much the same style as a ferromagnetic material becomes a magnetic dipole below its Curie temperature.) You will probably also appreciate that at a temperature slightly above the critical 125°C the crystal can be helped into its new form by an electric field, since this will tend to displace the ions. Even slightly below 125°C a field can urge the cations to move to the opposite ends of their new elongated cages. Consequently the polarization of a barium titanate crystal is very sensitive to applied field around 125°C. Compared with other field-induced displacements of charge, this sensitivity is very large, so the dielectric constant is extraordinarily high, typically several thousand. Figure 2.18 shows this

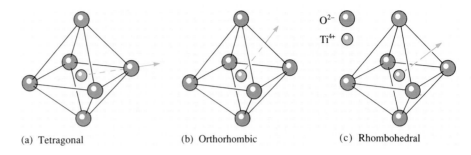

(a) Tetragonal   (b) Orthorhombic   (c) Rhombohedral

O^{2-}
Ti^{4+}

Figure 2.19 Displacement directions of titanium ions for the phase transitions of $BaTiO_3$

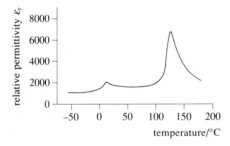

Figure 2.18 Dielectric constant of $BaTiO_3$ peaks at phase transition temperatures

and reveals similar but smaller peaks at lower temperatures which are the critical temperatures for further crystallographic phase changes, illustrated by Figure 2.19. The chemical origin of these distortions stems from titanium being a 3d transition element. You will remember that this means it has d-orbitals for electrons to form covalent bonds to its neighbours. The exact symmetry of non-directional ionic bonding is upset when these directional bonds come into action.

Several other combinations of oxides make technically useful perovskites. In the general formula ABO_3, where A and B are metals, the positive ions must have a total charge of +6; A and B must be of quite different sizes; and the smaller, higher-charged ion must be a transition metal. It is not essential, though, for all the ions represented by one letter to be the same. For example, when we look at ceramic dielectrics we shall explore substitutions for parts of the Ba^{2+} and for Ti^{4+} in barium titanate. Nor do the charges have to be +2 and +4. The combination +1 and +5 will do, and potassium niobate $KNbO_3$ exemplifies this. The ion radii are $K^+ = 0.133$ nm, $Nb^{5+} = 0.070$ nm so you see the sizes are very like those in $BaTiO_3$. Niobium is a member of the 4d transition series.

As ionic compounds, perovskites should display high d.c. resistivity. But they are mixed oxides and that makes it necessary to be very careful with their stoichiometry if they are to be good insulators. To make $BaTiO_3$, equal molar quantities of BaO and TiO_2 are mixed together as powders and fired to a high temperature. They react to form the new compound. If the proportions are not exactly 1:1 the mixture will still crystallize in the perovskite

structure but there will be some vacant sites, both cation and anion vacancies, which can support diffusion and hence ionic conduction. Even an error of 0.1% would create far more vacancies than the thermodynamic-equilibrium amount, so expectations of good insulation may be frustrated. Because the oxygen ions are nearer neighbours than the cations (see Figure 2.15 again) the activation energy for oxygen-ion diffusion is lower and conduction usually results from drift of the oxygen-ion vacancies. Another source of oxygen vacancies in technical quality perovskites is pollution by alumina from the balls used to grind the oxide powders. Since Al^{3+} is a small ion, it will fit the octahedral sites. But Al_2O_3, substituting for TiO_2, brings only $1\frac{1}{2}$ oxygen ions per cation, so the resulting crystal of perovskite structure will have several vacant oxygen sites. To compensate for these problems, a small amount of niobia Nb_2O_5 can be added. Niobium ions Nb^{5+}, being so close to the size of the titanium ions, are able to substitute onto the octahedral sites. But they bring $2\frac{1}{2}$ oxygen ions for every cation. Thus a small addition of niobia provides extra oxygen to fill up vacancies caused by inaccuracy of composition. You will meet further examples of such corrective measures later in this chapter and in Chapter 4.

2.3 Printed circuit boards

2.3.1 Choosing a board material

The circuit board must be an insulator, so polymers or polymer composite materials are appropriate. But a successful PCB substrate material needs many other virtues, chemical, thermal, mechanical and, as just discussed, dielectrical. Here is a list:

- Conducting tracks must adhere to it, but solder should not.
- It must not char during soldering.
- It must be proof against the chemicals used in processing.
- It should not absorb water (would spoil the insulation).
- It should be immune to damage by ultraviolet light.
- It should not suffer thermal damage such as distortion during soldering or in service.
- It should have appropriate thermal expansion characteristics. This could be important in determining how the assembly responds to stresses caused by temperature changes or for ensuring the components stay in place during soldering.
- Boards should be mechanically stiff and strong, even though quite thin, and they must be manufacturable as flat sheets, sometimes as large as 30 cm × 20 cm.
- The board material should be hard enough to be machined cleanly: tiny holes have to be drilled or punched without burring, smearing or clogging the drills.
- Low relative permittivity is preferred; see Exercise 2.6.
- Dielectric strengths of several kV/mm are specified. Even in low-voltage circuits, as conductors are brought ever closer together the dielectric limit might be approached. The specification usually imposes a safety margin against high-voltage surges in the circuit, such as might be produced by current interruption in inductive components. Apart from the need to withstand the surge itself, if the insulator does break down momentarily its subsequent quality may be seriously impaired.
- It must be cheap.

You probably know enough about polymers to go a long way towards a wise selection for this property suite.

EXERCISE 2.5   Here is a list of polymers that might be suitable for circuit boards.
LDPE; Nylon 6,6; PMMA; PTFE; PES (polyethersulphone); polyester resin; phenol-formaldehyde resin; epoxy resin.

Below (and on facing page) is a list of shortcomings. For each polymer above, select the shortcomings, if any, that could affect their selection for the application we have in mind.

(a) Thermoplastic with T_g too low for service.
(b) Thermoplastic with T_g too low to resist soldering temperature.

(c) Insulation and dimensional stability may be spoilt by water absorption.
(d) Risk of degradation by UV exposure.
(e) Would depolymerize at soldering temperatures.
(f) Difficult forming processes.
(g) Expensive.
(h) Poorly cross-linked thermoset liable to soften at high temperatures.
(i) Brittle.

Thermosetting resins offer the obvious solution to the problem of high solder temperature, provided they are adequately cross-linked. Phenol-formaldehyde resin and epoxy resin are satisfactory in all respects except for their brittleness. This is overcome by reinforcing them. 'Paxolin', paper reinforced phenolic resin, is cheap and finds wide application in domestic electrical goods. Glass fabric reinforced epoxy resin is perfect! It accounts for the majority of circuit boards for quality electronics. Both are made by impregnating the reinforcement with resin which is polymerized under pressure between hot, polished, steel plattens. Buried conductors can be included at this stage.

PES alone among the thermoplastics listed is suitable. It is one example of the modern thermoplastics which are stable to high temperature. These are hailed as new options for PCBs. Though their cost may be a drawback, they will permit three-dimensional circuit shapes to be injection moulded complete with component mounting holes and such 'furniture' as battery clips, so the final cost of the assembly may actually be less.

Dielectric properties influence circuit design at high frequencies. Conductors running close together form stray capacitors and signals may leak from one line to another if the capacitive reactance $1/(2\pi f C)$ between them is low. Clearly stray capacitances must be kept low. Having insulators with low dielectric constant certainly helps, but it's more important to keep critical conductors well apart. This can be a problem for a designer trying to achieve a compact circuit. Exercise 2.6 illustrates what can go wrong.

EXERCISE 2.6   A 200 μm wide conductor runs for 25 mm at 0.25 mm above an earth plane. The separating insulator has relative permittivity of 4.0. By modelling the arrangement as a parallel plate capacitor, plot a line on the axes provided as Figure 2.20 relating the leakage reactance to frequency. If the conductor carries signal to an amplifier of input impedance 10 MΩ, at about what frequency would the amplifier appear to be short circuited?

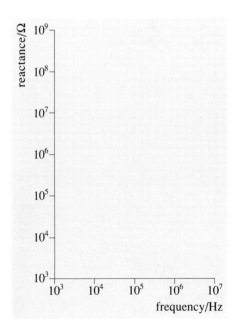

Figure 2.20

2.3.2 Laying the conductors

Laying down the conductive tracks of metal onto the substrate calls for some subtle materials science. Often the whole board is coated with metal and then unwanted parts are etched away. This is the so-called subtractive method. First the surface has to be thinly metallized by a chemical deposition. Palladium chloride absorbed into the surface catalyses the precipitation of nickel from a solution. That gives just enough conductivity to allow electroplating, and copper is laid to the standard thickness, $\approx 100 \ \mu m$. Copper-clad board is the usual input to the circuit assemblers.

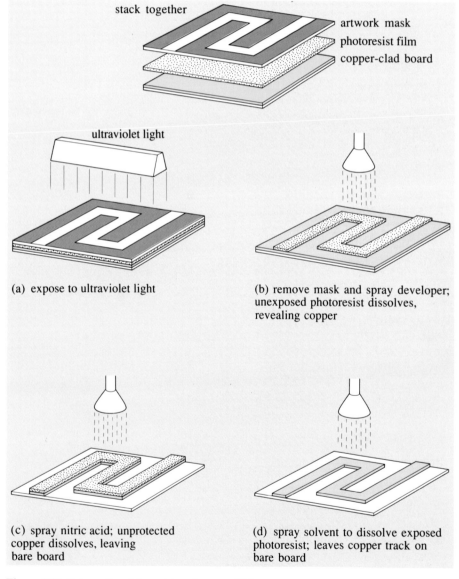

(a) expose to ultraviolet light

(b) remove mask and spray developer; unexposed photoresist dissolves, revealing copper

(c) spray nitric acid; unprotected copper dissolves, leaving bare board

(d) spray solvent to dissolve exposed photoresist; leaves copper track on bare board

Figure 2.21 Schematic of the subtractive metallizing process sequence

Cleanliness comes close to Godliness in circuit building so the processes are carried out in machines where air and treatment chemicals can be kept clean. Process times are decided by the rate at which boards pass through the machines. Figure 2.21 shows the sequence. The pattern of conductors and insulators for a circuit will have been drawn up as artwork (by a computer of course) and is brought to the production line as a 'mask', printed black at exact size on a clear plastic sheet. Let's have the insulating areas black. The copper-clad board is degreased and its metal oxide etched or scoured. It passes to a chamber where a layer of photoresist film is rolled onto the copper (see ▼ Chemistry of photoresists ▲). This happens in a vacuum so that air bubbles are not trapped. The mask is laid on top and a strong dose of ultraviolet light delivered. Where the mask transmits the radiation the photoresist polymer becomes insoluble in the 'developer'. But the unexposed areas, under the black of the mask, can be dissolved. The metal which we do not want is then dissolved in nitric acid. Yet another solvent then removes the polymer from the remaining conductors. Finally the conductors are tinned by a wave of solder so they can be stored without the copper tarnishing. 'Know-how' is the crux of success: correct ultraviolet exposure, the right solvent mixtures for the right time at the optimum temperature, adequate washing of acid and solder flux residues and cleanliness from dust are all vital. Notice also that the board polymer has to resist to chemical attack by a wide variety of chemicals used in the process.

SAQ 2.5   (Objective 2.4)
The subtractive process just described used a 'negative' photoresist (its solubility was *decreased* by the ultraviolet light) and copper clad board. Work out a sequence of operations for an 'additive' process which deposits copper only where it is needed on the substrate and which uses a 'positive' photoresist. (A 'positive' photoresist is one whose solubility *increases* under exposure to ultraviolet light.) What savings in materials can you see?

▼ Chemistry of photoresists ▲

Photoresists are polymer systems which change their solubility in specific solvents after exposure to ultraviolet light. There are two types. 'Positive' photoresists show *increased* solubility after ultraviolet irradiation; 'negative' photoresists have *decreased* solubility after exposure. In use, a film of photoresist is masked by a black and clear pattern so that selected areas of the circuit board are exposed to ultraviolet while others are screened from the light. Then the pattern is 'developed' by a solvent. If a positive photoresist has been used, the solvent strips the areas which were lit and leaves the screened areas alone. With negative resist the screened areas are dissolved and the exposed areas remain intact. The chemical principle in all cases is to provide a low-molecular-mass polymer and a photosensitive cross-linking agent. Where the polymer cross-links to become high-molecular-mass, its solubility is reduced. Evidently in negative resists the ultraviolet light provokes cross-linking while in the positive type it inhibits cross-linking. There are many formulations in each class but the chemical discussions which follow explain the action of nearly all of them. The first is a negative resist which uses an organic solvent developer.

Negative photoresists are mostly based on polyisoprene,

$$(-CH_2-C.CH_3 = CH-CH_2-)_n$$

This polymer can be further polymerized.

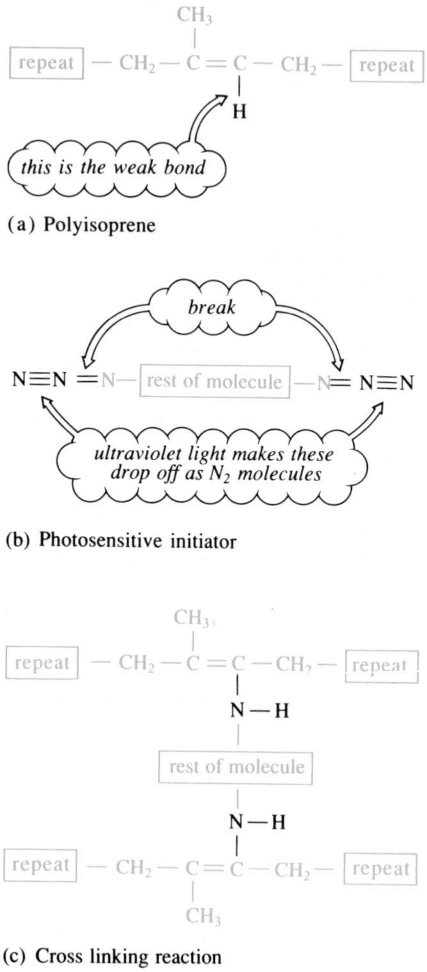

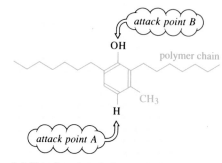

(a) Polyisoprene

break

N≡N ═N— rest of molecule —N═ N≡N

ultraviolet light makes these drop off as N₂ molecules

(b) Photosensitive initiator

(c) Cross linking reaction

Figure 2.22 Chemistry of cross-linking PIP

An initiator which cracks to give two free radicals on each of its molecules when irradiated by ultraviolet light acts to bridge between the polyisoprene molecules. Figure 2.22 shows the scheme. The molecular mass climbs from ≈ 60 000 to a few million and the film hardens and becomes insoluble in the developer which will dissolve the unreacted material. For your delectation you may like to know that the photosensitive ingredient is named 2,6-bis(para-azidobenzilidene)-4-methyl-cyclohexanone!

The second example is particularly versatile because it can exhibit both kinds of behaviour. A low-molecular-mass phenolic resin, the essential part of which is shown in Figure 2.23(a), is used in

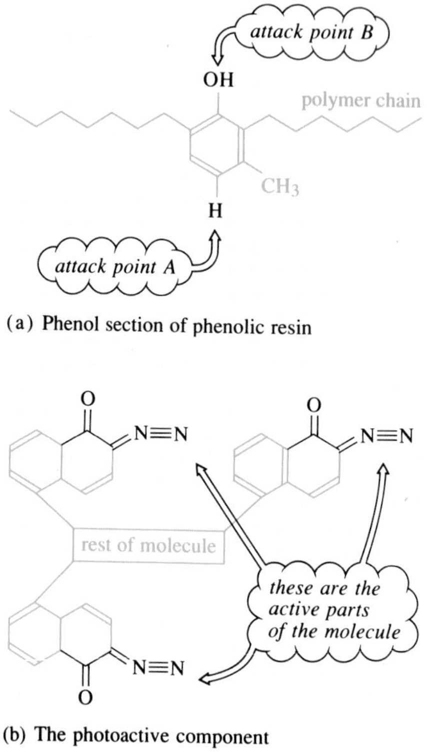

attack point B

OH

polymer chain

CH₃

H

attack point A

(a) Phenol section of phenolic resin

rest of molecule

these are the active parts of the molecule

(b) The photoactive component

Figure 2.23 Constituents of phenolic photoresist

conjunction with the chemical depicted in Figure 2.23(b). I won't attempt to name that stuff. Let's just call it PAC for 'photo-active component'. The developer is a 1% aqueous solution of sodium carbonate which adopts a pH of 11.8. Referring to Figure 2.23(b), we are only interested in the highlighted parts of the molecule. They are all the same, but it is important that there are three of them. One way or

another these are going to be the sites responsible for cross-linking the resin, so with three per molecule a solid network can be built.

We have three conditions to describe:

- the reaction of the unexposed mixture with the alkaline developer,
- what happens under ultraviolet light when moisture can get at the film,
- reactions after ultraviolet exposure under vacuum.

Let's take them in that order. In the first case the alkali causes a reaction between the ═N≡N group and the rings of the resin molecules (Figure 2.24). This cross-links the resin and renders it *in*soluble. We are therefore looking for a way of making the exposed parts soluble in alkali.

The action of ultraviolet light on the 'interesting' part of our PAC is quite dramatic (Figure 2.25a). In the presence of water (and the humidity of air is sufficient, as the material is a thin film with a large surface-to-volume ratio) the immediate photolytic derivative quickly reacts to form an acid (Figure 2.25b). Although this might cross-link the phenolic rings by attacking their —OH groups, this reaction is so slow that it can be dismissed. The acid dissolves in the alkali developer and so does the uncross-linked polymer: we have the result we wanted.

But in the absence of water, in vacuum, the photolytic derivative *does* react with the phenol —OH groups (Figure 2.25c), and successfully cross-links the polymer. Thus, through controlling the environment of the irradiation chamber, we can dictate whether the material shall be a positive or a negative resist. Let us think through the process sequences for the two actions.

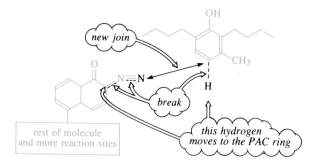

Figure 2.24 Action of alkaline developer on unexposed PAC. Alkali developer attacks point A of the resin. Then the resin reacts with unexposed PAC to cross-link the resin molecules

To use it as a positive resist, the areas to be dissolved are irradiated in air so that the acid derivative is formed. When developed in alkali, the irradiated areas dissolve while the non-exposed areas cross-link and remain intact. For negative-resist action, we keep the areas to be dissolved dark and expose under vacuum. The illuminated areas cross-link. Then admit air, remove the mask and re-expose the whole board. Those areas already cross-linked are not further affected, but the remainder undergoes the acid reaction and can then be dissolved off. This is the route used to make a solder resist which can be left on the board for environmental protection. Also, the resulting layer has good insulating and breakdown properties so this process is the key to building multilayer circuitry.

In Chapter 5 we shall find a need for photoresists in solution, but for PCB work a film is required. It is presented as a sandwich between protective films of polyethylene and 'Mylar' (a polyester) providing strength. The applicator machine strips off the polyethylene coat just before the board arrives. The Mylar film is left in place until the developer needs access to the resist.

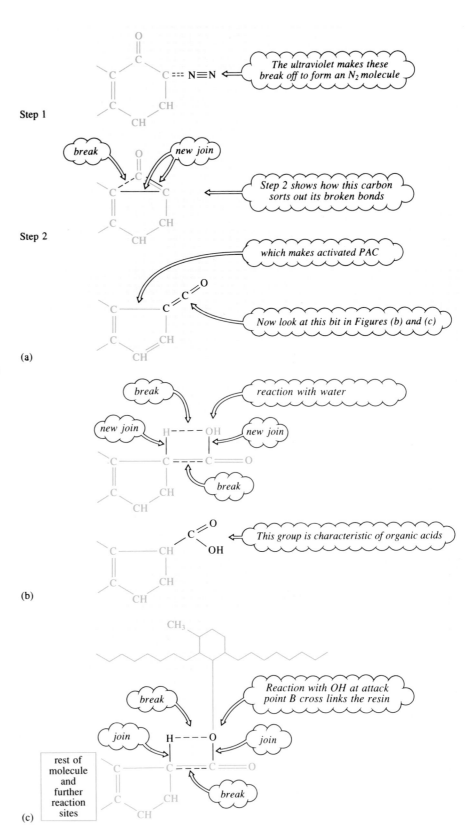

Figure 2.25 (a) (top) Ultraviolet light reorganizes the active parts of PAC.

Figure 2.25 (b) If water (H–OH) is around, the new end reacts quickly to make an acid which dissolves in the alkaline developer — preventing cross-linking of the resin

Figure 2.25 (c) In a vacuum (no H₂O) the activated PAC slowly reacts with site B of the resin molecules to cross-link them. The result is insoluble in developer

61

2.3.3 Putting the bits on

For placing and soldering devices onto the board, there is a choice between the older **through hole technology**, THT, and the newer **surface mount technology**, SMT.

In the older system, devices had wire connecting legs which passed through holes in the board and were soldered to the conductors on the back (Figure 2.2). Gravity holds all the bits in place on top of the board, while the solder is applied as a wave to the underside. Despite the simplicity of the method, there are problems:

(a)

- Hundreds of holes to be accurately drilled or punched requiring expensive machinery. Setting up the machinery for each job is costly.
- Devices are confined to one side of the board only.
- Buried layers in multilayer boards must be steered round the line of every hole unless a connection is wanted.
- Bent wires can miss the holes they should go through. This can cause assembly problems with many-terminal devices (such as 'chips').
- Each hole needs a ring of conductor for soldering, so the spacing limit between connections is $\approx$ 2 mm.

Surface mount technology avoids these problems. No holes are drilled, the solder 'pads' are solid and devices have either block ends or short, stubby J or L legs. You can see all this on Figure 2.26. Connections can be closer, allowing smaller chip carriers, and block-terminated devices are smaller too. Indeed, so small and tight is the spacing that the relative expansion between device and board could become a problem.

(b)

SMT has not only invited a whole new suite of device designs, it has called for a new approach to soldering. The bits have to be 'stuck' in place pending soldering and 'solder pastes' combining glue, flux and solder have been developed to do this. In automated assembly, which suits large-volume production of a single circuit design, automatic pneumatic dispensers put a measured speck of paste on each solder pad and the placement robot puts on the device. (Hand tools are available for short production runs where it is not worth programming the robots.) Although it is possible to turn the assembly device-side-down and pass it through a wave-solder machine, that is a crude and risky way of heating the solder paste. Instead, vapour phase heating can be employed, using a liquid with an appropriate boiling temperature. The fixed boiling temperature allows tight control of the process. Alternatively, a more sophisticated method is to use infra-red ovens giving nicely controlled rising temperature with time. This allows the organic glue to be driven off gently before the solder starts to flow.

(c)

The advantages of SMT outweigh its disadvantages, which will almost certainly diminish as the technology becomes more familiar and more widely used. By 1990, about a third of PCBs were being made in SMT, a market share which had grown from nothing a decade earlier.

Soldering is the most fault-prone step in circuit building. Basically there are two kinds of mishap: the solder fails to make electrical contact where

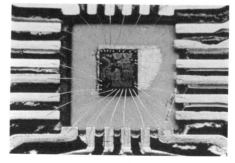

(d)

Figure 2.26 Terminal designs for SMT. (a) and (b) Legs. (c) Block ends. (d) Chip carrier

it should; or the solder makes electrical contact where it should not.

The first can happen if:
- The solder does not melt or misses the point of application.
- Surfaces to be joined are greasy or tarnished and do not 'wet'.
- A joint cracks during cooling owing to relative movement of the parts.

The second can happen if:
- Too much solder has been applied.
- A solder-wettable surface extends beyond the intended joint area.
- Movement of the circuit relative to the solder drags a 'wick' of solder from the join onto an adjacent site.

The three items in each category correspond interestingly: the first pair are processing-machine mistakes. The second are materials problems. The third items are due to movement. These similarities prompt similar solutions. The first pair are matters of machine setting and maintenance, though note that solder pastes for SMT need to have suitable flow and drying properties not to clog the fine nozzles through which they are applied. The second and third pair however can be looked at from a materials viewpoint.

We could:
- Cover all parts of the circuit except where we want solder with a 'solder resist'.
- Strip grease and tarnish from solder points with a 'flux' incorporated with the solder.
- Formulate solders so that they set suddenly, rather than via a pasty condition that persists over for a range of temperatures during cooling.

Solder resist is now usually applied as a photoresist film covering the entire circuit. The areas to be soldered are dissolved off, but the cross-linked polymer remains as a protective cover for the rest of the circuit during soldering and in service. The phenolic type discussed in 'Chemistry of photoresists' is favoured for this application. In the hardened form the polymer can withstand soldering temperatures. Also it has excellent electrical insulating properties and gives protection from water and abrasion to the circuit in service. The method of applying the film gives tight and hole-free coverage of the conductors; and the mask-and-dissolve technique permits accurate registration, so that the right parts are covered or revealed. Film resists are considered much better in these several respects than earlier screen printed types. The fundamentals of flux and solder formulations to solve the other problems are discussed in ▼Fluxes▲ and ▼Solders▲

The production technology which we have reviewed here grew to maturity in a mere 25 years or so. The industry is fiercely competitive, and has invested heavily in research and development, spurred by the goals of ever greater compaction, productivity and reliability. So although our main concern is with the materials, you should not overlook the industrial context of their development and use.

▼Fluxes▲

The twin roles of a solder flux are to strip tarnish from the metal to be soldered and to protect it from hot oxidation during soldering. Typically fluxes are based on resin formulations including some organic acids. These acids, although only weakly reactive compared with mineral acids, react fast enough with metal oxide tarnish at the high temperature of soldering. High-molecular-mass types will melt and cover the soldering site for a short while before burning off. If a stronger action is needed an ionic chloride, such as $ZnCl_2$, is included. The reaction

$$\boxed{organic}COOH + \boxed{metal}Cl$$
$$\rightarrow \boxed{organic}COO\boxed{metal} + HCl$$

provides a stronger acid to attack the tarnish. Now, however, there will be ionic residues on the surface which may cause short circuits so these need to be cleaned off.

▼Solders▲

Soldering uses metal as a 'melt-and-freeze' adhesive. As a liquid, the solder can take up any shape and wet the surfaces it has to join; when it freezes it has fixed the parts together. To a great extent, because of the nature of metallic bonds with their delocalized, shared electrons, any liquid metal can wet any other metal — provided both are truly clean. This is the secret of soldering. Getting a low melting temperature is important, and you'll remember from your studies of phase diagrams that that means going for 'eutectic' mixtures.

EXERCISE 2.7   Sketch a phase diagram for a binary alloy system which exhibits a eutectic (for example, the lead–tin system). What qualities of eutectic alloys could be relevant to soldering?

Complex mixtures of three or four elements provide eutectics at a range of temperatures lower than that of the Pb–Sn system. Given good temperature control these allow repeated soldering by using solders with progressively lower melting temperature.

2.4 Hybrid circuit materials

Hybrids are made by high-temperature ceramic processes, so none of the materials chosen for PCBs are any use at all. As in the previous section, we need to explore properties for both manufacturing and service functions.

2.4.1 The substrate

The substrate for a hybrid circuit has to meet most of the criteria listed for printed circuit boards and a few more. Chemical problems are unlikely to arise because ceramics are so inert, even in the high-temperature processes used, such as laying conductors. On the other hand, it is important that their coefficient of thermal expansion should match that of deposited components; and the brittleness of ceramic could cause problems both on the production line and in use. High thermal conductivity, if it can be arranged, will let heat escape from energy-dissipating components.

EXERCISE 2.8   Table 2.2 lists eight candidates for substrates and indicates whether the properties are wanted 'high' or 'low'. Highlight the top three for each property and decide a 'best material'.

Alumina is indeed a ubiquitous material for hybrid-circuit substrates, but not usually at high purity. A small fraction of a glassy phase which melts during firing is usually added to the alumina because it confers several advantages. The liquid encourages sintering to a low porosity at a reasonable temperature ($\approx 1600°C$). It also leads to a smooth surface

Table 2.2 Candidate substrate materials

Material	T_m	α	κ	ε_r	tan δ	dielectric strength	ρ	MOR
	°C	10^{-7} K^{-1}	W m^{-1} K^{-1}		10^{-3}	kV mm^{-1}	Ω m	MN m^{-2}
fused SiO$_2$	1700	0.5	1.2	3.9	0.3	15 to 25	10^{14} to 10^{18}	≈ 200
alumina	2020	60	33	9	0.3 to 2	10 to 15	10^{16}	500
Si$_3$N$_4$	1900	15	40	6	0.1	16 to 20	10^{14}	700
porcelain	1300	30 to 50	2	5 to 7	10 to 20	6 to 13	10^{14}	80
steatite	1350	60 to 90	6	6	1 to 4	8 to 14	10^{17}	150
Pyroceram 9606	≈ 1000	65	3.6	5.5 to 6.3	2 to 13	10 to 12	10^{12}	250
magnesia	2800	13	40 to 60	8	1	8 to 11	10^{14}	140
glasses	soften ≈ 600 to 900	8 to 13	3	4 to 8	0.5 to 10	8 to 13	$\approx 10^{12}$	≈ 100
ideal	> 1000	high (same as metals)	high	low	low	high	high	high

Key: T_m = melting temperature; α = thermal coefficient of linear expansion; κ = thermal conductivity; ε_r = relative permittivity; tan δ = dielectric loss tangent; ρ = resistivity; MOR = modulus of rupture (strength in a bend test)

▼Band casting▲

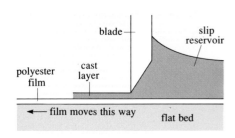

Figure 2.27 The band casting process

A slip is formed of the oxides with suitable organic solvents, surfactants and binders and is 'de-aired' under vacuum. It is spread onto a continuous plastic film, supported on a smooth flat steel bed. A 'doctor blade' accurately controls the spread thickness (Figure 2.27). The solvent is evaporated leaving the solid minerals held in the binder. This 'green' condition is a weak but leathery material which can be cut and punched to size. The firing schedule burns out the organics at low temperature and then moves up to high temperature for the sintering. Total shrinkage is about 30% so allowance has to be made to get tiles at specified dimensions. Flatness is only preserved through these changes if the tape is truly homogenous; process control of mixing and casting is important. Tiles are made in three standard thicknesses, $\frac{5}{8}$, 1 and $1\frac{1}{2}$ mm and areas up to 150 mm × 100 mm. Beyond this size it is too difficult to control distortion during firing.

finish, and it is to this glassy phase at alumina grain boundaries that the 'inks' adhere when they are fired. But this second phase may compromise the excellent properties of pure alumina, as revealed in Exercise 2.8. Silica and magnesia added in the ratio 2:1 react with the surfaces of the alumina particles at the high firing temperature to produce an intergranular liquid phase sufficiently viscous to avoid distortion of the slabs during firing. In preparing the material, the solid oxides are milled together for some hours. The resulting small particles have a wide size distribution so they pack efficiently. Both aspects help sintering to low porosity. The usual method of manufacture is ▼Band casting▲.

SAQ 2.6   (Objective 2.5)
Explain why a glass in the $SiO_2.Al_2O_3.MgO$ system is acceptable as an intergranular component of the substrate ceramic.

The continued use of alumina is assured, but it does have limitations:
- Large substrate tiles cannot be made because of firing distortion.
- Brittleness; a broken tile will cause havoc on an automatic line.
- Geometry; the finished circuit is bound to be a flat sheet.
- Power dissipation; thermal conductivity could be higher with advantage.

All of these limits would be pushed back by a substrate consisting of enamelled metal. But can there be an enamel which meets all the other demands? On the face of it the answer is no. Enamels are glasses, so are likely to soften when the glass-based inks are fired on. Also, to match the expansion coefficient of the enamel to that of the metal it is stuck onto, enamels tend to be high in alkali, which causes difficulties with resistivity and dielectric strength, especially in service at high temperatures. But the puzzle has been solved by ▼Keralloy▲. Just as thermoplastic PCBs open new design possibilities so too may metal-backed hybrid substrates.

▼Keralloy▲

A small UK company believed there must be a 'glass–ceramic' which could both fit a metal and meet the electrical specification. Their goal was to coat the cheap and ductile ferritic stainless steel used for such volume applications as car trim and cutlery. This has low thermal expansion (for metal) and a chromia passive surface to which the ceramic could adhere.

The idea of glass–ceramics is first to make a glass (amorphous structure) and shape it by any of the simple hot-working processes available for glasses. Then the material is crystallized by heat treatment, conferring the advantages of a crystalline ceramic (for example, strength and toughness). In this case the crystalline condition is favoured for its higher electrical resistivity. Actually there is always some glassy phase remaining after crystallization, so the resistivity of this phase had to be watched. But the glassy phase is an essential ingredient for getting the thermal expansions closely matched. The trick was to find a crystal phase with expansion coefficient exceeding that of the metal, and to find a glass with lower value. Then the proportions could be adjusted to get a composite of correct expansivity. Also the glassy phase allowed the new material to mimic the usual alumina/glass substrate composition.

To be accepted by circuit builders the new material must be compatible with existing component inks in three respects:

(a) After firing, inks on the new substrate must have similar electrical characteristics to inks fired on an alumina substrate.
(b) Components must adhere to the new substrate just as well as to alumina; in the event the adhesion proved superior.

(c) The new substrate must withstand the ink-firing schedules.

A silicate crystal with high enough thermal expansion was discovered. To coax a crystal into the glass-ceramic mode the usual approach is to move to the silica-rich side of the intended crystal composition and add a small proportion of weak-bonding alkali ions to encourage a mobile glass to form, from which crystals can be precipitated. But no alkaline glass could be tolerated here — it would be too conductive. *Large* divalent ions (Ba^{2+} or Pb^{2+}) both open up the structure to give a mobile glass and confer high resistivity. These big ions were unable to enter the crystal phase so they formed glass layers around the crystals as the devitrification proceeded. The proportions of the two phases could be easily controlled by the ratio of glass-making to crystallizing oxides.

The company still had much to learn, for example how to make the ceramic stick to the chromia layer on stainless steel and how to coat the steel with a thin uniform layer of glass powder and fire it. For the latter, electrocoating was the best method. The glass powder is made to carry an electrostatic charge and is sprayed onto the steel, held at earth potential. Electrostatic forces hold the powder in place and ensure even coating. The company also had to find a nucleating agent and a heating schedule to provoke crystallization. The final microstructure is shown in Figure 2.28. It consists of very small crystal grains (0.5 to 1 μm wide) with even narrower glassy interfaces. In this condition the material is stable to well above 1000°C so components printed with inks, which

0   1 μm

Figure 2.28 Keralloy glass-ceramic, final microstructure

mostly sinter at 850°C, can safely be built on.

In summary 'Keralloy — substrate of the future' (the company hopes) has all these advantages:

● It is completely robust both in the assembly plant and in service.
● There is no economic stricture on area.
● Fold-up geometries offer 'self containment' of circuits and other interesting topological possibilities.
● Thermal conduction from circuit to metal is fast, and heat extraction from the metal can be made efficient.
● Improved component adhesion means improved circuit reliability.
● Low porosity gives good breakdown performance.
● Fine surface quality prevents contamination causing inter-track short-circuits, improving reliability when components are tightly packed.

2.4.2 Components

▼Screen printing▲ is the process by which the geometry of conductors, resistors and insulating layers is defined for hybrid circuits. For each, an 'ink' has to be formulated having the necessary flow properties for printing, which binds together and to the substrate when dry and which contains ingredients for it to fuse to the substrate and to meet its service function. Basically, therefore, four ingredients are required: fluid, glue, glass and 'functional material'. The first two are fugitive: the fluid will evaporate to dry the ink immediately after printing and the glue must be burnt off in the early stages of firing. Only the last two will survive. Firing sinters them to a continuous solid, but in the ink they must be powders so they can flow. The mixture is proportioned to have a treacle-like viscosity that prints well.

Recipes for the fugitive parts are numerous. One example uses a modified cellulose polymer dissolved in a not-too-volatile solvent to provide the fluid and the glue. Polymers have characteristic non-Newtonian flow properties: they flow more easily as stress rises. This helps the ink get through the screen mesh, and then to stay put. A non-foaming detergent is included to suppress agglomeration of the powders and to aid wetting of the substrate.

▼Screen printing▲

The ancient craft of screen printing provides a simple way of printing several colours with accurate registration. Designs are delineated on fine-weave cloths as sealed and open areas. These are then used as masks. They are placed onto the surface to be printed on and ink is spread across. The ink prints through the open zones only. With a different mask for each colour, a picture is soon built up.

As adapted for the electronics industry, much of the craft element of the process may have gone, but it's still a process where everything has to be just right to get good results. This means such factors as: the right mesh at the right tension; the right blocking agent and the right ink; the right squeegee to drive the ink through and the right amount of spring-back to lift the screen off the work.

Screens are usually nylon or steel mesh, and mostly plain weave. They are tensioned into a frame and coated with a photoresist in emulsion form. After exposure with artwork in place to define the required pattern the resist is developed and the unhardened parts dissolved away.

Alternatively patterns can be defined by etching or electron-beam machining holes into metal foils which may be used directly as screens or reinforced with a mesh.

The printing action is shown in the picture sequence of Figure 2.29. Notice how the pressure of the squeegee brings the mesh into contact with the substrate so that the edges of the pattern are accurately defined. This also arranges that the amount of ink delivered is controlled by the thickness of mesh. The mesh must then spring back by at least its own thickness so it lifts clear of the print, and the ink is then expected to level itself by its surface tension. To make all this happen the screen tension, its spacing from the substrate and the squeegee pressure have to be mutually adjusted, and parallel alignment is critical. The speed and 'angle of attack' of the squeegee together with physical properties of the ink determine how the ink flows. When everything is working perfectly, layers from 25 µm up to about 150 µm thickness are quickly, reliably and very consistently printed. Integrity of the thinnest deposits depends much on the surface quality of the substrate and the upper limit reflects the tendency of the wet ink to slump when the screen has left it.

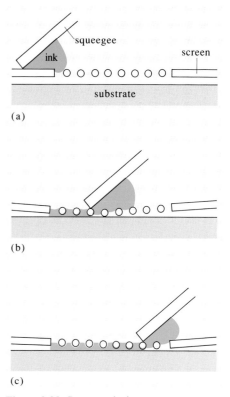

(a)

(b)

(c)

Figure 2.29 Screen printing sequence

The glasses used are commonly members of the lead borosilicate family, designed to meet several compromises. They must:

- react with the substrate glass phase to ensure adhesion,
- below T_g have thermal expansion similar to that of alumina to avoid thermal stresses developing upon cooling,
- be good insulators (note the lead borosilicate family contains no alkali ions, which are relatively mobile),
- sinter correctly.

This last point is also a compromise. The viscosity/temperature characteristic is crucial. A glass will sinter readily if it becomes very fluid, but of course it may then flow across the surface of the substrate and reduce the thickness and ruin the definition of the component. On the other hand, slow sintering gives poor adhesion and, with conductors, the glass gets incorporated in the metal, increasing its resistivity and spoiling its solderability.

The functional ingredients, those which decide whether the component is a conductor, a resistor or an insulator, need separate discussion.

Conductors

EXERCISE 2.9   The circuit conductors will be metal of course. Can you suggest some candidates? Remember they have to survive subsequent high temperature processing.

Silver is the cheapest noble metal, has low resistivity (0.002 Ω square^{-1}, see ▼Sheet resistance▲), sinters beautifully and is easily soldered onto. But it has one devastating snag: a nasty habit of migrating away from where it has been put. It dissolves quite readily in glasses, so it diffuses during high-temperature processing or under persistent electric fields in service (especially in humid environments). It also tends to dissolve in solder. Alloying silver with palladium or platinum removes these problems but resistivities are higher, 0.01 to 0.04 Ω square^{-1}. The choice has depended on the relative prices of Pd and Pt.

▼Sheet resistance▲

Consider the conduction of an electric current through the object depicted in Figure 2.30. If the material resistivity is ρ the resistance of the piece is

$$R = \frac{\rho l}{A} = \frac{\rho l}{wt}$$

Now let's imagine the conductor to be assembled from a number of squares of width w so the length is nw. Then

$$R = \frac{\rho nw}{wt} = \frac{\rho n}{t}$$

The ratio ρ/t is a description of this particular conductor, what it is made of and how thick it is. This is called the 'sheet resistance' R_{sheet} of the conductor. It will be measured in Ω square^{-1} since $\rho/t = R/n$.

Ink manufacturers can describe their products by a nominal sheet resistance because each ink fires to a known

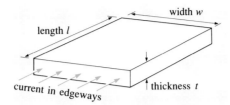

Figure 2.30 Dimensions specified for sheet resistance deduction

resistivity and has a recommended print thickness. The *ratio* of length to width of a resistor is thus defined as soon as the resistance value is chosen. The size of each square still has to be decided on power-dissipation grounds. With current I flowing in each square the power per square is $I^2 R_{sheet}$ and the designer has to ensure the square is big enough to dissipate that power without undue temperature rise.

Resistors

Metal particles dispersed in glass will make a resistive phase and its resistivity will depend on the 'touchiness' of the particles. But this is not a practical way to make resistors: the dependence of the sheet resistance on the concentration of metal particles is far too steep to be controllable (see Figure 2.31). Until some particles touch, the resistance is infinite; once they all touch the resistance is short circuited.

Semiconductor-oxide dispersions have much flatter dependencies (Figure 2.31 again). Of the three examples shown, ruthenium oxide is believed to diffuse into the glass and make it semiconductive, the palladium-oxide–silver composition uses silver links between the oxide particles, and bismuth ruthenate probably reorganizes during firing to form a continuous phase of semiconductor. This last material is the basis of

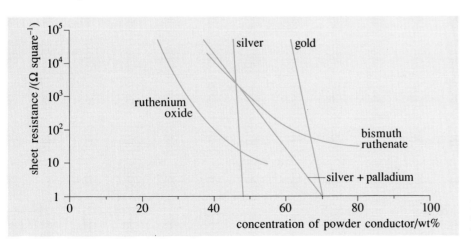

Figure 2.31 Resistance square^{-1} for different resistor formulations of the same thickness

Birox resistors, which are now perhaps the most popular, giving consistent values and good stability with age. Although the sheet resistance expected from an ink laid at standard thickness is specified, the actual resistance of a deposit may have to be adjusted after firing by trimming the dimensions using a laser beam. The resistance can be monitored and compared with a goal during the trimming so that precise values can be achieved. Figure 2.32 shows favoured trimming patterns. Similar processes are used for making ▼Discrete resistors▲.

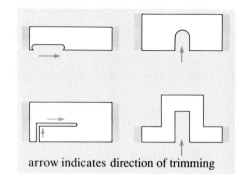

arrow indicates direction of trimming

Insulators

Insulators are required for separating layers of circuitry, for overglazing the completed work and for building capacitors. (We shall look at capacitor dielectrics in Section 2.5.) For the first two applications, a low-loss, low-dielectric-constant, high-resistivity material is required.

Figure 2.32 Trimming patterns for resistors on hybrid circuits

▼Discrete resistors▲

Two design criteria for resistors, resistance and power dissipation, are illustrated in Exercise 2.10.

EXERCISE 2.10 Choose power ratings for the resistors in the circuits of Figure 2.33(a) and (b). In Figure 2.33(a), note that the current into the base of the transistor can be neglected.

In power circuits, where heat dissipation dominates the architecture, large, wire-wound resistors may have to be used and attention given to ventilation and heat sinks. A metal substrate such as Keralloy may find favour. But low-power resistors, which can be very small, are now mostly made as films on ceramic.

The two technologies, known as **thick film** and **thin film** use semiconductor oxide or metal respectively. Thick film resistors are made as we have described for the hybrid-circuit components. Many resistors are printed onto a ceramic sheet which, after firing, is diced and each resistor provided with solderable terminations. To get the desired sheet resistance from a highly conductive metal means that the metal layer must be very thin — much less than a micron. Such thin films can be laid down by evaporating metal onto a substrate. Dicing and termination steps are the same. Laser trimming to specific values is used. For the traditional cylindrical format of resistors, thin ceramic rods are coated with either resistive medium. The rod is cut into short lengths and terminals put on. Then

the laser cuts a spiral track to increase the number of squares between the terminations. Of course the cut can be feedback-controlled to get a tight tolerance on the resistance value. All discrete resistors must have an outer coat of insulation to protect the element from its working environment.

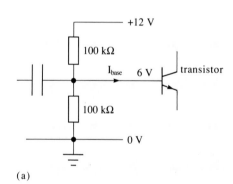

(a)

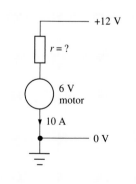

(b)

Figure 2.33

Substantial crystal content within the glass favours these properties and provides a substrate suitable for further print work. Glass-ceramics offer one route (as you saw in 'Keralloy') but loading the ink with a ceramic powder, such as alumina, is simpler. It is so important to avoid pinholes in these layers that it is usual to 'double print', the chances of coincident pinholes being remote.

SAQ 2.7   (Objective 2.5)
What aspects of hybrid circuit technology make it suitable for building capacitors direct onto the substrate?

2.5 Capacitors

The function of capacitors in circuits is often to block d.c. while allowing a.c. to pass. In such applications precise values are rarely important. In tuned circuits and filters, precision and stability are important. Since most of the volume of a capacitor is its dielectric, deciding how to make small capacitors is a materials-property competition. Remembering from 'Dielectric definitions' that

$$C = \frac{\varepsilon_0 \varepsilon_r A}{d}$$

the general trends of geometry and materials selection to make a high-capacitance device are to have a thin insulating layer (d small) of a material which has high relative permittivity (ε_r large). Polymer and ferroelectric ceramic dielectrics now offer more compact designs than were possible with early dielectric materials such as waxed paper, electrical porcelain or the sheet mineral mica, which had modest ratios of ε_r/d. Comparing capacitors of similar specification (Figure 2.34) you can see that capacitors are rather smaller than they used to be ▼Capacitor construction▲ outlines modern methods of assembly and the design exercise in Section 2.5.1 guides you from the defining formula towards real options.

2.5.1 Capacitor design exercise

The sequence of questions in SAQ 2.8 to 2.19 leads through the major considerations of capacitor design. They expose the relative merits of polymer and ceramic dielectrics and contrast their performance with one of the earlier materials. Objectives 2.6 and 2.7 will be met if you work this design exercise satisfactorily.

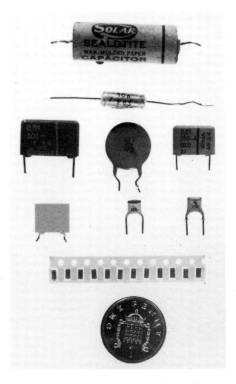

Figure 2.34 Selection of capacitors, all about 0.01 μF. Above the coin is a packaged row of surface-mount ceramic capacitors. The row above them contains ceramic capacitors. In the centre of the picture is a single-layer disc ceramic capacitor. The rest are polymer, apart from the vintage paper capacitor at the top

Table 2.3 is data for the exercise for three typical dielectrics, a ceramic, a polymer and a paper.

Table 2.3 Dielectric data

Material	ε_r	1000 tan δ at 1 MHz	ρ_{dc}	Max. freq.	Dielec. Str.	Min. thick.
			Ω m	MHz	kV mm^{-1}	μm
X7R ceramic	1800	60	10^{12}	25	10	25
PET polymer	3.2	10	10^{16}	2	30	2
Waxed paper	2.3	2.4	10^{10}	1	16	125

SAQs 2.8 to 2.19 (Objectives 2.6 and 2.7)

SAQ 2.8 'Capacitance' and 'working voltage' characterize a capacitor. What do they mean? What other qualities matter to its performance? Try for four items.

SAQ 2.9 How do high capacitance and high working voltage have conflicting design implications?

SAQ 2.10 Suppose the goal in designing a capacitor is to get the required capacitance into the smallest volume possible. Neglecting the volume of the conductors, express the volume $\mathscr{V}$ of the device in terms of A and d. How is the volume for a given capacitance related to the dielectric thickness? What is the designer's response?

SAQ 2.11 Let's get numerical and compare designs for some 0.01 μF capacitors. This value is at around the cross-over between ceramic and polymer being chosen. The voltage rating would probably govern the choice of dielectric. Calculate the volume of dielectric in a 0.01 μF capacitor made with the minimum thickness of X7R ceramic. Would a much larger device result from a PET dielectric with $d = 2$ μm? How would the paper dielectric fare? Use the approximate value $\varepsilon_0 \approx 9 \times 10^{-12}$ F m^{-1}.

SAQ 2.12 What conductor areas must be disposed onto these volumes of dielectric?

SAQ 2.13 What do you think of these capacitors as engineering propositions? How do they fit circuit dimensions?

SAQ 2.14   Apart from enabling capacitors to be smaller, what other capacitor property, listed in the answer to SAQ 2.8, will benefit from the dielectric volume being as small as possible? Why?

SAQ 2.15   Our calculations demonstrate that the extremely thin dielectrics proposed may not be necessary in order to make capacitors of modest value which are small enough for modern circuitry. Which of the factors in the answer to SAQ 2.8 will benefit from increasing the dielectric thickness?

SAQ 2.16   Calculate the d.c. resistances of the three devices.

SAQ 2.17   Now let's look at the dielectric losses. How do we find the equivalent a.c. resistance in parallel with C which accounts for the loss and what is the power loss when a voltage of r.m.s. value V_{rms} is applied? Choosing 1 MHz as the test frequency, work out the equivalent a.c. resistance for each capacitor, and the power dissipations per unit volume when V_{rms} is 10 V.

SAQ 2.18   If we do take advantage of the minimum practical thicknesses, how will the dielectric strengths of the capacitors compare?

SAQ 2.19   Using the ranking 1 = best, 3 = worst, enter X7R, PET and paper into Table 2.4, for each virtue listed.

Table 2.4

Virtue	Rank		
	1	2	3
small size			
small area			
large d.c. resistance			
small dielectric loss			
great dielectric strength			

This exercise has shown the major design factors of capacitors in terms of three representative materials. Actually the choice is much wider; Table 2.5 reviews the field and ▼Ceramic dielectrics▲ and ▼Polymer dielectrics▲ look at the materials cited. ▼Capacitor construction▲ outlines modern methods of deploying them.

Table 2.5 Properties of dielectric materials

Material	ε_r	1000 tan δ at 1 kHz	$\dfrac{1}{C}\dfrac{dC}{dT}$ % K^{-1}	ρ_{dc} Ω m	Max. frequency MHz	Dielec. strength kV mm^{-1}
Ceramics						
X7R	1800	20	±0.1	10^{12}	25	10
Z5U	6000	30	±0.1	10^{11}	2	10
NPO	60	1	0.003	10^{12}	70	10
Polymers						
PET (polyester)	3.2	4	0.02	?	2	30
PC (polycarbonate)	2.8	2	?	10^{16}	1	18
PS (polystyrene)	2.5	0.1	0.001	10^{18}	2500	20
PP (polypropylene)	2.3	0.5	?	10^{17}	?	20

▼Ceramic dielectrics▲

The designations of commercial ceramic dielectrics, Z5U, X7R, NPO used in Table 2.5 are codes for performance rather than for particular formulations. Especially, they refer to the temperature sensitivity of the dielectric constants. Here we shall discuss only Z5U and X7R which are based on barium titanate. The origins of barium titanate's high dielectric constant were revealed in 'Perovskites' but composition changes bring its magic temperature into a practical range for service and make the effect much less temperature sensitive.

The charge balance of the crystal will remain unaltered if different divalent and quadrivalent positive ions are substituted for Ba^{2+} and Ti^{4+}. But if the replacement ions are not the same size, the amount of 'rattle' at which the crystal can sustain a given structure will change. Substituting larger 4+ ions causes the main phase change to occur at lower temperature. The influences of some possible options are seen in Figure 2.35.

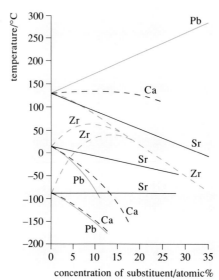

Figure 2.35 Phase transition temperature versus concentration of substituents

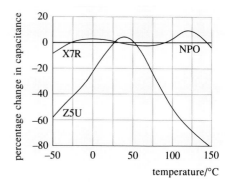

Figure 2.36 Percentage change of capacitance with temperature for threee dielectric materials, relative to capacitance at 25 °C. The change is due to variation of ε_r with temperature

---

SAQ 2.20   (Objectives 2.2 and 2.3)
What is the logic, in atomistic terms, of this last assertion? Here are hints on how to think it through:

(a) What will larger 4+ ions do to the octahedral cages?
(b) What happens as a consequence to the cuboctahedral cages?
(c) How does the entropy of the divalent ions in these spaces change?
(e) Hence, what temperature is needed to switch the free energy equation?

In connection with Figure 2.18 we saw that, in addition to the phase change at 125°C, barium titanate has a phase change at room temperature. The effect on this lower-temperature phase transition of substituting zirconium for titanium is to raise the transition temperature. Thus with about 15% of $BaZrO_3$ alloyed in to make a *homogenous* solid solution into $BaTiO_3$, the transitions are pinched together near room temperature. Moreover if $CaZrO_3$ is the alloying oxide, the solid solution has some enlarged and some unchanged cuboctahedra, and among each kind some contain large Ba^{2+} ions and others smaller Ca^{2+} ions. The resulting confusion broadens the temperature range over which an electric field can shift the ions around.

In essence this is how to make a Z5U dielectric. The temperature profile of its dielectric constant is shown in Figure 2.36.

Z5U still has a distressingly sharp temperature response. X7R sacrifices some ε_r but wins a lower temperature coefficient (see Figure 2.36 again). One formulation creates a heterogenous microstructure of barium titanate crystals whose surfaces are heavily doped with niobium ions. A small quantity of niobium oxide, Nb_2O_5, is added to the precalcined barium titanate before sintering. When powders are sintered to continuous ceramics the crystals have to grow to occupy the spaces between the particles. During this time the niobium ions are incorporated into the newly grown structure. With charge +5 and radius 0.070 nm they take up octahedral sites in the perovskite structure, so effectively we are substituting Nb_2O_5 for some of the TiO_2 at the surfaces of the grains. The final microstructure will show a profile of niobium concentration varying from a rich concentration at the surface of each grain to virtually zero at the core. Niobium, adjacent to zirconium in the periodic table of the elements, produces a similar reduction of the temperature for the polarization transition. So if we imagine a cooling grain, its core will polarize at 125°C and the transition will spread outwards as the temperature falls. With 1% niobium distributed in this uneven way the permittivity versus temperature curve is fairly flat over a wide range of

Notice that each niobium ion brings $2\frac{1}{3}$ oxygen ions to the structure, an increase over the two O^{2-} ions which accompany every Ti^{4+} ion. So it is to be expected that the newly grown material will have only a very low population of oxygen vacancies. Processes relying upon oxygen ion diffusion will therefore be slowed considerably. Two consequences flow from this observation. First, ionic conductivity is suppressed and we may hope that the dielectric will display a good d.c. resistivity as its application demands. Actually, because the substitution also upsets the charge balance of the compound this needs more careful thought and we will return to the topic in Chapter 4 when conduction phenomena are taken further. The second effect of filling oxygen vacancies is that the growth of grains, which requires atoms to be shunted around, will be inhibited. A beneficial fine grain-size can be produced if a fine powder is used to start with.

In large grains (Z5U) the tetragonal crystallographic change can occur in different directions in different parts of the grain. In a changing field these boundaries move to bring the displaced ions onto axes more nearly parallel to the field. Distortions of the structure at 90° to one another raise considerable strain energy in the grain, so moving the boundary involves energy dissipation. Smaller grains cannot support such an internal structure so this loss mechanism is eliminated giving X7R ceramics lower tan δ than Z5U. (This is a preliminary skirmish with the idea of 'domains' of polarization, which get a lot of attention in the magnetic context of the next chapter.)

▼ Polymer dielectrics ▲

Polymers, in common with all materials, can show polarization due to displacement of their electrons. Additionally many polymer molecules contain polar groups such as

$$\overset{+}{C}=\overset{-}{O} \text{ or } \overset{+}{C}-\overset{-}{Cl}.$$

Displacement of these by applied fields gives further scope for polarization.

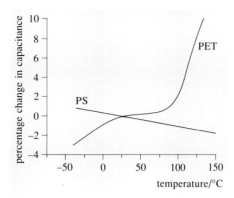

Figure 2.37 Percentage change of capacitance with temperature for PS and PET dielectric relative to capacitance at 25 °C, at 1 kHz. The change is due to variation of ε_r

EXERCISE 2.11   How would these two sorts of polarization be affected by

(a) the closeness of polymer dielectric's glass transition temperature to its operating temperature,

(b) frequency of the applied field?

Neither of the polarization mechanisms of polymers gives the very high ε_r values seen for the ferroelectric ceramics, so their use as dielectric materials in capacitors is dominated by the processing factor: can a really thin film of good quality be made? The four most commonly used, PET, PC, PS and PP, have this feature, allowing defect-free films down to about 2 µm thick, though the ultimate is rarely used because ceramics can give better ε_r/d ratios for low-voltage capacitors. Table 2.6 and Figure 2.37 relate molecular structure to dielectric properties for a polar and a non-polar polymer.

Polystyrene, PS, has no polar groups so its dielectric constant reflects the displacement of electrons relative to nuclei when a field is applied. Electrons, with their very low mass, can 'keep up' with high frequencies so ε_r is the same at 1 kHz and 1 MHz. PET (polyester) has a higher dielectric constant at 1 kHz because the polar parts of its molecules respond to the field. But this

contribution collapses at higher frequencies; molecular displacements can no longer keep up with the rapid oscillations of the field. Temperature effects are present in PET since temperature is associated with molecular motion. Above T_g, PET has a further increase of ε_r but becomes very 'lossy' (large tan δ). Of course temperature has no significant effect on the internal electron dynamics of molecules so PS is temperature stable. PS is the chosen dielectric for capacitors of which good frequency and temperature stability is demanded.

SAQ 2.21   (Objective 2.2)
PP serves at temperatures above its T_g. Why does this not affect its dielectric constant? What can you remember of the environmental degradation susceptibilities of PP which might influence its success as a dielectric film?

Table 2.6

Material	T_g	polar?	ε_r			1000 tan δ	
	°C		1 kHz $< T_g$	1 MHz $< T_g$	1 kHz $> T_g$	60 Hz $< T_g$	60 Hz $> T_g$
PET	70	yes	3.2	≈ 2	> 4	2	≈ 20
PS	105	no	2.5	2.5	2.5	0.1	0.1

▼Capacitor construction▲

The problem to be solved is how to stack a large area of thin material into a more equiaxed shape. The terms ε_r and d dictate the conductor (and dielectric) area; the mechanical properties of the materials influence the forms of construction that may be used. Thus there are many design solutions to the problem.

Flexible dielectrics allow the 'Swiss-roll' method of packing. This method was used with paper dielectrics and is nowadays used with polymer dielectrics. The process is simplified if the polymer tape is metallized (by evaporation or by chemical deposition) before winding (Figure 2.38a). Just two tapes are fed to the machine with the necessary offset to allow connection to the terminals. However the roll form gives the device significant inductance L, limiting its frequency range. (Electrically speaking, capacitance and inductance are 'opposites'. Each counteracts the other to an extent determined by the frequency. A capacitor C with an inductance L has its capacitive effect cancelled by its inductive effect at its self-resonant frequency, $\sqrt{LC}/2\pi$.)

The 'concertina' format, Figure 2.38(b), has less self inductance but requires a slightly more complicated machine to make it. A single tape, metallized on both

faces, is used. The whole device is fitted with terminals and encapsulated for protection.

Rigid dielectrics call for other methods of assembly. A fairly primitive option is to stack thin dielectric and metal sheets alternately. Much of the success of multi-layer ceramic capacitors, as well as their limitations, can be attributed to automating this lay-up process in the way shown in Figure 2.39. The dielectric material is produced as a calcined powder and presented to the assembly plant as a tape held together by an organic binder. It is hopelessly weak and requires delicate handling. Nevertheless it can be fed to a screen printing station where a silver-palladium alloy 'ink' is applied, similar to those used for putting conductors onto hybrid circuits. Successive layers of tape and ink are stacked, and the job turned through 180° between ink layers to offset the metal. When the designed stack has been made, the whole is compressed between platters, diced into individual chips and then fired to drive away the volatile components and sinter the dielectric ceramic. Here is where the process gets tricky. The dielectric binder and the fugitive ingredients of the ink have

to evaporate without internal pressure delaminating the stack. The secret is to lose the organics from the ink first so that porous channels are formed which conduct

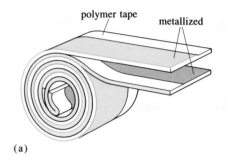

(a)

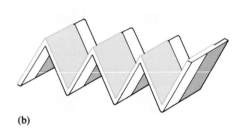

(b)

Figure 2.38 (a) 'Swiss-roll' construction using two tapes, each metallized on one surface and with metallization offset to one edge. (b) Concertina construction using film metallized on both surfaces but offset from opposite edges

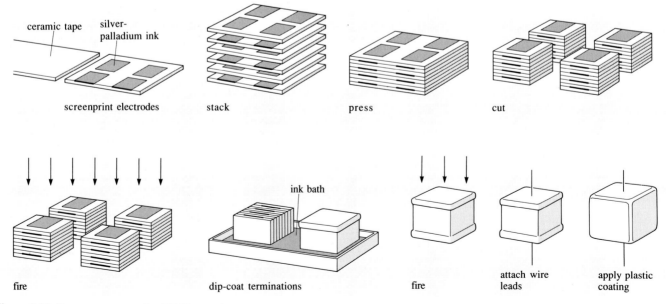

Figure 2.39 Process sequence for MLC capacitors

vapour from the ceramic parts to the outside. A gentle temperature rise is used during 'burn out' to hold the rate of generation of vapour down to a manageable level. The ceramic sinters typically at about 1300°C. After the rough edges of the now solid ceramic chips have been cleaned up, electrodes are fitted either by electroplating or by firing on a metal-loaded glass (of the sort used for hybrid circuit conductors). Somewhere along this complex line of production there's a good chance that something will go wrong so every single item, perhaps ten million a day, is tested. The 30% rejects must be weeded out to protect the reputation of the good remainder. These are regarded as precious metal scrap.

Capacitors would be significantly cheaper if less palladium could be used, but high sintering temperatures dictate a high proportion of palladium to avoid the metal melting. Liquid-phase sintering in molten glass at lower temperature is one possibility. The glass must have high resistivity and low melting temperature; it must interact sufficiently with the barium-titanate based dielectric to promote sintering yet not spoil the dielectric properties. One patent specifies a glass containing only 5 weight% of SiO_2 together with 23% Bi_2O_3, 5% B_2O_3, 1% Al_2O_3 as network modifiers and 36% CdO, 25% PbO and 5% ZnO to provide divalent ions. The heavy-metal oxides (Cd, Pb and Bi) give low melting temperature and the 'rules' for ensuring high resistivity have been followed. Capacitors can be sintered at 1100°C allowing an electrode alloy with only 15% palladium.

The above techniques are for 'assembled' capacitors, where the dielectric and plates start as distinct items to be assembled. But other sorts of capacitor are 'grown' rather than assembled. One ceramic variety avoids precious-metal electrodes by sintering the ceramic with no metal in place. Instead, layers of a different ceramic are designed to have connecting pores after firing. This structure can be impregnated with a cheap, low-melting-temperature metal such as lead to provide electrodes. Another 'grown' type is the electrolytic capacitor, in which a *very* thin dielectric layer is deposited electrochemically at the interface of a metal and a paste electrolyte. This method can provide large capacitance values. Finally, capacitance in ICs is provided by carefully designed p–n junctions, as you will discover in Chapter 5.

With size often the predominant influence on the choice of capacitor in modern circuit building, ceramic dielectrics dominate the small-value, low-voltage section of the market while polymers are favoured for higher values at higher voltages. Tables 2.7 and 2.8 give you some flavour of the choice from a recent catalogue. The prices are pence each for quantities of 1000; our analysis has not gone into all the factors influencing cost so these figures may not be seen as directly competitive.

Table 2.7 lists some ceramic capacitors. The first pair shows the competition for tiny surface mount capacitors. Below them, the table compares the three ceramic dielectrics and shows the extra cost in both pence and volume of the temperature-stable NPO ceramic with its lower ε_r. Temperature stability is also a virtue of polymer dielectrics. When constructed for close tolerance, they are best for filter applications. NPO ceramic, formulated to access this niche, has some size advantage.

Table 2.7   Some multilayer ceramic 0.01 µF, 100 V capacitors

Maker	Type	Dielectric	Volume/mm³	Tolerance %	Price/p
AVX	A	X7R	4.5	10	5.4
Mullard	A	X7R	4.5	10	3.4
AVX	B	Z5U	37	10	14
Mullard	B	Z5U	37	20	6
AVX	B	X7R	82	20	8.9
Mullard	B	X7R	82	20	9.1
AVX	B	NPO	145	5	43
Vitramon	B	NPO	145	5	54

Key: Type A = bare chip; B = encapsulated block, side wires.

Table 2.8 shows the extra size we had expected of polymer dielectrics, and this is exaggerated by their being designed for higher voltages. The wide variety of prices reflect many other aspects of quality than those tabulated and, presumably, the production quantities are strongly influential. What costs, apparently, is the skill in providing a high-voltage performance in a smaller volume as the Rescam PET products show. Finally if you must replace that old paper one in your 'wireless' set with something similar, you'll have to pay!

Table 2.8 Some polymer 0.01 µF capacitors

Maker	Type	Dielectric	Voltage/V	Volume/mm³	Tolerance/%	Price/p
Mullard	C	PET	250	620	10	9.3
Rescam	C	PET	250	125	20	73
Rescam	C	PET	500	266	20	84
Mullard	B	PET	400	600	20	3.3
Suflex	C	PS	160	1600	10	18.5
Mullard	C	PC	400	620	10	12.4
Mullard	B	PC	400	585	10	5.2
Mullard	B	PP	1500	4440	10	23
Rescam	C	Paper	500	745	20	780

Type B = block, side wires; C = cylinder, axial wires.

Objectives for Chapter 2

You should now be able to do the following:

2.1 Differentiate between printed, hybrid and integrated circuit technologies. (SAQ 2.1)

2.2 Describe the role of free and bound charges in determining the electrical response of materials through conductivity and permittivity. (SAQ 2.2, 2.3, 2.20, 2.21)

2.3 Discuss dielectric media in terms of polarization, loss and breakdown. (SAQ 2.20, 2.21)

2.4 Describe (with scientific rationale) the design and manufacture of a circuit board for discrete components. (SAQ 2.5)

2.5 Describe (with scientific rationale) the design and manufacture of a substrate for hybrid circuitry. (SAQ 2.6)

2.6 Discuss the design specification for discrete capacitors. (SAQs 2.8 to 2.19)

2.7 Explain the physical design of a range of types of resistor and capacitor. (SAQ 2.7 and SAQs 2.8 to 2.19)

2.8 Define and be able to recognize: loss tangent, relative permittivity, sheet resistance.

Answers to exercises

EXERCISE 2.1
See Table 2.9
With an activation energy of 0.1 eV, significant fractions of ions will be available to diffuse even at room temperature — and certainly at the higher temperature. A modest binding energy of only 1 eV locks virtually all ions, even at 1000 K.

Table 2.9

E_a/eV	$\exp\left(\dfrac{-E_a}{300k}\right)$	$\exp\left(\dfrac{-E_a}{1000k}\right)$
0.1	0.02	0.31
1.0	10^{-17}	10^{-5}

EXERCISE 2.2
(a) The polymer contains very few ions, but the open structure of an amorphous polymer should provide plenty of places for the ions.

(b) A crystalline oxide is made of ions; but a crystal has few vacancy defects.

(c) Window glass has an open, amorphous covalent network loaded with a fair population of cations.

EXERCISE 2.3 With anions of radius R_O just touching, the diagonal centre-to-centre distance across an octahedron is $(2\sqrt{2})R_O$. The gap between the anion surfaces along this axis is therefore $2R_O(\sqrt{2} - 1)$. The cation must be bigger than this so it can touch the oxygen ions without them touching one another. The critical minimum radius for the cation is therefore $R_{cat} = 0.414 \times 0.14 = 0.058$ nm.

EXERCISE 2.4 Pythagoras' theorem applied to the triangle of Figure 2.16(c) gives

$$2x^2 = (2R_O + 2R_{Ti})^2$$

$$x = (R_O + R_{Ti})\sqrt{2}$$
$$= (0.14 + 0.068)\sqrt{2} \text{ nm}$$
$$= 0.294 \text{ nm}$$

The appropriate radius for an ion to just fill the 12 cornered cage defined in Figure 2.17 is $x - R_O = (0.294–0.14)$ which is 0.154 nm.

EXERCISE 2.5 In turn, the suggested polymers all have problems. LDPE suffers from items (a) and (d). Nylon 6,6 from

items (b), (c) and (d). PMMA from items (b) and (e). PTFE from items (f) and (g). PES from item (g). Polyester resin from item (h). Phenol-formaldehyde resin and epoxy resin from item (i).

EXERCISE 2.6 The necessary formula for C is in 'Dielectric definitions'. Reactance Z is

$$Z = \frac{1}{2\pi f\, C} = \frac{d}{2\pi f \varepsilon_0 \varepsilon_r A}$$

For the board in question,

$$Z \times f = \frac{\text{spacing}}{2\pi\varepsilon_0\varepsilon_r\,(\text{length} \times \text{width})}$$
$$= \frac{250 \times 10^{-6}}{2\pi\varepsilon_0 4\,(25 \times 200 \times 10^{-9})} \Omega\text{ Hz}$$
$$= 224 \times 10^9\ \Omega\text{ Hz}$$

On the log–log axes provided a straight line results passing through, for example, 224 kΩ at 1 MHz and 224 MΩ at 1 kHz (see Figure 2.40). The amplifier will be effectively short circuited when the reactance is about the same as the amplifier's input impedance. Now, $Z = 10$ MΩ at 22.4 kHz, which is only at the top end of the audio spectrum! At digital data processing frequencies, typically tens of MHz, this stray capacitance is effectively a dead short across the amplifier. Nor is there useful improvement available by

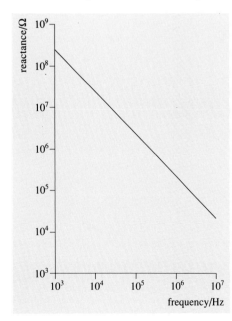

Figure 2.40

choosing an insulator with lower relative permittivity: 1.0 is the minimum (air). Clearly the earth plane must not be put near this signal lead.

EXERCISE 2.7 Figure 2.41 shows the Pb–Sn phase diagram with the eutectic composition at 38% Pb. The significant features of eutectic alloys are first, that they are the lowest melting alloy of the system; second, because they melt at a single temperature, an exact working temperature can be specified and the sharp freezing point ensures rapid solidification without the parts moving due to contractions; third, the two phases with compositions defined by the ends of the eutectic tie line must form simultaneously throughout the liquid, so a fine grained microstructure with good strength results.

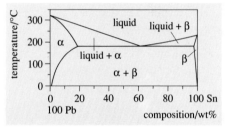

Figure 2.41 Pb–Sn phase diagram

EXERCISE 2.8 Alumina scored six times (α, κ, tan δ, dielectric strength, ρ and MOR). Though not the best in any one category, it has the best overall suite of properties.

The entries in Table 2.2 were chosen to be reasonably cheap, a very important property; but even when compared with fancier materials, alumina still comes out as the best all-rounder.

EXERCISE 2.9 Obviously a low-resistivity metal must be chosen. But the high-temperature process will oxidize copper, the cheap choice, so hybrids are forced to employ noble metals such as silver, gold, platinum and palladium — and they are all expensive. The high-melting temperatures of the latter two imply that they will not sinter from powder to solid very easily.

EXERCISE 2.10 The base current of the transistor in Figure 2.33(a) is negligible, so each 100 kΩ resistor has a 6 V drop across it. Power $= V^2/R = (36/100\,000)$ W, which is 0.36 mW.

The other circuit is rather different; 10 A must flow through r because the motor needs that current to do its job. Also the resistor must drop 6V. So $r = V/I = 0.6$ Ω and the power dissipated is $VI = 60$ W.

EXERCISE 2.11

(a) Polarization by electron displacement

will be insensitive to temperature. That due to molecular movements will be sensitive however. Above T_g polymer molecules become free to wriggle, so the contribution of polar group movement is enhanced. However, these movements are then coupled to thermal agitation, so dielectric losses can be expected to rise.

(b) The very low inertia of electrons allows them to follow much higher frequencies than the more massive molecules. Higher relative permittivities, due to molecular orientations, may be expected to fall off at high frequency.

Answer to self-assessment questions

SAQ 2.1   Capacitors. Printed circuits use discrete capacitors soldered in position. The same products might also be used in hybrid circuits, although this technology can be used to build up capacitors directly on the substrate. Integrated circuits can include internal capacitors or else can be used in combination with discrete components.

Resistors. Printed circuits use discrete soldered resistors. Hybrid circuits use relatively resistive sections of connecting track as resistors. Integrated circuits may have resistances 'grown' into them during manufacture.

Inductors. Printed and hybrid circuits use discrete inductors soldered in position. Identical products might be used in either technology. (Integrated circuits generally do not include inductors.)

Transistors. Printed and hybrid circuits use the same discrete components soldered in place. In integrated circuits they are built in.

SAQ 2.2   The electrical properties of salt water have a strong bearing on radio communication and submarine power transmission.

You should have thought about some of the following. In aqueous sodium chloride solution, the dominant free-charge carriers are positive and negative ions. When electric fields are steady or slowly varying, the ions can respond in much the same way as electrons in metals and the properties of sea water are those of a poor conductor. At higher frequencies (well beyond radio-wave frequencies), the ions may be inertially restrained from keeping up and oscillating in phase with electric fields. The charge carriers might as well be tied down as they are in dielectrics such as (cold) ionic solids.

SAQ 2.3   A high dielectric constant means a high permittivity relative to vacuum —

and that means it's easily polarizable. The charges on the molecules (or indeed the molecules themselves) of a high permittivity material are easily displaced by external electric fields.

Comment: In fact many insulating solids hold their charges quite tightly and have dielectric constants around 2 due to electron displacement only. Where molecular dipoles can be oriented (as in some polymers) or where some ions can be displaced (as in some glasses), values range up to 10. In special cases where many ions are displaced in concert (perovskites) permittivities orders of magnitude greater are found.

SAQ 2.4   Given the instantaneous applied voltage $v = V \sin \omega t$, the instantaneous current is

$$i_C = C\frac{dv}{dt} = \omega CV \cos \omega t$$

$$= \frac{V}{1/\omega C} \cos \omega t$$

Since

$$\sin\left(\omega t + \frac{\pi}{2}\right)$$
$$= \sin \omega t \cos \frac{\pi}{2} + \cos \omega t \sin \frac{\pi}{2}$$
$$= \cos \omega t$$

the required result is proven.

The equivalent equations for a resistor subject to that alternating voltage are

$$i_R = \frac{v}{R} = \frac{V}{R} \sin \omega t$$

By comparison we see that $1/\omega C$ plays the equivalent role to R in relating the magnitude of current and voltage, while the trigonometric functions describing their oscillation are 'in phase' for resistors and $90°$ out of phase for capacitors.

SAQ 2.5   The steps are:

1 Lay down the thin nickel metallization.
2 Apply positive photoresist and mask insulating areas of the board.
3 Expose to ultraviolet light, remove mask and 'develop' the resist.
4 The conductor areas are revealed; plate copper electrolytically.
5 Strip remaining resist and give short nitric acid wash to remove nickel from insulating zones.
6 Tin (that is, coat with solder) the conductors.

Note: 1 must be done before 2 and 3 or all separate conductor zones have to be individually contacted for electroplating. During 5, because the copper conductors are much thicker than the nickel they will not be significantly eroded while the nickel is being dissolved. The material saving is in the copper: the subtractive process wastes all the copper put on the areas which are to be insulating.

SAQ 2.6   The formulation is designed for high resistivity following 2 out of 3 of the propositions of 'Making an insulating glass'. Al_2O_3 acts like B_2O_3 as a network modifier and the added ions are divalent Mg^{2+} for strong bonding and immobility in electric fields. Large ions are not chosen as they would reduce the glass viscosity at the sintering temperature so risking distortion of the sheets.

SAQ 2.7   You may have noted some of the following

(a) The use of ceramic as the main insulator means that it is also available for use in capacitors.

(b) The resistive and conducting inks are primarily an insulating printable medium with additions to sustain conduction, so the prospects for printing plate/dielectric sandwiches for capacitors are good.

(c) Screens and inks for printed components permit the dimensional tolerance required. Thickness of insulator, because of its connection with breakdown strength, is probably the most crucial.

SAQ 2.8 The functional characteristic is the capacitance C, expressed in farads, though microfarads ($\times 10^{-6}$) down to picofarads ($\times 10^{-12}$) are the practical ranges. As we have seen, this represents an a.c. reactance of $1/\omega C$. The other part of the specification is the working voltage which, with some safety margin, is the maximum r.m.s. voltage which it can withstand. Above this level, dielectric breakdown may occur.

Other performance factors, which are to do with both material and geometry, are:
- the d.c. resistance,
- the dielectric loss,
- the maximum working frequency,
- the temperature-variation of capacitance,
- overall size.

SAQ 2.9 High capacitance comes from conductors with large area, separated by a thin layer of insulator with a high dielectric constant. High working voltage calls for a thick layer of insulation so that the maximum voltage gradient that the material can withstand is not exceeded. Here is the fundamental conflict of capacitor design.

SAQ 2.10 The volume of the dielectric is
$$\mathscr{V} = A \times d$$
so
$$A = \frac{\mathscr{V}}{d}$$
Since $C = \varepsilon_0\varepsilon_r\mathscr{V}/d^2$, then
$$\mathscr{V} = \frac{C d^2}{\varepsilon_0\varepsilon_r}$$

Thus the designer will strive for the thinnest possible dielectric if minimum volume is the goal.

SAQ 2.11 For the ceramic
$$\mathscr{V}_{ceramic} =$$
$$\frac{10^{-2} \times 10^{-6} \times (25 \times 10^{-6})^2}{9 \times 10^{-12} \times 1800}\, m^3$$
$$= 0.038 \times 10^{-8}\, m^3$$
$$= 0.38\, mm^3$$

The polymer capacitor will be smaller by a factor of

$$\left(\frac{25}{2}\right)^2 = 156 \text{ times}$$

because of its thinner dielectric, but bigger by a factor of
$$\frac{1800}{3.2} \approx 560 \text{ times}$$

by virtue of its poorer dielectric constant. The outcome is 3.6 times larger, but the volume is still only
$$\mathscr{V}_{polymer} = 3.6 \times 0.38\, mm^3$$
$$\approx 1.4\, mm^3.$$

The connections will make the whole device larger than these volumes. Proportionally the ceramic is going to lose some of its advantage.

Comparing the paper dielectric with the ceramic, both multipliers are upwards, by a factor of about 780 for the effect of ε_r and 25 for the extra thickness. The final volume is $\mathscr{V}_{paper} = 780 \times 25 \times 0.38\, mm^3$
$$= 7400\, mm^3$$
$$\approx 7\tfrac{1}{2}\, cm^3$$

which is very much larger.

SAQ 2.12 The conductor areas are $A = \mathscr{V}/d$, giving:
- Ceramic area
$$= \frac{0.38 \times 10^{-9}}{25 \times 10^{-6}}\, m^2 = 0.15\, cm^2$$
- Polymer area
$$= \frac{1.4 \times 10^{-9}}{2 \times 10^{-6}}\, m^2 = 7\, cm^2$$
- Paper area
$$= \frac{7400 \times 10^{-9}}{125 \times 10^{-6}}$$
$$= 590\, cm^2$$

SAQ 2.13 The tiny area of the ceramic allows a single layer to form a worthwhile capacitance (see the single-layer ceramic capacitor in Figure 2.34). However, the proportions of the polymer are awkward: a volume of 1.4 mm³ and a plate area of 7 cm². For the paper capacitor things are even worse. Fortunately, the low dielectric constant of both these materials is combined with a high degree of mechanical flexibility and the 'Swiss-roll' construction can be used.

SAQ 2.14 The net dielectric loss is roughly proportional to the volume of dielectric, so it will be reduced if the dielectric volume is reduced.

Loss arises from the alternation of the polarization in the dielectric not being perfectly 'elastic'. The amount of loss is related to the number dipoles being switched, and hence to the amount of dielectric used.

SAQ 2.15 The d.c. resistance of the device will be increased. Also if the maximum working voltage is restricted by dielectric breakdown, a thicker dielectric will allow a higher voltage rating because the field strength (volts per metre) in the dielectric will be reduced.

Comment: Limits to thinness are to do with the practicalities of making very thin films without holes, and to do with handling during manufacture, when dielectrics may be damaged.

SAQ 2.16 You need the formula $R = \rho l/A$ with the dielectric thickness d being the length l of the resistive path. We have for the ceramic
$$R_{dc} = \frac{10^{12} \times 25 \times 10^{-6}}{0.15 \times 10^{-4}}\,\Omega$$
$$= 1.7 \times 10^{12}\,\Omega$$

whereas the polymer provides a d.c. resistance of
$$R_{dc} = \frac{10^{16} \times 2 \times 10^{-6}}{7 \times 10^{-4}}\,\Omega$$
$$= 2.8 \times 10^{13}\,\Omega$$

Paper performs less well:
$$R_{dc} = \frac{10^{10} \times 125 \times 10^{-6}}{590 \times 10^{-4}}\,\Omega$$
$$= 21\, M\Omega$$

The polymer wins by a good margin, though the ceramic gives a comfortably large value. The paper is barely adequate. Probably processing vagaries in bulk resistivity or on surface leakage will dominate d.c. leakage of modern capacitors; values of d.c. insulation resistance of around $10^{10}\,\Omega$ are claimed in catalogues.

SAQ 2.17 From 'Tan δ' we know

$$\tan \delta = \frac{1}{\omega C R_{ac}}$$

so

$$R_{ac} = \frac{1}{\omega C \tan \delta}$$

The power dissipated is

$$\text{power} = \frac{V^2_{rms}}{R_{ac}}$$

You can confirm this equation is consistent with the deduction in 'Tan δ' that

$$\text{energy lost per cycle} = \frac{V^2 \pi}{\omega R_{ac}}$$

The equivalent a.c. resistances of the capacitors are as follows.
For the ceramic,

$$R_{ac} = \frac{1}{2\pi \times 10^6 \times 10^{-8} \times 0.06} \Omega$$

$$\approx 265 \, \Omega$$

For the polymer,

$$R_{ac} = \frac{1}{2\pi \times 10^6 \times 10^{-8} \times 0.01} \Omega$$

$$\approx 1600 \, \Omega$$

For the paper

$$R_{ac} = \frac{1}{2\pi \times 10^6 \times 10^{-8} \times 0.0024} \Omega$$

$$\approx 6600 \, \Omega$$

The power dissipations become:
For the ceramic

$$\frac{V^2_{rms}}{R_{ac}} = \frac{10^2}{265} \text{W} = 0.38 \text{ W}$$

For the polymer

$$\frac{V^2_{rms}}{R_{ac}} = \frac{10^2}{1600} \text{W} = 0.063 \text{ W}$$

For the paper

$$\frac{V^2_{rms}}{R_{ac}} = \frac{10^2}{6600} \text{W} = 0.016 \text{ W}$$

The differences are exaggerated when the volumes of the dielectrics are considered. The powers dissipated *per unit volume* are

$$\frac{0.38}{0.38} \text{ W mm}^{-3} = 1 \text{ W mm}^{-3} \text{ (ceramic)}$$

$$\frac{0.063}{1.4} \text{ W mm}^{-3}$$

$$= 45 \text{ mW mm}^{-3} \text{ (polymer)}$$

$$\frac{0.016}{7400} \text{ W mm}^{-3} = 2.2 \text{ μW mm}^{-3} \text{ (paper)}$$

The significant losses in the ceramic may be sufficient to influence the running temperature of the device and hence both its capacitance and voltage rating.

SAQ 2.18 A thickness of 25 μm is 0.025 mm, so the ceramic capacitor should withstand (10 × 0.025) kV or 250 V, whereas 2 × 10^{-3} mm of PET can withstand (30 × 0.002) kV or 60 V. The thick paper can withstand (16 × 0.125) kV = 2 kV. A fully dense ceramic would have much higher breakdown voltage ($\approx$ 25 kV mm^{-1}) but the fabrication is liable to leave porosity in the material. They are often rated at only 4 kV mm^{-1}. The polymers can give their best performance more reliably and are used in higher-voltage applications even though a thicker than minimum film is required.

Table 2.10

SAQ 2.19 See Table 2.10.

SAQ 2.20 Putting larger 4 + ions into the octahedral cages will push the oxygen ions further apart. The twelve-cornered cuboctahedral cages formed by corner-sharing octahedra will therefore be enlarged. So the 2 + ion in this cage in its cubic form will have greater entropy (more spatial freedom) than before. In the free energy equation the $-T\Delta S$ term therefore becomes dominant at a lower T and the switch to cubic structure as T rises happens earlier.

SAQ 2.21 PP ought not to contain any polar groups, so its dielectric constant, depending on electron displacement only, should not be affected by temperature. However the C—H bond on the atoms linked to three other C atoms is easily broken by ultraviolet irradiation or by heat. The subsequent oxidation of this stable free radical introduces

$$\overset{+}{\diagup}C = \overset{-}{O}$$

Given that a very thin film is to be made which will present a large surface area for reaction, unwise processing conditions (too hot, in air) could then produce a rather lossy dielectric film.

Virtue	Rank		
	1 (best)	2	3
small size	X7R	PET	paper
small area	X7R	PET	paper
large d.c. resistance	PET	X7R	paper
small dielectric loss	paper	PET	X7R
great dielectric strength	paper	X7R	PET

Chapter 3 Something about magnets

3.1 Describing magnetic materials

3.1.1 Thinking about magnets

EXERCISE 3.1   (Revision)

(a) Figure 3.1 shows typical *B–H* curves for hard and soft magnetic materials. Which is which?

(b) What is hysteresis loss and how is it associated with the curves in Figure 3.1?

(c) What are domains and how do they affect the magnetism of a bulk sample?

(d) Distinguish between hard and soft material in terms of domains.

(e) What happens at the Curie temperature of a magnetic material?

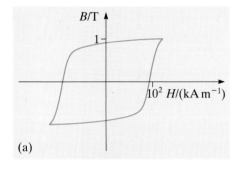
(a)

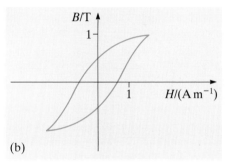
(b)

Figure 3.1 Schematic *B–H* curves

Let's begin with Exercise 3.1 and some casual observation and speculation. If you can, get a piece of magnetic recording tape or a floppy disc and look at its active area. What can you see? Just a brown, shiny surface — quite featureless. In Figure 3.2 a scanning electron micrograph (SEM) reveals a little more detail. At low magnification there is still little to pick out. But look at the sub-micron scale. The coating is a mat of short 'sticks'. I suppose they are particles really. Remember, this is magnetic tape, so can we speculate on what makes it magnetic? Each particle must be a tiny magnet. Remember, too, that this is a well developed product so the size and shape of those tiny magnets must have been specified or selected as suitable. There must be reasons for the choice. We'll be in a position to think about the shape later, but we can immediately think about the scale of the particles. I suppose that larger particles might make a recording sound 'grainy', rather as a photograph looks 'grainy' when examined on a small enough scale. Now, with 0.1 µm particles I could easily record things on the tape which change over about 1 µm — say a sine wave with 1 µm wavelength. Also, my tape

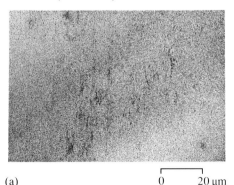
(a)   0   20 µm

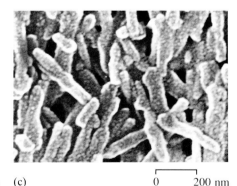
(b)   0   2 µm

(c)   0   200 nm

Figure 3.2 Micrographs of recording tape

recorder moves the tape at about 1 cm s^{-1} so the 1 μm wave would get carried past the playback head at a rate of $10^{-2}/10^{-6} = 10^4$ cycles per second, or 10^4 Hz. Interestingly, a piano runs out of keys at about half of this frequency and the best human ears don't work much beyond twice it. So the particle size is not unreasonable.

Why don't they pack the particles even closer together? Of course, two magnets will only sit comfortably close if paired oppositely side by side or else maybe nose to tail (see Figure 3.3). And magnetic fields somehow 'prefer' to be in magnetic material (which must be something to do with being in a low energy state). Perhaps closer packing would confine the field too closely in the material for it to be useful. We need to know more.

But what about domains? How do they fit into this? Do I see lots of tiny, apparently magnetic particles when I look at a lump of iron? Figure 3.4 shows a micrograph of a cheap permanent-magnet ferrous alloy. You've probably seen other micrographs of magnetic material. Clearly this is different from the tape. We can't see domain structure this way because we're looking at physical microstructure. (In fact the magnetic microstructure generally lies within the physical structure.) There are distinct particles in Figure 3.4, but many of them are actually non magnetic. Domains cannot 'ignore' non-magnetic inclusions, so perhaps these particles are there to impede changes in the domain structure. (We wouldn't expect such particles in soft magnetic material like silicon iron would we?)

I have already asked more questions than I have answered, so let's take stock and see where we can go for answers. There is magnetism in the particles on the tape, and generally in permanent magnets and in soft magnetic alloys, and someone has had a go at manipulating it. We'll look at the range of available materials in the next subsection, which starts by reminding you of the relevant jargon. We will also check there that we know the basic science before we have a look at motors and transformers. Controlling domains is at the heart of optimizing magnetic materials; we will look into how we can do this as we go along. The final sections return to magnetic recording (but not just tapes); this allows us an even closer look at some useful magnetic materials.

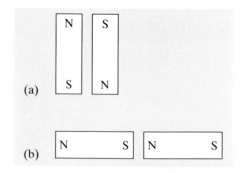

Figure 3.3 Favoured positions for a pair of magnets

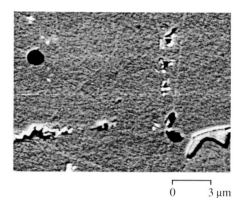

0   3 μm

Figure 3.4 Micrograph of ferrous (magnetic) alloy

3.1.2 Talking about magnetic materials

To discuss magnetism, you've first got to be able to get your tongue and your brain around some symbols, words and phrases in which subtle changes have profound effects. Watch out for μ (Greek mu), μ_B and μ_0 which are all different quantities with traditionally the same symbol. Beware too of variations like ferromagnetic and ferrimagnetic, where the prefix hardly changes but the material does! You should have been in parts of this minefield before, but just to refresh your memory, try ▼A little electromagnetism▲

▼A little electromagnetism▲

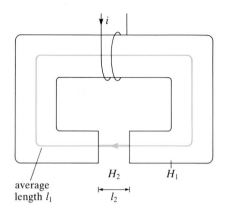

Figure 3.5 Ampere's law

$$H_1 l_1 + H_2 l_2 = 2i \qquad (3.1)$$

EXERCISE 3.2 If a field of H_2 exists in the air gap in Figure 3.5 when $i = 0$ because the material is a permanent magnet, obtain an expression for the field in the magnet.

Even more familiar to you should be Faraday's law: The e.m.f. induced in a loop of wire is equal to the rate of change of (magnetic) flux through it. So N turns of wire on a coil through which the rate of change of flux is $d\Phi/dt$ have a total voltage of

$$\varepsilon = N d\Phi/dt$$

across them. The following will allow you to exercise your electromagnetism.

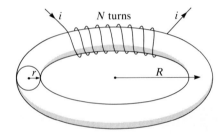

Figure 3.6 Electromagnet for Exercise 3.3

I'm going to assume that you've come across basic electromagnetism elsewhere in your studies and that here I need only remind you of some useful ideas, in addition to what was said in Chapter 1.

Can you recall Ampere's law for deducing the strength of a magnetizing field in and around magnetic material (see Figure 3.5)? For our purpose it is sufficient to state the law thus: Following the path of magnetizing fields which form a closed loop in sections, the sum of $H \times l$ in all the sections is equal to the total current enclosed by the loop. In Figure 3.5 therefore

EXERCISE 3.3 Figure 3.6 shows N turns of wire carrying a current i around a toroid (or doughnut) of magnetic material in which the flux density B ($= \Phi/\pi r^2$) increases in direct proportion to the magnetizing field, that is $B = kH$. Radii R and r are defined in the figure. Complete the following expressions with appropriate algebra.

(a) The 'Ampere's law path' at radius R encloses a total current of Ni so

$$2\pi R H =$$

(b) The mean flux density B is therefore
$$B =$$

(c) Hence the mean flux Φ ($= \pi r^2 B$) changes if the current i does, and the rate of change is

$$\frac{d\Phi}{dt} = (\qquad)\frac{di}{dt}$$

(d) The induced e.m.f. across the windings V is

$$V = (\qquad)\frac{di}{dt}$$

where the bracketed quantity is called the inductance of the coil.

We must rummage right down into the atoms if we are to find the source of magnetism in materials; and it is the co-operation of vast numbers of atoms one with another, through the intricate behaviour of electrons, which finally gives rise to the magnetic effects we detect on scales of microns to millimetres and upwards. Yet we don't detect the strong phenomena of magnetism in all matter but only in the elements iron, cobalt and nickel (and one or two rare ones), in most of their alloys and in many transition-metal oxides. We shall be looking for what it is that these materials have in common. Much of our discussion of the technologically important magnetic phenomena will be on a scale comparable with the metallurgists' optical micrographs. Even at this level we shall see distinct and useful differences between these materials which we can exploit in a variety of applications from power transformers to floppy discs.

Table 3.1   Magnetic terms and symbols

Quantity	Symbol	Remarks	Typical figure
Saturation flux density or induction	B_s	The maximum contribution to the magnetic flux density from the material — all atomic moments maximally aligned.	
Remanence	B_r	The magnetic flux density in the material when the net magnetizing field strength H (due to external currents and the material) is zero.	
Coercivity (and intrinsic coercivity)	H_c (and H_{ic})	The magnetizing field strength required to reduce the net flux density B (or else the flux density due to the material alone, B_i) to zero.	
Relative permeability	μ_r	$\mu_r = B/\mu_0 H$. The ratio of flux density to net magnetizing field strength.	
Differential permeability	μ	$\mu = \dfrac{1}{\mu_0}\dfrac{dB}{dH}$. Note: this is not the same quantity as $\mu_0\mu_r$.	
Saturation magnetization	M_s	$B_s = \mu_0 M_s$.	
Hysteresis loss or energy loss per cycle	W_h	A frequently used figure of merit for soft material: the energy cost of one cycle of magnetization.	
Energy product	$(BH)_{max}$	A frequently used figure of merit for permanent magnet materials indicating capacity to support external fields.	

Table 3.1 and the text which follows present some of the quantities which are used to characterize magnetic materials. I expect you have come across them before. (The blank column in Table 3.1 is part of an exercise which you should attempt at the end of this section.) Soft magnetic materials are comparatively easy to magnetize to saturation and are likewise easily remagnetized in the reverse direction; they find applications in transformer cores, inductors, electromagnets and elsewhere. Hard magnetic materials are useful as permanent magnets in motors, actuators and fastenings and also as a storage medium on tapes and discs.

3.1.3 B, H, M and permeability

The *concentration* of a *magnetic* field is measured by the density of magnetic field lines, or by the flux density B. The *strength* of a *magnetizing* field H is measured in terms of equivalent electric current (per metre of magnetizing field line) which could be used to generate the field H. A magnetizing field H is associated with real currents in wires (such as in a coil) and also with the ends of magnetized material (that is the poles of a magnet). The flux density B is associated with a field H, which in a vacuum is

$$H = B/\mu_0$$

(where μ_0, the permeability of free space, is a constant equal to $4\pi \times 10^{-7}$ henry per metre).

When some material is present we can introduce a relative permeability μ_r so that the same flux density B would be produced by a magnetizing field now given by

$$H = B/\mu_0\mu_r$$

For most non-magnetic materials μ_r is independent of B and is very close to unity. Diamagnetic substances have μ_r less than one, whereas paramagnetic substances have μ_r just greater than one. Figure 3.7(a) shows these different cases. Figure 3.7(b) and (c, black curve) show B as a function of H for soft and hard materials. Clearly we cannot determine a single value for the relative permeability since the quantity μ_r is

$$\mu_r = \frac{B}{\mu_0 H}$$

and it has a different value for different flux densities. Indeed for some values of B there are two values of H and so two values of μ_r. Nevertheless, $(\mu_r)_{max}$ is sometimes used to describe the maximum μ_r on the increasing H portion of a B–H loop (lower path of loop) and serves to indicate the 'softness' of a material: a high $(\mu_r)_{max}$ means *large* flux density for little magnetizing field, that is softness.

When magnetizable materials are present, the flux density B is due both to magnetizing fields H and to the magnetization of the material (M). Since the relative permeability μ_r is not so convenient, an alternative relationship is this one:

$$B = \mu_0(H + M) \tag{3.2}$$

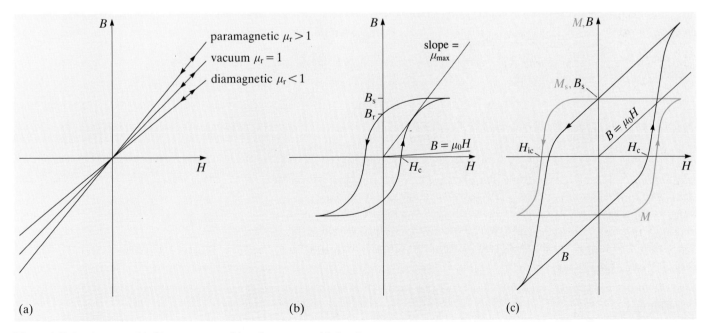

Figure 3.7 *B*–*H* curves. (a) Non magnets, (b) soft magnets, (c) hard magnets

Here the idea of relative permeability is set aside in favour of a more explicit account of the source of the flux. Now the complicated behaviour is in the variation of magnetization M with magnetizing field H. The response of a magnetic material to a magnetizing field is a characteristic of the material, and is often summarized in the M–H curve (Figure 3.7c, green), or loop, but you are probably more familiar with the B–H loop.

The source of magnetism is within the atoms of the material and once each atom has contributed, no further magnetization can be achieved. The material is said to be saturated. Thus the B–H curve at large magnetizing fields approaches the limit

$$\begin{aligned} B &= \mu_0 H + B_s \\ &= \mu_0(H + M_s) \end{aligned}$$

in which B_s is the constant saturated flux density arising from the material.

Once a material is magnetized into saturation, simply reducing the magnetizing field H to zero leaves a remanent flux density B_r. This is exploited in permanent magnets. In order to reduce the flux to zero it is necessary to apply a demagnetizing field (negative H), the specific value being termed the coercive field or **coercivity** H_c. Soft materials must have low coercivity, giving little opposition to changes in magnetization. In permanent magnet material (hard) the coercive field may reduce the net flux B to zero but it may still leave most of the material magnetized as before (that is $M = M_s$). In this case it is simply that the demagnetizing currents give rise to a field which exactly cancels the material's magnetization, without reducing it at all. The intrinsic coercivity H_{ic} is that field required to reduce to zero the intrinsic magnetization M.

EXERCISE 3.4   If the magnetization M of a material simply increases in direct proportion to the magnetizing field H, what can you say about its relative permeability?

3.1.4 Energy loss and energy product

The B–H curve exhibits **hysteresis**. That is, increasing H produces values of B different from those where H is decreasing. Notice the arrows in Figure 3.7. Some of the microscopic processes of magnetization are not thermodynamically reversible. As a result, energy is expended in completing one circuit of a B–H loop. In fact the area enclosed by a B–H loop is equal (in SI units) to the energy per unit volume which must be expended in traversing the loop. See ▼ B–H loops and energy ▲.

For soft magnetic materials in a.c. applications (such as mains transformers) we require small area B–H loops since each time the cycle is completed a 'loop area' of energy is expended (irreversibly, as heat). The appropriate material property is the energy loss per cycle, per cubic metre of substance.

▼ B–H loops and energy ▲

We can work out the energy expended in taking a specimen of magnetic material round its B–H loop by using straightforward arguments from electromagnetism. Consider a toroid (or doughnut) of magnetic material with a single turn of wire around it. See Figure 3.8(a). Now, if a current i flows in the wire, Ampere's law tells us that the associated magnetizing field H in the toroid is

$$H = \frac{i}{2\pi R} \qquad (3.3)$$

where R is the mean toroidal radius. If the current flows for a time Δt, the associated step in flux density ΔB means that the magnetic flux through the wire loop changes by ΔBA in a time Δt and Faraday's investigations lead us to conclude that the voltage induced across the wire of the coil is

$$V = \frac{A\Delta B}{\Delta t} \qquad (3.4)$$

But to supply a current i against a voltage V requires a power (energy per second) P given by

$$P = iV \qquad (3.5)$$

Equations 3.3, 3.4 and 3.5 can be used to account for the power supplied to change the flux density by ΔB (in time Δt):

$$P_{\Delta B} = 2\pi RH\, A\,\frac{\Delta B}{\Delta t}$$

The energy expended is $P_{\Delta B}\Delta t$ and the magnet volume is $2\pi RA$, so the energy ε supplied per unit volume for a change ΔB would be

$$\varepsilon_{\Delta B} = H\Delta B$$

and for a change from $-B_r$ to B_{max} we could integrate the energy expression:

$$\varepsilon_1 = \int_{-B_r}^{B_{max}} H\mathrm{d}B.$$

On the B–H curve in Figure 3.8(b) this is the area between the rising B curve (from a to b) and the B axis. This energy has been put into the effort of magnetizing and into the magnetic field itself. The energy to reduce B to $+B_r$ is negative (B decreases) as some energy stored in the magnetization is given back. This is the path b to c on Figure 3.8(b). Energy is stored in the magnetic field and some can be recovered by reducing the magnetization. The net expenditure of energy per unit volume in going from $-B_r$ to $+B_r$ is thus the area between the curve abc and the B axis.

What is the net total energy cost if the flux density is returned to $-B_r$?

It's the area of the loop multiplied by the magnet volume.

(a)

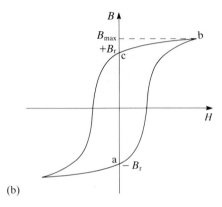

(b)

Figure 3.8 (a) A toroidal electromagnet and (b) its B–H curve

With hard magnetic materials for permanent magnets, cycling from one magnetization to the reverse and back is generally not a service requirement. Instead, it is customary to catalogue the maximum instantaneous value of the product BH for the portion of the B–H curve where B is positive but H is negative (demagnetizing). This is shown in Figure 3.9. In fact for permanent magnet material $(BH)_{max}$ would be somewhat less than a quarter of the cyclic energy loss; it has a special significance in optimising the design of a working magnet (see ▼$(BH)_{max}$▲).

There is a useful class of magnetic material which is hard but is repeatedly cycled in use. This is the stuff used for magnetic recording on tapes and disks. (Iron oxide is an example.) In this application the small volume of material makes the energy costs associated with recording secondary to quality of reproduction; energy loss data are not usually quoted.

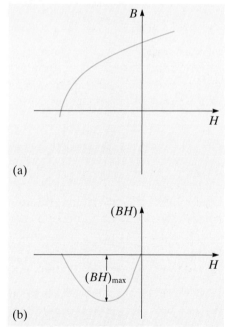

(a)

(b)

Figure 3.9 Part of B–H curve and (BH) product

▼(BH)_max▲

Figure 3.10(a) shows the doughnut magnet of Figure 3.8 magnetized to saturation and then sawn into two equal halves. With a little electromagnetism we can work out what happens to the flux density when the two halves are separated. When the magnet is intact, if there are no currents enclosed by the magnetic circuit around the doughnut, we have $H = 0$ and $B = B_r$. Now the magnetic circuit (across both gaps and around both halves of the magnet) still encloses no current. Ampere's law would tell us that

$$2H_m l_m + 2H_g l_g = 0 \qquad (3.6)$$

where H_m and H_g are the magnetizing fields in the magnet and in the gap respectively. If the gaps are small compared with the thickness of the section, the flux density will be the same in both magnet and air gap, say B. Furthermore, in the air B and H_g are approximately related by

$$B = \mu_0 H_g$$

Putting $H_g = B/\mu_0$ in Equation (3.6) gives

$$B = -\frac{\mu_0 H_m l_m}{l_g}$$

This straight-line equation must be satisfied simultaneously with the B–H-loop relation for the material. Figure 3.10(b) sketches the graphical solution. Notice that the magnets are a little demagnetized, but we now have a magnetic field, and so energy, outside the magnet and available to do work. How much energy depends upon both the flux density and the magnetizing field strength in the air gap, and on the volume of the gap, so we can write the available energy as

$$W_g \propto BH_g l_g$$

Substituting for H_g from the Equation (3.6) gives

$$W_g \propto BH_m l_m$$

For a given magnet, l_m is fixed and the gap energy is maximum when (BH_m) is maximum. Designs which operate at $(BH)_{max}$ make best use of available magnetic energy.

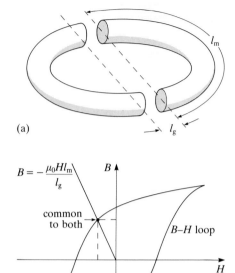

(a)

(b)

Figure 3.10 A magnet with an air gap

Tables 3.2 and 3.3 collect data for a selection of soft and hard materials. Remember that the magnetic properties of materials depend critically on the manufacture and previous history of the specimen. The values in Tables 3.2 and 3.3 should therefore be taken as typical only.

In due course we will go down almost to the atomic level to see what separates hard and soft materials on this scale, and therefore to see how we can control these qualities. We finish this section by noting the types of material which appear in Tables 3.2 and 3.3.

3.1.5 Elements, alloys and oxides

The state of spontaneous magnetism is not restricted to a few pure elements (Fe, Co, Ni, Gd) but is exhibited by certain alloys and compounds of these elements and also certain alloys and compounds of some other transition and rare earth metals.

Alloys of iron and nickel form a range of magnetic materials in which the atoms of the component elements interact strongly so that the magnetism and the electrical conductivity are sensitive to composition (Figure 3.11). The interaction is such that iron–nickel alloys having about 27% nickel are not ferromagnetic at room temperature. In contrast, iron and cobalt atoms interact in such a way that a 50% alloy exhibits about 12% more magnetism than pure iron. This alloy has the highest saturation density of all.

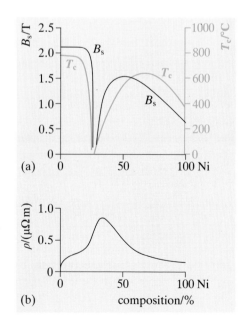

(a)

(b)

Figure 3.11 Magnetic properties of Ni–Fe alloys

Table 3.2 Magnetic properties of some soft magnetic materials

Material	T_c/K	B_s/T	H_c/A m^{-1}	μ_r	$\rho/\mu\Omega$ m	W_H/J m^{-3} cycle^{-1}
pure iron	1043	2.2	4	2×10^5	0.1	30
mild steel	1000	2.1	143	2×10^3	0.1	500
silicon (3%) iron	1030	2.0	12	4×10^4	0.5	30
78 permalloy (Fe 21.5 Ni 78.5)	800	1.1	4	10^5	0.2	
supermalloy (Fe 16 Ni 79 Mo 5)		0.8	0.16	10^6	0.6	
ferroxcube 3 (Mn Zn ferrite)	570	0.25	0.8	1.5×10^3	10^6	13
amorphous iron boron silicon	630	1.6	—	$>10^5$	10^3	15

Table 3.3 Properties of some hard magnet materials

Material	Composition (balance Fe)				T_c/K	Remanence B_r/T	Coercivity H_c/kA m^{-1}	Maximum B × H $(BH)_{max}$/kJ m^{-3}	Comments
	% Al	% Ni	% Co	% Cu					
alnico IV H	12	26	8	2		0.6	63	13	isotropic
alnico V	8	13.5	24	3	1160	1.35	64	44	anisotropic-columnar crystals (aligned by magnetic annealing)
barium ferrite	BaO (Fe$_2$O$_3$)$_6$				720	0.4 / 0.2	264 / 135	28 / 8	anisotropic isotropic
samarium cobalt	SmCo$_5$				1000	0.85	600	140	
neodymium iron boron	Nd$_2$Fe$_{14}$B				620	1.1	890	216	low T_c (420 K)
γ iron oxide	γFe$_2$O$_3$					0.21	25		acicular particles
magnetite (iron ferrite)	FeOFe$_2$O$_3$				850	0.27 / 0.27	25 / 43		acicular particles 100 nm film

Alloying with a magnetically inert metal will principally dilute the magnetism of a pure element. But, transition-metal alloys are notoriously fickle so it's no surprise that stranger things happen. For example, stainless steels include a large fraction of chromium and nickel, which can lead to an austenitic (FCC) phase at room temperature which is 'non-magnetic'.

Naturally occurring magnetism (in rocks and stones) is predominantly an ionic phenomenon, since ferromagnetic materials are not generally found as pure elements. (Iron is found as sulphide or oxide.) The mineral magnetite FeO.Fe$_2$O$_3$, also known as lodestone, has been exploited in navigation for centuries because of its spontaneous magnetization. It is one of a group of so-called ferrimagnetic materials known generally as **ferrites**. **Ferrimagnetism** is exhibited by many transition-metal oxides; the detailed atomic-scale mechanism differs from that of ferromagnetism but at the domain level there is little apparent difference. As we shall see presently, some ferrites are particularly attractive for high-frequency applications (where electrically insulating material is essential), although the inevitable dilution of magnetic atoms by oxygen in the oxide limits the achievable saturation flux density to less than 0.5 T, or so.

EXERCISE 3.5   Use the data from Tables 3.2 and 3.3 to fill in the 'typical figures' column in Table 3.1, giving answers for soft and/or hard where appropriate.

SAQ 3.1   (Objective 3.1)
Sketch the B–H loop for a sample of barium ferrite and on the same axes, sketch that of manganese-zinc ferrite. Pay attention to their respective coercivities, permeabilities, remanences and loop areas.

3.2 Permanent magnet motors

Magnetic materials are used in many electrical and mechanical devices, such as motors, transformers, transducers, catches and so on. In this section and the next we will look at just two examples of magnetic material in service. One example uses hard material (permanent magnets for d.c. motors) and the other uses soft material (transformer cores). Look out for how in each example the design considerations are bound up with the shape and property requirements of the magnetic material. Remember that similar observations will apply in other applications.

Electric motors with permanent magnets (p.m.) are used in a wide range of applications such as those illustrated in Table 3.4. In a p.m. motor, the current injected into the motor windings creates a magnetic field. The interaction between this field and that of the permanent magnets produces the force which drives the rotation.

SAQ 3.2   (Revision)
A conductor carries a current i perpendicular to a field B. What is the force per unit length of conductor?

▼P.m. motors▲ will remind you of the general form of the motor in question. In designing such a motor we must examine some of the factors which bear on the selection of the permanent magnets — namely demagnetization, efficiency and magnetic flux density.

▼**P.m. motors**▲

Figure 3.12 shows a schematic p.m. motor. A commutator arranges for currents always to be down and up as shown while the rotor spins. The rotational force arises from rotor coil currents flowing perpendicular to the field of the permanent magnets. Can you see that the motor shown will run clockwise? In practice the rotor contains several turns of wire.

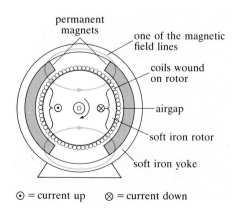

⊙ = current up   ⊗ = current down

Figure 3.12 A p.m. motor

Table 3.4 Applications of permanent-magnet (p.m.) motors

Area	Examples	Power level
industrial servo-drives	lathes, winding gear	kW
domestic equipment	audio turntable, model railway engines	W
automotive accessories	starter motors, windscreen motors	W–kW
computer peripherals	disc drives	W
transportation	propulsion units	kW–MW
aerospace	flight control actuators	W–kW

3.2.1 Demagnetization

We have already seen that with an air gap, magnetic field is forced to be outside the magnetizable material (that is, in the gap) and this costs energy. As the field is squeezed out, the magnet starts to demagnetize itself to provide the energy, so that values of B and H correspond to a point with negative H, such as P_1 in Figure 3.13(a). Bigger gaps produce more demagnetization (recall '$(BH)_{max}$'). In addition, it turns out that motor currents produce additional demagnetizing fields so that B and H may be shifted further, to P_2 say. Thus there are two sources of demagnetizing influence:

(a) Air gaps in the magnetic circuit.
(b) Currents being enclosed by the magnetic circuit.

A suitable magnet for a p.m. motor should retain sufficient magnetization during operation and withstand all demagnetizing influences, returning to the same state of magnetization whenever the motor is switched off (that is, the B–H curve between P_1 and P_2 must be reversible).

Figure 3.13(b) shows a magnetic circuit equivalent to that in a motor, although rearranged for convenience — notice that the circuit drawn has the following important features:

(i)   A permanent magnet forms part of a magnetic circuit.
(ii) The magnetic circuit has a short air gap.
(iii) The magnetic flux encloses current flowing in a coil, and the direction of the current is to demagnetize the magnetic circuit.

The circuit in Figure 3.13 (b) is supposed to mimic aspects of the *magnetic* behaviour of the motor — at this stage I don't need the complication of a spinning rotor, but the current in the coil can take account of the rotor current.

3.2.2 Efficiency

The power developed in a p.m. motor depends on a number of quantities, which we can assess as follows. First if a force F moves a distance Δx (in the direction of the force) in a time Δt then the associated rate of work (that is, the power) is

$$P_m = \frac{F\Delta x}{\Delta t}$$

This is the mechanical power output. Remember that the motor relies upon the force on several current carrying conductors (wires of length l which are part of the coil) in a magnetic field B so we can write the force per conductor as

$$F = Bil$$

Thus the output power per conductor will be:

$$P_m = Bilv \tag{3.7}$$

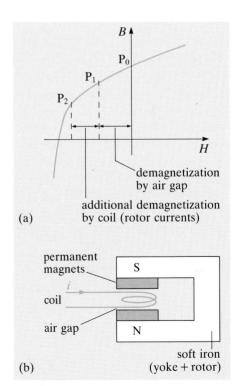

(a)

(b)

Figure 3.13 (a) *B–H* loop for the permanent magnet. (b) An equivalent magnetic circuit for a p.m. motor (where v is the conductor speed $\Delta x/\Delta t$). Of course, excluding superconductors, real conductors have resistance and so may dissipate power through ohmic heating. This power loss must also be supplied. For a conductor of resistivity ρ and cross section A, the power loss per conductor is

$$P_e = \rho\frac{1}{A}i^2$$

The total input power, ignoring all other losses, is the sum of mechanical power and ohmic heating power, so the efficiency (output/input) is

$$\eta = \frac{P_m}{P_e + P_m} = \frac{1}{\dfrac{P_e}{P_m} + 1}$$

The highest efficiency will have the ratio P_e/P_m as small as possible. How does this influence the design? Well, look at the ratio P_e/P_m:

$$\frac{P_e}{P_m} = \frac{\rho i}{BvA}$$

Efficient motors will have currents i small, but field B and rotor velocity v large (large diameter or high r.p.m.).

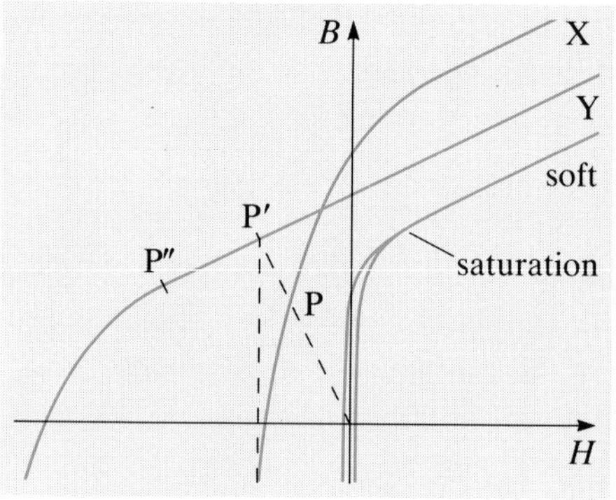

3.2.3 Magnetic flux density

Efficient p.m. motors will require magnets able to sustain a large flux density B. Is there a limit to how large a value of B could be used if available? There is indeed a limit. If you refer back to the magnetic circuit in Figure 3.13 you'll see soft iron is used to contain the flux in regions not required to be a permanent magnet. The soft yoke will have a very high permeability (lots of B for very little H) but saturation of the iron may severely limit the effectiveness of the magnet. (There may be mechanical constraints which prevent us from attempting to change the soft material, so let's stick with soft iron.)

EXERCISE 3.6   Figure 3.14 shows demagnetization curves for two hard magnet materials, X and Y, together with part of the B–H loop of the soft yoke parts of a p.m. motor. Why can't I simply make magnets from material Y for a motor design which specifies material X?

EXERCISE 3.7   Figure 3.15 shows demagnetization curves for p.m. materials (there are of course many others). Comment on the suitability of each one for a p.m. motor magnet. (Refer to Table 3.3 for other data.)

Figure 3.14 Demagnetization curves for two p.m. materials with a B–H loop for a soft material

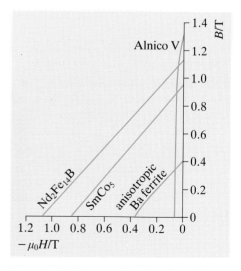

Figure 3.15 Demagnetization curves for various materials

3.2.4 Comparison of two p.m. materials for a motor

Suppose I have a 750 W, 2000 r.p.m. d.c. motor with a ferrite magnet. What scope is there for economy if I change the magnet material to neodymium–iron–boron? The motor power due to each conductor of length *l* carrying a current *i* (Equation 3.7) is

$$P_m = Bilv$$

Now, the new material will allow me to increase the flux density *B*, provided the soft iron does not saturate. So we could decrease *l* and get the same output from a shorter (and hence lighter) motor. But I could also consider using thinner magnets to get the *same* flux density as before. In this case the rotor diameter could be increased (and still fit inside the old case) so that at the same r.p.m. the conductors move faster (*v* increases) so again I could shorten the motor. Then again, I could increase *B* just a little and still end up with thinner magnets so that the diameter of the whole motor could be decreased, since a smaller conductor speed (*v*) could be accommodated.

Which should I choose? There is no easy answer. Remember that I ought also to worry about how hot the magnets get as the two materials are differently affected. So, having identified the potential for radical redesign leading to a lighter, more compact motor, I can hand the problem over to a computer-aided designer. In fact, D. Howe of Sheffield University has established that the following changes are feasible simultaneously.

- Decrease motor length by 30% (volume decreases by 30%).
- Decrease magnet volume by 80% (thinner and also shorter because of the above).
- Decrease volume of copper conductors by 25%.

These are achieved with a doubling of the operating flux density. Notice that the net weight saving will be dominated by the motor volume change and will be around 30%.

3.3 Transformer cores

Transformers with magnetic cores are used widely in several areas — usually, but not always, in connection with the supply of electrical power. Table 3.5 gives some examples.

Table 3.5   Transformer applications

Area	Examples
50 Hz Industrial/domestic power	(i) Step up for transmission at 11–400 kV (ii) Step down from 400–132–33–11–0.415 kV for distribution
Radio frequency power	Solid state power amplifier output stage
Aerospace power systems	400 Hz a.c. power distribution in aircraft
D.C. supplies for electronics e.g. at 24, 18, 12, 6 V d.c.	Mains-fed power supplies Switched mode power supplies
Signal processing	Pulse/h.f. transformer

The role of the core of a transformer is to couple magnetic flux from one circuit into another and it does this by providing the flux with a relatively easy path. In power applications, energy is transferred through the core by an oscillating magnetic field. The minimization of energy losses in the magnetic materials is of paramount importance. In this section I want to examine the factors which influence the selection of core material for a power distribution transformer. ▼General principles of transformers▲ is a reminder of how transformers work.

▼General principles of transformers▲

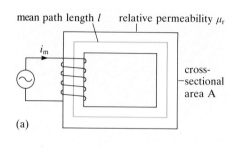

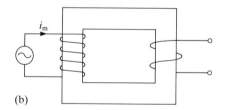

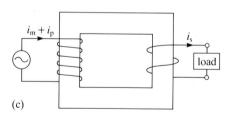

mean path length *l*   relative permeability μ_r

cross-sectional area A

(a)

(b)

(c)

load

i_m

i_m

$i_m + i_p$

i_s

Figure 3.16 Transformer core plus coils

This discussion is intended to help you appreciate the specifications of transformer cores. I want you to realize why transformer cores for power distribution must have high permeability and low loss.

Let's invent the transformer from scratch. Figure 3.16(a) shows a coil wound around

a loop of soft magnetic material (the core). If the coil is connected to an alternating voltage source, can we say how the current drawn from the source will vary? There will be an alternating current i_m, say. Next, Ampere's law tells us that the alternating current in the coil will produce an alternating flux Φ in the core. Faraday's law tells us that the alternating flux in the core would produce an alternating e.m.f. (a voltage) in the turns of any coils threaded by it, including the one whose current is producing the flux. If the total e.m.f. induced in our coil does not exactly match what we are supplying then more or less current will be forced to flow. Suppose the supply voltage fractionally exceeds the e.m.f. Fractionally more current gets into the coil and more flux ensues, and in turn more e.m.f. is generated. A dynamic balance is maintained.

Figure 3.16(b) introduces a second coil. An e.m.f. will also be produced across the turns of this coil if the core flux is changing. Each turn of the coil has $d\Phi/dt$ generated across it (Faraday's law), so N turns will have $Nd\Phi/dt$ across them. We call this coil the **secondary winding** and the first one the **primary winding**.

EXERCISE 3.8
Explain, with words only, how the current which flows in the primary coil is determined.

In a distribution transformer, as consumers use power the secondary circuit must in turn drive currents (using the e.m.f caused by flux changes), so Figure 3.16(c) is now

appropriate. But currents in the secondary will contribute to the flux in the core. Hence the new e.m.f. of the primary will not match the voltage source unless the primary current adjusts to restore the core flux and so bring back the dynamic balance between primary e.m.f. and source voltage. As power is drawn out of the secondary, it is funded by the supply of an equivalent amount to the primary. The basis of power transformer action is therefore:

1 The primary coil magnetizes the core to ensure that flux changes generate an e.m.f. which just matches the applied voltage. The associated current is called the magnetizing current, i_m.

2 The secondary coil has an e.m.f. induced in it.

3 If current flows in the secondary, it starts to upset the primary balance, so an extra component i_p of primary current flows to compensate and so retain the balance.

4 Power supplied by the secondary, say $V_s i_s$, must be supplied to the primary, say $V_p i_p$. In practice, of course, some of the power supplied to the primary ends up as waste heat, so the secondary doesn't get it all.

EXERCISE 3.9
What can you say about the magnetizing current required for a high permeability core as compared with that for a low permeability core?

3.3.1 The economics of power transformers

A power transformer costs money to buy, install and run. Although it has no moving parts to wear out and can be protected from serious faults, there is a running cost associated with power lost as a result of its operation. You probably know that electricity is transmitted at high voltage in order to reduce power losses (hence the need for transformers in the supply network). What are the sources of power loss in a transformer? There are two sorts of loss: magnetic and resistive. The magnetic losses are of most concern to us here, but the resistive component has a strong bearing on the complete design.

Let's deal first with resistive power loss. Transformer coils have some resistance, as have all practical conductors, and we must keep an eye on it. Resistance is a combination of a materials property (resistivity ρ) and a product property (dimensions l, A); it is $\rho l/A$. Also, we must keep an eye on the current i within the resistance. Together resistance and current produce 'waste' heat at the rate or $i^2\rho l/A$ (in watts, if l and A are suitably dimensioned). When the transformer is in the so-called 'on load' condition, currents flow in both the secondary (i_s) and primary (i_p) coils. Don't forget though that a magnetizing current (i_m) must also flow in the primary, even when the secondary is 'off load' (from 'General principles of transformers'). Here then is our first core specification: to minimize magnetizing current i_m and its associated heat loss, the core should magnetize as readily as other conditions allow; it must be of high permeability.

The core itself gives rise to two magnetic losses. The first is hysteresis loss. Each time the core is taken round its B–H loop, magnetized one way, then the other and back again, energy is lost reshaping magnetic domains. Here is our second core specification: the B–H loop of the core must be narrow; it must have low energy loss per cycle.

The second magnetic loss also arises from the changing flux in the core. Just as there is Faraday induction of e.m.f. in the coils, so we get Faraday induction of e.m.f. in the core material. It causes circulating currents within the core material (more heat loss). ▼Eddy currents▲ examines them further. A third specification is for a core design and material which keeps down the losses due to eddy currents.

The core material is responsible for much of the weight and volume of the transformer. It must contain flux Φ, but it can do this only up to its saturation limit on the flux density (B_s). The larger B_s, the smaller the core's cross-sectional area A which is needed to contain the peak flux and so the less core (and coil) material is required. Heavy transformers cost more to buy and to install, than lighter ones.

▼Eddy currents▲

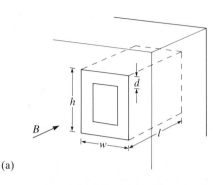

(a)

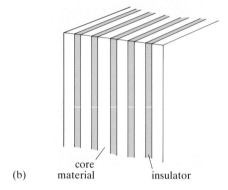

(b) core material   insulator

Figure 3.17 Eddy current paths and laminated cores

Figure 3.17 shows a portion of magnetic material in which a flux density of B is varying. I want to look at currents induced within it and the associated power loss. Faraday's law will help. Any shell of material such as that shown (thickness d, height h, length l and width w) encloses a flux which changes and hence generates an e.m.f. Provided d is small, the flux threading through the shell is roughly

$$\Phi = Bhw$$

and its rate of change is

$$\frac{d\Phi}{dt} = hw\frac{dB}{dt}$$

So the e.m.f. induced in the shell loop is

$$\varepsilon = hw\frac{dB}{dt}$$

The resistance of a path around the shell is

$$R = \frac{\rho 2(h + w)}{ld}$$

and so there must be a current in the shell of

$$i = \frac{\varepsilon}{R}$$

with an associated power loss ΔP_e (i^2R or ε^2/R):

$$\Delta P_e = \frac{(hw\,dB/dt)^2 ld}{2(h + w)\rho}$$

I can consider all the material to be made up of a set of concentric shells so that we have just considered one of several nested rings. Each ring has its so-called eddy current and an associated loss. We don't need to do any more algebra though to be able to conclude how to control eddy currents. (Notice that eddy currents themselves produce magnetic fields, which are directed against the field causing them, so any detailed calculations must use the net magnetic field $\mathbf{B}$ taking into account the eddy currents.)

How then can losses associated with eddy currents be made low? Look at the form of ΔP_e. We can't do much about dB/dt and l, which may be fixed by other needs, and the shell thickness d is just an arbitrary fraction of h and w. However, if h or w were small (tall, thin or short, wide loops) and if the resistivity ρ were kept as large as possible, ΔP_e would be kept low.

This is why iron transformer cores are laminated, with alternate sheets of iron and insulator as illustrated in Figure 3.17(b). The insulation constrains eddy currents to circulate within one sheet and so to adopt paths which are tall and thin (w small compared with h). The higher the resistivity of the iron, the better.

The total cost of a power transformer can thus be assessed as the cost of:

    purchase and installation
    + off-load loss (resistive + hysteresis + eddy current)
    + on-load resistive loss.

In a complete network with several transformers, the two running-cost elements are comparable with the maintenance costs of the cabling.

To summarize. The core design should aim to:
- maximize permeability and saturation flux
- minimize hysteresis and eddy current losses.

EXERCISE 3.10   Give two reasons why a transformer core gets hot, even when its not actively delivering power.

3.3.2 Material for power transformer cores

Exercise 3.11 lists four candidate materials for a power transformer core. I want you to have a go at selecting one, on the strength of what we have discussed. But to appreciate how refinements can be made, and have been made, we must return to the microscopic scale of domains in order to see what governs magnetic properties.

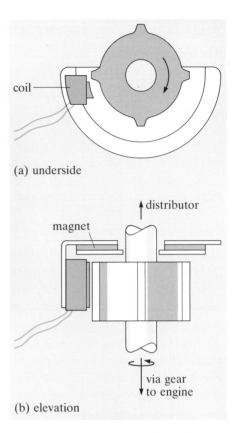

(a) underside

(b) elevation

Figure 3.18 Reluctance transducer for SAQ 3.4

> **EXERCISE 3.11**   Which of the following materials are unsuitable for a power transformer core? Appropriate date can be found in Table 3.2.
> (a) Ductile 0.3 mm sheets of 3% SiFe.
> (b) MnZn ferrite formed and sintered to order.
> (c) 30 μm thick ribbon, 30 cm wide, of amorphous iron boron silicon.
> (d) Dry air which has infinite resistivity and a constant permeability ($\mu_r = 1$).

Summary of Sections 3.2 and 3.3

● Magnetic materials may be hard or soft. Within these two categories, choice of material is determined by design factors such as cost, size, ease of manufacturing, service conditions and so on.

● Magnets in p.m. motors must provide the required magnetic flux efficiently and in the face of demagnetizing influences arising from air gaps and currents.

● Power transformer cores must magnetize with very little magnetizing current up to as high a flux as possible. Cyclic operation of transformers incurs hysteresis and eddy current losses in the core material.

> **SAQ 3.3   (Objective 3.2)**
> In electric motors, magnetic flux is generally guided to the air gap between stationary and moving parts by soft magnetic material. What sort of specification would you expect for its saturation flux density, permeability, resistivity, hysteresis loss, energy product?

> **SAQ 3.4   (Objective 3.3)**
> Figure 3.18 shows a simple 'breaker-less' spark timing system used in some electronic ignition systems for petrol engines. Instead of contact breaker points, an electronic switch (normally closed) is triggered open to interrupt a current in the conventional 'ignition' coil. The trigger is provided from a 'reluctance transducer' in the base of the distributor. As the lug on the shaft passes close to the core of the sense coil, magnetic flux is coupled through the coil, inducing a voltage. Discuss the selection of material for the magnetic components.

3.4 Magnetism and microstructure; domains

Ferromagnetic and ferrimagnetic materials both demonstrate the effect of spontaneous interaction of atomic magnets. Although different in atomic detail, both types of solid comprise regions of local net magnetism, called **domains.** At the domain level, ferromagnets and ferrimagnets behave in much the same way. To go further, we must look for the source of magnetism in atoms, so that we can understand how large numbers of atoms interact to form the domains. (An analogy would be to move from a discussion of microstructure to a discussion of chemical bonding.)

3.4.1 The quantum mechanical atom

The modern model of an atom evolved from a need to account for the characteristic light emissions from excited atoms. The model arises from quantum mechanics, in which electrons are considered more like waves than particles, being smeared out around the nucleus. From the mathematics come four **quantum numbers** which uniquely identify each possible electron state for an isolated atom. If we wish to visualize the mathematics, we must consider states in which electrons can reside, arranged in shells around the nucleus. The earlier image of electrons orbiting a nucleus is somewhat naive in detail but remains a useful concept. It is as if the electrons actually whizz around their quantum mechanical orbitals rather than just becoming smeared out mathematical abstractions. Indeed, the electrons have angular momentum as if they were circulating, so that they are in a way like electric currents in loops of wire. In fact, their 'orbiting' motion does produce a magnetic effect. Furthermore **electron spin** also produces angular momentum, and it too produces a magnetic effect. Spin is the more important source of magnetism in solids as the basic orbital effects get cancelled out in bonding. Each electron spin behaves like a tiny bar magnet, and when two spins are paired opposites their spins and their magnetism exactly cancel. So, it is only when there are unpaired spins that atomic magnetism stands out. (▼Atomic magnets and spectra in magnetic fields▲ looks at some evidence.)

The fundamental quantity of atomic magnetism is the so-called **Bohr magneton**:

$$\mu_B = 9.27 \times 10^{-24} \text{ A m}^2 \text{ (or J T}^{-1})$$

which is associated with an isolated electronic spin. Quantum mechanics won't allow us simply to add the magnetic effects of each electron in an atom. A more careful combination is called for. However in the strong, technologically useful, magnetism of solids even quantum mechanics is compromised and it is often almost as if individual electron contributions were simply summed.

If unpaired spins are a source of magnetism, transition metals (unfilled 3d shell) and rare earth metals (unfilled 4f shell) should be good candidates with their part-filled shells, especially since all shells fill by having as many unpaired spins as possible. In solids, the constraints of bonding may affect the situation further.

We already have enough theory to test against real materials. Try to follow the next example.

▼Atomic magnets and spectra in magnetic fields▲

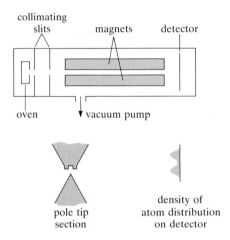

Figure 3.19 Stern and Gerlach's experiment

Stern and Gerlach in 1922 provided an elegant demonstration of the origin of magnetic phenomena in the quantum mechanical atom. They wanted to show that if quantum mechanics forbids all but specific (quantized) orientations of atomic magnets, with respect to an external field, then by using magnetic forces a spatial separation of atoms would be possible according to the quantized states available. (At the time the quantum mechanical description of atoms was incomplete and electronic spin had not been established. Nevertheless it is convenient to describe the results in terms of spin.)

Stern and Gerlach introduced a beam of silver atoms into a magnetic field with strong spatial field gradients (Figure 3.19). Silver ($1s^2 \ldots 4d^{10}5s^1$) has a single 5s electron which must lie spin up or spin down in any external field. The core ($1s^2 \ldots 4d^{10}$) is fully paired and has no net magnetism.

Magnets experience forces in external fields which tend to twist them into alignment. But in atoms, electronic spin (the atomic magnet) cannot be other than parallel or antiparallel. If the external field is not uniform but has a gradient, there are translational forces as well (this is what makes one magnet move towards another, and what makes steel screws jump onto the pole of a magnet). By careful design of the pole pieces of their magnet, Stern and Gerlach were able to produce a field which changed enough to produce an observable effect. A low-density beam of atoms from an oven was injected into the non-uniform field in a vacuum chamber. Atoms would have to orient spin up or spin down, and if there were no collisions that's the way they would stay. The cold detector plate therefore collected atoms in two distinct blobs, some atoms having been pushed one way by the non-linear field, the rest — being oppositely oriented — having been pushed the opposite way. This experiment gave direct evidence of atom sized magnets.

A second effect which gives an indication that the atomic magnet is a valid picture comes from light emissions from excited atoms. Emission spectra for low-pressure gases are made up of narrow bands of frequency, characteristic of each atomic species. Light is emitted when electrons fall down from excited states to lower energy levels. Each transition has an energy separation which determines the frequency of the emitted photon, and sets of transitions are unique to atoms of a given atomic species (Figure 3.20a). Atomic energy levels can contain two electrons each (one spin up, one spin down), but when there is a magnetic field present it turns out that we should split the level into a single electron state higher in energy and a second single electron state lower in energy. Thus, transitions to or from the split state have slightly different energies which result in a splitting of the spectral lines (Figure 3.20b). (The whole story is even more complicated.) The fine structure of energy levels revealed by magnetic fields is referred to as the Zeeman effects and is attributable to the quantum mechanical way electrons behave within atoms.

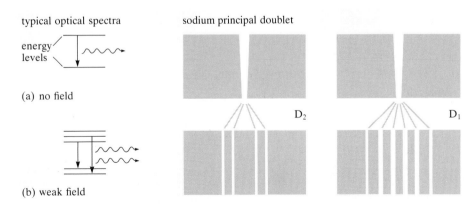

Figure 3.20 Atomic spectra. (a) Without magnetic field. (b) With magnetic field

EXAMPLE   Iron saturates at a flux density of 2.18 tesla. How many electrons per atom appear to be involved?

ANSWER   A little clear thinking is called for. Let's work up from the atom. The Bohr magneton μ_B is the amount of magnetism per electron. If n electrons on the atom contribute to the magnetism, then

amount of magnetism per atom $= n\mu_B$ (A m^2/atom)

Now keep an eye on the units. Suppose there are N atoms m^{-3}, then

amount of magnetism per cubic metre

$$= Nn\mu_B \left(\frac{\text{atom A m}^2}{\text{m}^3 \text{ atom}} \right)$$

$$= Nn\mu_B \text{ (A m}^{-1})$$

This is what we called the **magnetization** M or rather the saturation magnetization M_S. Notice the units. They are the same as those of H. So to get flux density units we need to multiply by μ_0, whereupon we find

$$B_s = \mu_0 Nn\mu_B \text{ (tesla)}$$

For iron $N = 0.85 \times 10^{29}$ atom m^{-3}, $\mu_0 = 4\pi \times 10^{-7}$ H m^{-1}, $\mu_B = 9.3 \times 10^{-24}$ A m^{-2}. We have been told that $B_s = 2.18$ T. From the above equation

$$n = \frac{B_s}{\mu_0 \mu_B N}$$

$$= \frac{2.18}{4\pi \times 10^{-7} \times 9.3 \times 10^{-24} \times 0.85 \times 10^{29}}$$

$$= 2.19 \text{ electrons per atom.}$$

Comment.   The answer looks plausible because iron does have unpaired d-shell electrons. Apparently in the solid not all these electrons are always available for magnetism.

SAQ 3.5   (Objective 3.4)
Look up the electronic structure of the transition metals Sc to Zn. Inferring the electronic structure of the ions by first removing 4s electrons, and then 3d as necessary, suggest which of the following you would expect to give rise to the strongest ionic magnets.

Ti^{4+}, V^{4+}, Sc^{3+}, Ti^{3+}, V^{3+}, Cr^{2+}, Mn^{2+}, Fe^{3+}, V^{2+}, Cr^{2+}, Mn^{2+}, Fe^{2+}, Co^{2+}, Ni^{2+}, Cu^{2+}, Zn^{2+}.

▼Exchange and superexchange▲

Ferromagnetism owes much to the bonding of the transition metals atoms involved. The picture we have is of electrons time-sharing between delocalized metallic bonding states and localized covalent-type bonds and orbitals. In iron, on average about two of the six electrons from the atomic 3d shell contribute to its magnetism. The rather mysterious quantum-mechanical coupling between atoms which we call the exchange interaction follows from the swappings and wanderings of electrons between neighbouring atoms, which keep atoms aware of the orientations of each other's spin. But coupling is not guaranteed: it depends on how the atoms are arranged in space — where and how close. It turns out that only iron, cobalt and nickel of the transition series spontaneously couple with spins aligned to form ferromagnetic metals.

In contrast, ferrimagnetism is a property of the oxides of several transition metals (these oxides are insulators). Here it is transition-metal ions which are magnetic. Fe^{3+} for example, has the structure $1s^2 \ldots 3p^6 \, 3d^5$ and is able to show off five unpaired 3d-shell electrons. The magnetic coupling between neighbouring atoms is fundamentally different since any magnetic ion has non-magnetic negative oxygen ions next door. Nevertheless, the negative ions around each positive ion are blessed with the quantum-mechanical ability to couple spins between shared positive ions in a so-called **superexchange interaction**. This effectively promotes oppositely aligned ionic magnets on alternate sites. In 'antiferromagnets' the oppositely aligned magnets exactly cancel and no magnetism results. However, in ferrimagnets, the oppositely aligned magnets are of unequal strength and do not cancel; there is net magnetization in the material.

3.4.2 Weiss's precedent

Now we have to try to appreciate why some solids are magnetic and others are not. The approach to ferromagnetism suggested by Weiss considered atoms in a solid coupled together in some way or another that energetically favours in a solid spontaneous alignment of the atomic magnets.

The modern theory now explains this as follows. If two neighbouring atomic magnets are at some angle θ, there is a quantum mechanical energy of interaction (the **exchange energy**) given by

$$W_{ex} = A\,(1-\cos\theta) \tag{3.8}$$

The constant A is essentially a materials constant which measures the strength of the effect, rather like a bond energy. It is positive for ferromagnetic and ferrimagnetic coupling between adjacent atoms (see ▼Exchange and superexchange▲ and ▼Single and mixed cubic ferrites▲).

> **SAQ 3.6 (Objective 3.5)**
> How is it that pure iron (and generally its alloys) is magnetic, and most transition metal ferrites (for example, $Fe^{3+}(Mn^{2+}Fe^{3+})O_4^{2-}$) are also magnetic, yet zinc ferrite, $Zn^{2+}(Fe^{3+}Fe^{3+})O_4^{2-}$, is not?

Let's return to the microstructural scale. Suppose a lump of iron is cooled from above its Curie temperature T_c (about 1043 K, or 770 °C for iron) to room temperature, in zero magnetic field. What magnetization will it have? (Remember that above T_c, thermal energy prevents any long-term alignment of atomic magnets.) On average, it has none, although it is made up of many cells of magnetization, apparently randomly oriented (see Figure 3.21). Within each cell or domain, atomic magnets will be aligned so that in the central region, the domain is magnetized to saturation. At the boundaries between domains a transition of atomic magnetization from one domain to the next takes place over many atomic layers (as shown in Figure 3.21). ▼Hearing and seeing magnetic domains▲ will remind you how we know.

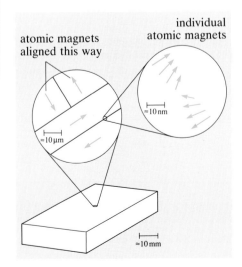

Figure 3.21 Unmagnetized iron

▼Single and mixed cubic ferrites▲

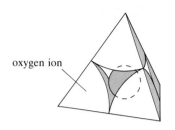

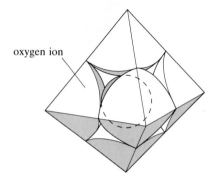

Figure 3.22 Tetrahedral and octahedral cages

Table 3.6 Expected and observed magnetism in ferrites

Ferrite	Spinel structure	Formula magnetism		Saturation flux density at 290 K
		expected	observed	
Mn	Inverse	$5\mu_B$	$4.6\mu_B$	0.5 T
Fe (magnetite)	Inverse	$4\mu_B$	$4.1\mu_B$	0.6 T
Co	Inverse	$3\mu_B$	$3.6\mu_B$	0.5 T
Ni	Inverse	$2\mu_B$	$2.3\mu_B$	0.3 T
Cu	Inverse	$1\mu_B$	$1.3\mu_B$	0.2 T
Zn	Normal	—	0	—

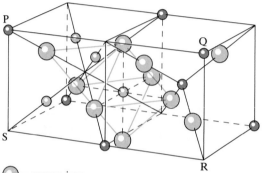

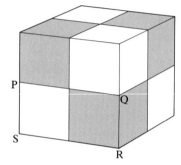

= oxygen ions

= metal ions on tetrahedral (A) sites

= metal ions on octahedral (B) sites

Figure 3.23 Magnetite/inverse spinel

The crystal structure of a cubic ferrite such as magnetite ($Fe^{2+}Fe^{3+}_2O^{2-}_4$) is as follows. First we consider components of the network of oxygen ions. These are co-ordinated groups of four or six oxygen ions around so-called tetrahedral and octahedral holes. We can go on to see the whole oxygen-ion network by considering the simplified pattern of regular tetrahedra and octahedra formed by lines joining ion centres in these two groupings. See Figure 3.22.

When regular octahedra are joined edge to edge, the only spaces they leave are exact regular tetrahedra between their faces and this is how things are in ferrites. Since octahedra have eight faces and tetrahedra four, there will be two tetrahedra for each octahedron.

Now these co-ordinated groups of like ions around holes must be held together by some oppositely charged ions which occupy some of the holes, with overall charge neutrality. In magnetite it is found that one eighth of the tetrahedral holes (so-

called A sites) house Fe^{3+} ions, while half of the octahedral cages (so-called B sites) are empty, and the remainder contain Fe^{2+} or Fe^{3+} in equal proportion. It turns out therefore that half of the Fe^{3+} are on B sites and half on A sites, see Figure 3.23. We show this in the chemical formula by bracketing B sites: $Fe^{3+}(Fe^{3+}Fe^{2+})O_4^{2-}$

EXERCISE 3.12
(a) Confirm that the number of Fe^{3+} ions on B sites equals the number of Fe^{3+} on A sites in a cubic ferrite.
(b) Using the electronic structures of Fe^{3+}, Mn^{2+}, Fe^{2+}, Co^{2+}, Ni^{2+}, Cu^{2+}, Zn^{2+} from SAQ 3.5, work out the magnetism expected of each ion (in units of μ_B).

The mineral spinel has the related structure suggested by the formula $Mg^{2+}(Al^{3+}Al^{3+})O_4^{-2}$, with trivalent ions on B sites only. Magnetite is said to have **inverse spinel** structure.

In ferrites, superexchange interaction ensures that A and B site ions are antiparallel. Therefore in one formula unit of magnetite we have $5\mu_B$ (A site) antiparallel to $(5 + 4)\,\mu_B$ (B site) which means Fe^{3+} ions on the two sites cancel each other but B site Fe^{2+} ions remain unmatched. The material is ferrimagnetic. Now, if another divalent transition metal ion is substituted for Fe^{2+} we get another amount of unmatched magnetism. Table 3.6 shows expected and observed magnetism in a formula unit of some single ferrites (that is, one divalent-ion types).

▼Hearing and seeing magnetic domains▲

Notice that in fact zinc ferrite is a normal spinel. The A site is therefore non-magnetic (Zn^{2+}) and so superexchange cannot couple A and B sites, so we do not expect any magnetic co-operation between ions. There is no magnetism at all.

The behaviour of zinc is important in oxide alloys such as the mixed manganese–zinc ferrites. Since the single ferrites are respectively inverse and normal spinel, a fraction x of zinc ferrite alloyed with manganese ferrite will have a formula

$$Zn^{2+}_x\ Fe^{3+}_{1-x}\ (Mn^{2+}_{1-x}\ Fe^{3+}_{1+x})O^{2-}_4$$

When x is zero, the B-site magnetism is $10\mu_B$ (5 each from Mn^{2+} and Fe^{3+}) and the A-site is $5\mu_B$ (opposed). As x increases, the B-site magnetism remains unchanged (Fe^{3+} ions are just as magnetic as the Mn^{2+} that they replace), the A-site moment goes down (Zn^{2+} with no magnetism replace Fe^{3+} ions) and the net magnetic moment rises. But as the zinc fraction approaches unity we know that superexchange switches off so the overall trend with x shows a maximum. In practice, mixed inverse spinel–zinc ferrites show a maximum magnetization around 50% composition (see Figure 3.24).

In 1919, Barkhausen listened for domains, the idea of local co-operative interaction to form domains having been proposed by Weiss in 1907. Barkhausen connected an amplifier and headphones system to a coil wound around a ferromagnetic sample. Sudden changes in magnetization as domains grew and rotated in response to an external magnetizing field produced audible clicks as a result of voltages induced in the coil.

By 1931, attempts were being made by Bitter to render domains visible. He used a technique similar to that which uses iron filings to trace magnetic fields. Ferromagnetic powders, and soon afterwards colloidal suspensions, were used to try to trace out magnetic discontinuities where domain boundaries intersect the surface. These **Bitter patterns** were true domain pictures only when a sufficiently fine colloidal suspension was used on surfaces which exposed particular crystal planes. Ferromagnetic colloidal particles suspended in a liquid are 'free' to move in response to magnetic fields, accumulating in areas of non-uniform (spatially changing) field, such as along domain boundaries. In practice particles of about 10 nm, aggregating up to 100 nm, are used. We shall see later that these particles, like those on the tape in Figure 3.1, will themselves be single domains.

The **Kerr effect** makes domains visible by using subtle differences in the reflective properties of magnetized surfaces. When plane polarized light is directed at a ferromagnetic surface, two adjoining domains magnetized in opposite directions tend to rotate the plane of polarization in opposite senses. When the reflected light is viewed through a slice of polaroid (analyser), then an orientation of polaroid can be found which transmits maximum light from one of the domains. Reflected light from the other domain, being oppositely rotated will not be transmitted. Thus one domain appears lighter and the other darker. The angle through which the polarization is rotated is the **Kerr polar angle** θ_k. In practice it is a fraction of one degree, but this is readily observed. See Figure 3.25. Transient and oscillatory behaviour is easily followed with this technique.

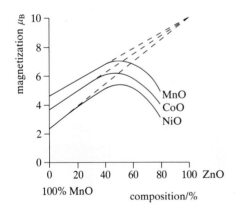

Figure 3.24 Magnetism in mixed ferrites

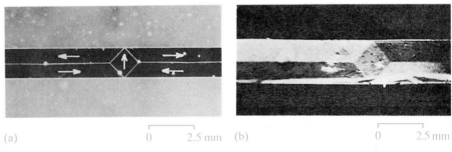

(a)　　　　0　　2.5 mm　(b)　　　　0　　2.5 mm

Figure 3.25 Domains in an iron whisker revealed by the Bitter technique (negative image) and the Kerr effect

3.4.3 Introducing the energy budget of domains

In this subsection we will broach a question which is crucial to the exploitation of magnetic materials. What determines the size and shape of magnetic domains? The importance of this question can be illustrated by two examples.

• Permanent magnet motors develop a torque in direct proportion to the strength of magnet involved. Careful exploitation of magnetic materials may lead to more powerful or more efficient motors, or to savings in weight of conventional designs, or may even lead to novel designs not previously feasible. 'Strong' permanent magnets require a domain *structure* which is resilient to demagnetizing influences.

• Suppose that a transformer is used to step the voltage from a power station up to 400 kV for transmission, and that another is then used to step it down to 415 V for distribution to consumers. If each transformer gives a power loss of 0.5%, the overall efficiency is 0.995×0.995, or slightly over 0.99. The power loss is almost 1%. Such losses are expensive and wasteful (1% of your bill may not be much but what about 1% of a whole nation's bill?). If the magnetic power loss can be reduced by better use of materials then it should be. Energy loss during magnetization is in part a consequence of associated changes in the domain *structure*.

To comprehend the scope for controlling domains, we must look at the various sources of energy in magnetic materials. This we do in Section 3.5. The final magnetic condition will be an equilibrium of forces arising from these energy sources. (If you like thermodynamics, you can think of minimizing energy — it's the same thing.) We will think about this equilibrium in terms of an 'energy budget' in Section 3.6.

3.5 Magnetic energies

3.5.1 Thermal energy

Thermal energy can destroy magnetic order altogether, swamping the magnetic interactions between atoms. The **Curie temperature** (T_c) marks the onset of thermal disorder. Above T_c, atoms are still magnetic but they behave independently, resulting in a feeble magnetic effect (**paramagnetism**).

3.5.2 Exchange energy

The fundamental source of magnetic interaction on the atomic scale is the exchange interaction. The magnitude of the co-operation between adjacent atoms (A in Equation 3.8) can be approximated by the thermal energy necessary to destroy it, kT_c. This energy is called the **Curie energy**. Table 3.7 gives values for ferromagnetic transition elements. I have also stated these energies in electron volts for easy comparison with the energies of chemical bonds, visible light photons, work functions, Fermi energies, band gaps and so on, which are all a few electron volts.

Table 3.7 Values of kT_c (Curie energy) for ferromagnetic transition elements

Element	Curie temperature	Curie energy	
		kT_c/joules	(kT_c/e)/eV
Fe	1043 K	1.4×10^{-20}	0.09
Co	1404 K	2.0×10^{-20}	0.12
Ni	631 K	0.9×10^{-20}	0.06

3.5.3 Two magnetic anisotropy energies: K and λ

The crystal structure and K

The crystal structure of magnetic materials puts some demands on the disposition of electrons. The electron spins are not entirely free to act independently of their host atom; and the regular atomic order within a crystal exerts some influence on the magnetic order, giving rise to what is called **magnetocrystalline anisotropy**. For example, cobalt has a hexagonal close packed structure which leads to one preferred direction of magnetization in a crystal (the so-called easy axis), perpendicular to the hexagonal planes. (This direction in HCP crystals is often called the c-axis.) Additional energy is required for other directions of magnetization. The cubic structure of iron gives three easy axes along cube edges, but in nickel the four body diagonals are preferred.

We shall shortly assume that there is an average fixed energy charge K (joules per cubic metre) for magnetization not-aligned with a preferred (or easy) direction. This K is called the **anisotropy constant** and is typically 10^5 J m^{-3}.

Atomic spacing, magnetostriction and λ

If the atomic spacing in a material changes, for example as a result of an applied stress, the strength of the exchange interaction between adjacent atomic magnets may change. Certain directions of magnetic orientation may become more favoured. The converse effect is also found: magnetization of the material in one direction may strain the crystal lattice. This effect is called **magnetostriction** and is another source of magnetic anisotropy. The connection between lattice distortion and magnetic orientation is very important. It allows anisotropy to be induced by rolling or other stress-building operations, and this anisotropy may counter or supplement the crystal's inherent magnetic anisotropy (that is, its magnetocrystalline anisotropy).

Magnetostriction may cause tension (for example in iron) or compression (cobalt and nickel), depending upon the direction of magnetization and the anisotropy of the elastic moduli of the material — the lattice distorts more or less easily in different directions. The strain induced by the magnetizing of a multidomain sample will eventually saturate when the domains have so reoriented that the magnetization is saturated. A typical saturated strain λ_s is 10^{-6} to 10^{-5}. The magneto-elastic energy stored is around 10 J m^{-3}, usually much less than for crystal anisotropy. The domain nature of magnetism allows a degree of physical control over magnetostriction; Figure 3.26 shows some different cases.

Magnetostrictive effects are also very sensitive to composition, so alloys can be readily prepared to produce a wide range of behaviour. Figure 3.27 illustrates the potential for chemical control of magnetostriction. (▼A note on transformer hum▲ looks at a practical instance of magnetostriction.)

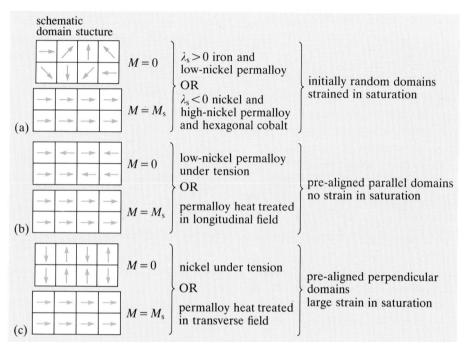

Figure 3.26 Controlling magnetostriction with domain structure

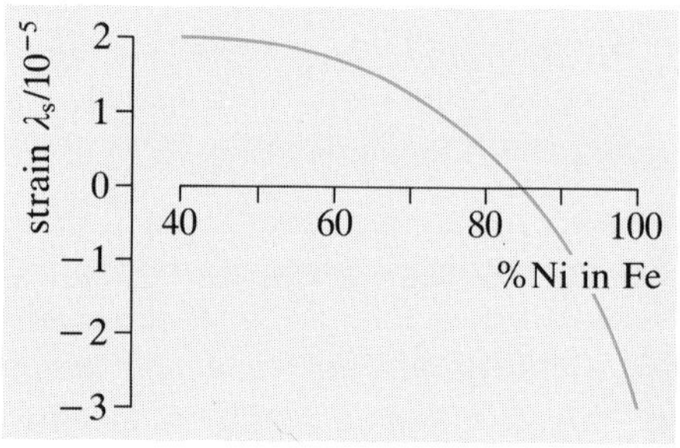

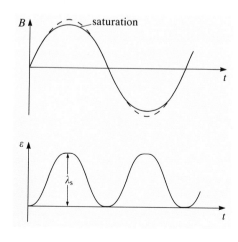

Figure 3.27 Controlling magnetostriction with composition (Ni–Fe)

▼A note on transformer hum▲

When a transformer core is magnetized first one way then the other, some domains will rotate into and out of alignment at the extremes. Consider the simplified picture of Figure 3.26(a). Each time saturation is reached, the axial magnetostrictive strain will be greatest, returning to zero as the axial flux passes through zero. Figure 3.28 shows schematically how the dimensional changes correspond with magnetic flux changes. If the flux cycle repeats 50 times per second (mains frequency), the flux passes through zero (reverses) 100 times per second and so the length of the bar in Figure 3.26(a) would change at this rate, vibrating surrounding material (oil coolant or air or mechanical supports) at 100 Hz. Because neither the flux nor the strain are likely to be pure sine waves, the sound

waves launched by this vibration will have harmonics at 200, 400, 800 Hz and so on. These frequencies correspond to a slightly flat A in the UK. (In the USA, where the mains frequency is 60 Hz, they would correspond to a slightly sharp A.) Magnetostriction is the major source of hum in conventional transformer cores.

EXERCISE 3.13 In general, transformer hum is not desirable because it wastes power (although it is not a major loss) and because it is a nuisance. Suggest a scheme which might exploit magnetostrictive effects.

Figure 3.28 Flux and strain variations

3.5.4 Magnetostatic energy

Here I want to illustrate how magnetic potential energy is associated with having a magnetic field coming out of a magnetic material.

Consider first the potential energy of a system comprising a water well and a bucket on a rope; see Figure 3.29(a). If one kilogram of water is wound up to one metre above the water level, the potential energy of the system (*mgh*) is about ten joules. The bucket could be allowed to return under gravity producing ten joules of rotational energy in the winding machinery.

Consider now a toroidal iron magnet, which is magnetized to saturation. The magnet has been cut twice to make two equal halves. Figure 3.29(b) shows the two halves being separated by a winch similar to that in the well, but in a horizontal plane to avoid complications of gravity. Once again potential energy is stored, this time as magnetostatic energy in the magnetic fields between the 'poles'. If the restraining force supplied by the winch is released, the two halves are free to move to a lower energy state by coming together again. As with the well and bucket, there is a transfer of energy into rotation of the winch. Creating magnetic field outside the iron requires work, that is, energy. The more field there is outside a magnet the larger the **magnetostatic energy** associated with it.

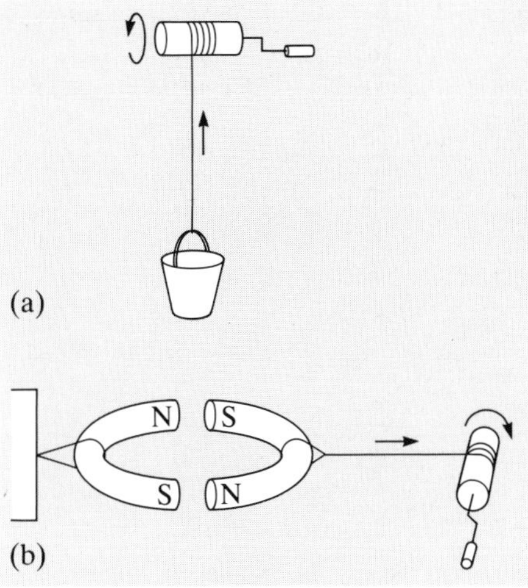

Figure 3.29 (a) Potential wells and (b) potential magnets

Summary of Sections 3.4 and 3.5

● Magnetism in solids comes chiefly from unpaired electron spins. Fe, Co, Ni are ferromagnetic solids. Transition metal ferrites (except Zn) are ferrimagnetic solids.

● Magnetic alloys of ferromagnetic metals and ferrimagnetic oxides can be prepared with a range of properties.

● Below the Curie temperature, domains of co-aligned atoms are the equilibrium state. Thermal exchange, magnetocrystalline (K), magnetoelastic ($\frac{1}{2}E\lambda_s^2$, E=Young's modulus) and magnetostatic energies are involved.

3.6 The energy budget

3.6.1 Why have domains at all?

EXERCISE 3.14   In a boundary region between domains of different orientation, some atomic magnets are not on easy axes. There is an energy penalty for this. Using Figure 3.30 explain why a single crystal of freshly smelted iron has no net magnetization.

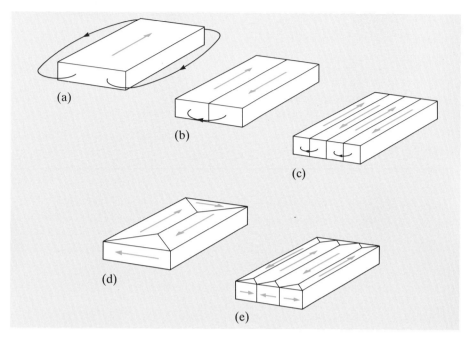

Figure 3.30 Saving energy with domains

How can domains save energy? Having just two domains in place of one (Figure 3.30b) would clearly save magnetostatic energy, at the expense however of introducing a boundary region. In the boundary there is an energy cost owing to the disruption of the alignment favoured by the exchange interaction between atoms (crystal magnetic anisotropy). More domains will save more magnetostatic energy but cost more wall energy. Even more magnetostatic energy can be saved by having the closing of magnetic flux inside the material, as shown in Figure 3.30(d). The energy cost might now include some more anisotropy energy if the closure domains are not magnetized along an easy axis. However iron is cubic, with easy axes for magnetization along cube edges, so Figure 3.30(d), with its large closure domains, is not unlikely. Observations on a single-crystal iron whisker confirm this (Figure 3.25). On the other hand, cobalt, a uniaxial material (one easy axis) may be energetically more stable with many smaller closure domains (Figure 3.30e).

As magnetic material cools slowly, domains nucleate and grow because this corresponds to the lowest energy state — a thermodynamic equilibrium. We must now look further into what equilibrium sizes and shapes can result.

3.6.2 The size of magnetic microstructure

In iron, the boundary regions between oppositely magnetized domains is around 10–100 nm wide. Why? Since iron prefers to be magnetized along cube axes, why not switch direction from one plane to the next?

Suppose the direction of magnetization did switch from one plane to the next. There is an energy requirement to hold this arrangement against the forces of the exchange interaction (which want everything parallel). This must be balanced against the (anisotropy) energy cost of not keeping the magnetization along their easy axes, and clearly the more atoms in the wall (the region of transition), the higher the cost. On the other hand it turns out that the exchange forces favour many atoms with a small angle between adjacent atoms (rather than a few with large angles). Consequently, a compromise can be reached. This argument is given a little mathematical backing in ▼Between domains; width and energy▲.

The domain boundary (wall) has a surface energy associated with it just as the free surface of a solid or liquid has a surface energy. The value of the surface energy is a material property (although it is not very precisely defined). The wall energy is saved if the wall is removed, or destroyed by two oppositely twisted walls coalescing. In a perfect crystal the wall will be free to move within the bulk with no energy penalty.

So much for the domain wall width. What determines the size of the domain itself in an unmagnetized sample? I indicated in Figure 3.30 that magnetostatic energy can be saved by fitting closing domains into the material. The final state will as usual be the result of a balance between competing effects. Let's consider a cubic material such as iron so that we can leave out crystal anisotropy since then easy directions for closure domains are always available. The competition is now between domain-wall energy, which favours few domains, and stresses arising from the magnetization in closure domains, which favour many small closure volumes. ▼Within domains; thickness▲ introduces the mathematics, and leads to the important result that domain size depends upon specimen size!

3.6.3 Single-particle domains

Since domain boundaries have finite thickness, it is not surprising that very small particles can exist as single domains in which every atomic magnet points in the same direction. In 'Within domains; thickness' it became clear that if we consider just wall energy and magnetoelastic energy, we expect to see single domains in particles smaller than 100 μm. In fact the opposition to a single domain is from not only magnetoelastic effects but also the magnetostatic energy which arises when closure domains are squeezed out. As a result, it is not until less than about 50 nm that single domains are energetically favourable in iron. As with wall thickness and domain width, the dependence on material properties is not strong, so the figure of 50 nm is characteristic of a wide range of materials. Our simple estimate of domain-wall width is comparable with this anyway, so we should have extra exchange forces to contend with if we required narrower walls. In needle shaped (acicular) particles, magnetized along the length, the magnetostatic energy is much less and larger single domains are possible (see Figure 3.2). See ▼Small domains and superparamagnetism▲

▼Small domains and superparamagnetism▲

The smallest useful domain will be one in which all the atomic magnets can hold their 'agreed' direction in the absence of an external field. If we get too small a group of atoms then one lump (or phonon) of thermal agitation energy (kT) might be enough to turn all the atoms, as a block if necessary, into some other direction, in spite of, say, the crystal anisotropy. We can estimate the smallest useful size by comparing thermal energy with the anisotropy energy of a volume l^3. When

$$l^3 K = kT$$

thermal randomizing of the domain orientation is likely. At room temperature (300 K) with the typical figure for K of 10^5 J m^{-3}, we find

$$l \approx 3 \text{ nm}$$

Notice again how insensitive this figure is to the specific material property. This time it's a function of the cube root. Just to make sure we don't underestimate this limit, let's be a little more cautious. For the particles used on recording tapes and discs, we'd like to know how small they can be before they are useless. Since the energy kT is some kind of average lump of energy, let's insist that particles are resilient to ten times the average thermal energy. Also, K may not be quite so large — we could reduce it to 10^4 J m^{-3}. The new limit for domain size is then 15 nm. Any smaller and we can't expect the domains to remain magnetized in any one direction for long. Such small particles are said to be **superparamagnetic** because each one behaves like an atom in a paramagnetic substance (see Section 3.1.3). (Look again at the scales in Figure 3.2.)

EXERCISE 3.15 A solid is made from superparamagnetic particles of iron by compacting and sintering. Discuss its domain structure and magnetic properties.

▼Between domains; width and energy▲

Consider the boundary between oppositely magnetized domains. Across the boundary, the atomic magnets progressively change their orientation in the way shown in Figure 3.31(a). The exchange energy of a pair of atomic magnets varies with angle as shown in Figure 3.31(b), and we can approximate the relationship of Equation (3.8) for small angles θ to

$$W_{ex} = \frac{A\theta^2}{2}$$

for any pair of atomic magnets out of alignment by θ (θ in radians), where A is the Curie energy, kT_c.

Now if there are $N+1$ atomic layers in the wall, then there are N pairs in each line of atoms (Figure 3.31c), each pair consisting of an atom and its next-door neighbour. If the spacing between atoms is a, there are $1/a^2$ such lines of atoms per square metre (Figure 3.31d). The N pairs share the angle change of π radians (180°) equally. If N is large, θ (which is π/N) is small, giving the exchange energy cost per square metre of boundary σ_{ex} as

$$\sigma_{ex} = \frac{A}{2}\left(\frac{\pi}{N}\right)^2 N \frac{1}{a^2} \text{(J m}^{-2})$$

We must also pay the anisotropy cost of not having all atoms aligned along easy directions. This is K (joules per cubic metre of boundary region). Suppose we have 1 m² of boundary. Its depth is Na, so its volume is $1 \times Na$. The anisotropy energy cost per square metre σ_A is

$$\sigma_A = KNa \text{ (J m}^{-2})$$

These are the two most significant contributions to the energy budget. Figure 3.31(e) shows the variation of energy with the number of atomic layers in the wall. We need to know the number of layers for which the total energy is a minimum. The minimum occurs when

$$\frac{d(\sigma_{ex} + \sigma_A)}{dN} = 0$$

or

$$-\frac{A}{2}\left(\frac{\pi}{a}\right)^2 \frac{1}{N^2} + Ka = 0$$

Therefore, we can deduce that

$$N_{min} = \frac{\pi}{a}\sqrt{\frac{A}{2Ka}}$$

and the wall thickness δ ($= N_{min}a$) is

$$\delta = \pi\sqrt{\frac{A}{2Ka}} \text{ (m)}$$

and the equilibrium surface energy of the wall σ_w ($= \sigma_{ex} + \sigma_A$, with $N = N_{min}$) is

$$\sigma_w = \pi\sqrt{\frac{2A\,K}{a}} \text{ (J m}^{-2}) \qquad (3.9)$$

Notice that the wall thickness and energy depend only on the square root of material properties a, A and K. For iron ($K \approx 0.5 \times 10^5$ J m^{-3}, $a \approx 0.3$ nm, $A = 1.4 \times 10^{-20}$) δ is about 70 nm, there are about 230 atoms in the wall and the wall energy is about 0.007 J m^{-2}.

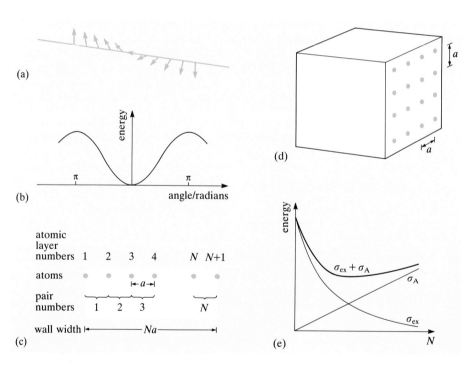

(a)

(b)

(c)

(d)

(e)

Figure 3.31 Domain wall structure. (a) Orientation of atomic magnets across domain wall. (b) Exchange energy vs angle. (c) Numbering of layers and pairs. (d) Layer separation and cross-sectional area of wall. (e) Energy vs N

▼Within domains; thickness▲

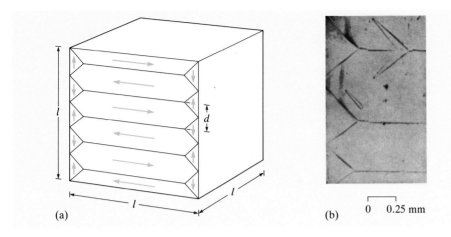

(a)

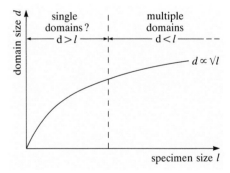

(b)

| 0 | 0.25 mm |

Figure 3.32 Typical domain structure in iron. (a) Schematic. (b) Observed Bitter pattern

Consider the domain structure in a single-crystal cube of iron (Figure 3.32a). Domain-wall energy is chiefly due to the horizontal walls, having area approximately l^2 with l/d such walls. Ignoring the smaller wall area due to closure domains, we have energy in the walls given by

$$W_{\mathrm{w}} = \frac{\sigma_{\mathrm{w}} l^2 l}{d}$$

where σ_{w} is the surface energy for domains in iron. This is opposed by magnetoelastic stresses in the end domains. Since the bulk remains at width d, the magnetostrictive strain λ_{s} is opposed by elastic forces (bonds) and so magnetoelastic stress builds up an energy ε per unit volume. (The size of ε depends upon the Young's modulus E

of the material.) Elastic energy W_{E} is thus ε times the total volume of the end domains:

$$W_{\mathrm{E}} \approx \frac{\varepsilon l^2 d}{2}$$

As with the wall-thickness calculation, one energy term increases with the dimension of interest (d here), and the other decreases with the same dimension. There is equilibrium when

$$d = \sqrt{\frac{2\sigma_{\mathrm{w}} l}{\varepsilon}}$$

The elastic energy density ε can be shown to be given by $E\lambda_{\mathrm{s}}^2/2$, so

$$d = \frac{2}{\lambda_{\mathrm{s}}}\sqrt{\frac{\sigma_{\mathrm{w}} l}{E}}$$

Taking values for iron we get

$$d \approx 0.3\sqrt{l\,(\text{in mm})}\ \text{mm}$$

So a 1 mm cube would have 0.3 mm-wide domains (see Figure 3.32b). This example shows that if the energy is dominated by magnetoelastic and domain-wall contributions, then the sample size has some bearing on domain structure. Even when other energy terms must be considered, some sample-size dependence will remain. Figure 3.33 shows the variation of domain size with specimen size and reveals an interesting limit — for this simplified case one domain will more than fill the sample ($d > l$) when l is less than about 0.1 mm. In fact the calculation is no longer strictly valid since we must reintroduce magnetostatic energy if field is forced outside the material, but it indicates an interesting possibility: single-domain particles.

Figure 3.33 Domain size d vs specimen size l

3.6.4 Polycrystalline materials

What can we say about domains in polycrystalline materials? Figure 3.34 shows a logarithmic length scale with the characteristic distances of magnetism. Now, within one crystal of polycrystalline material, in the unmagnetized state, there may be many domains and if crystal axes are randomly distributed then domains see the crystal limits as the shape within which energies should be minimized. However magnetostatic energies may be lowered around a crystal if neighbouring crystals are conveniently oriented to carry the magnetic field outside it. Also, if some degree of grain orientation exists then domains may extend over several crystals. The equilibrium state is of course decided by energy minimization.

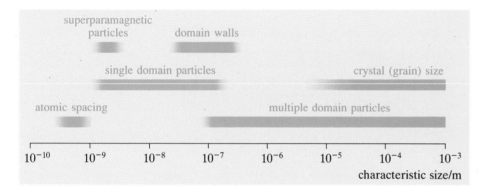

Figure 3.34 Scales in magnetism

SAQ 3.7 (Objective 3.6)
Specify the properties required of a material that is to have large domains with narrow boundaries (look at our algebra on domain dimensions).

3.7 Magnetization, hysteresis and domains

In this section we will revise the processes of magnetization and magnetic properties at the domain level.

In an unmagnetized sample of polycrystalline magnetic material, domains will be oriented randomly overall, but within each crystal the domains are likely to be magnetized along easy directions. You should know what happens to domains when a magnetizing field is applied, so rehearse this before trying Exercise 3.16.

EXERCISE 3.16 In Figure 3.35, there are two green routes, one describes a 'minor' hysteresis loop (from f to f′ and back) and the other a spiral path which ultimately returns the material to the unmagnetized state. Give a brief domain-based explanation of each.

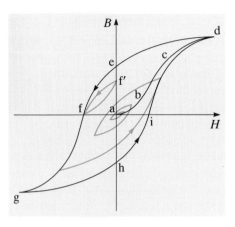

Figure 3.35 Schematic magnetization curve and hysteresis loop

3.7.1 Remanent flux

The visualization of domain patterns provides important clues as to the possibility of controlling the magnetic flux remaining in a material in the absence of magnetizing field (that is, when $H = 0$). Between saturation and remanence, a general relaxation of domain magnetization to the nearest easy axes takes place. Suppose by dint of careful preparation a given material has easy axes only perpendicular to the direction of saturated domains (as in Figure 3.26c). What remanence might we expect? The answer is that we may end up with zero net flux at zero magnetizing field; hysteresis in this case may even be defeated (▼Inducing anisotropy▲). Of course the price to be paid is that the material is being magnetized along a non-easy direction. This requires a lot of magnetizing field (H) for a little magnetic flux density (B), so the material has low permeability.

Now the same material might be differently compounded so that an easy axis of each domain coincides with the magnetizing field. In this case, saturation will be achieved by growth of favourable domains without the need to rotate any domains from their preferred magnetic orientations. The remanent state might be no different from the saturated state, giving one suitable characteristic for a permanent magnet. Alternatively, we might simply ensure that the material has lots of easy axes, so that the magnetization direction is never far from a domain's preference.

▼Inducing anisotropy▲

Virtually zero remanence might be achieved by preparing a material from elongated single-domain particles aligned with the long axis perpendicular to the proposed magnetization direction. The magnetostatic energy component will be a minimum if each particle is magnetized along its length, so the long axis of the particle is the easy direction. Figure 3.26(c) shows the saturated and remanent conditions for a group of particles. In being driven to saturation, the domain magnetizations must rotate away from the preferred (easy axes); there need be no domain wall motion. In pure particles there may be no losses associated with rotation and the process would therefore be reversible.

Stress-induced anisotropy (by cold rolling and then annealing, a process of grain orientation) exploits magnetoelastic coupling to produce similar effects in multidomain crystals. A 50% FeNi alloy (called isoperm) owes its constant permeability and near zero remanence to this. Isoperm has a maximum relative permeability of 100 and an average of 90.

Maximum remanence can be achieved using the same elongated, single-domain particles (or some similar anisotropy-induced preference), this time co-aligned with the magnetizing field (Figure 3.26a). Such particles may remain in a stable equilibrium up to quite large demagnetizing fields. Some anisotropic permanent magnets owe much of their strength to this effect (for example barium ferrite, Table 3.3).

In low-carbon silicon iron for transformer cores, dramatic improvements in permeability can be achieved by ensuring that a large number of crystals are aligned with what will be the magnetization direction. This is also done by cold rolling to build up dislocations and then annealing at relatively high temperature. The iron recrystallizes with a high degree of grain orientation, owing to the directional dependence of crystal growth rate. During recrystallization, dislocation-free crystals grow throughout the material, replacing the dislocation-riddled crystals formed during rolling, In iron this leaves a **texture** with easy axes parallel to the roll direction.

Some of the high permeability NiFe alloys (permalloys) achieve their special property after being annealed in a magnetic field. In this process, small magnetostrictive stresses are built in on cooling from the anneal temperature giving preferred axes in the direction of the field during annealing.

3.7.2 Permeability

What does the domain picture suggest for a high permeability? If we don't want to spend a lot of time rotating misoriented domains, then as in the quest for high remanence we will be interested in easy axes parallel to the desired magnetization. This is achieved in polycrystalline iron alloys by means of texturing (as described in 'Inducing anisotropy'). In anisotropic (for example, grain-oriented) high-permeability materials, reverse magnetization is achieved by the nucleation and growth of 180° (reverse) domains without the need for rotation from easy axes. If such a material is to be used where repeated cycling from one magnetization to the reverse is inevitable (such as in transformer cores), then the domain walls once nucleated must also be free to move throughout crystals without hindrance (see ▼Moving domain walls▲).

A second approach suggested by the domain picture favours less anisotropic materials. Several of the Ni–Fe alloys which are the basis of many high permeability materials have both low crystal anisotropy and low magnetostriction, with correspondingly broad-walled, wide domains. On magnetizing, easy enhancement of the external field may occur by rotation of atomic magnets within the boundaries, narrowing the wall structure, as well as by the usual growth of aligned domains.

▼Moving domain walls▲

We saw earlier that domain boundaries extend over hundreds of atomic layers. Any factors which disrupt order within the wall region will impede its motion. There are three main causes of disruption:

• Non-magnetic impurities. These will introduce local irregularity and stress in the otherwise smooth transition of atomic-magnet alignment.
• Internal stresses arising from any lattice defects (dislocations).
• Crystal boundaries.

The initial growth of domains is locally arrested whenever a wall encounters one of these disruptive features. At larger magnetizing fields, the obstacles are overcome.

Figure 3.36 shows a domain wall distorted as it passes a non-magnetic inclusion.

EXERCISE 3.17　Why are transformer steel laminations annealed after being stamped from sheet?

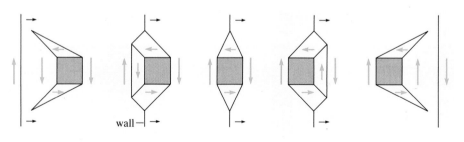

■ non-magnetic impurity

Figure 3.36 Domain wall pinning

EXERCISE 3.18   Confirm the association of broad domain boundaries and large domains with low anisotropy constants (λ_s and K). (Refer back to 'Between domains; width and energy' and 'Within domains; thickness' in Section 3.6.)

This strategy for high permeability suggests we might try to avoid easy axes altogether. That means no anisotropy, perhaps no crystal structure at all. Keep an eye open for amorphous materials! Without strongly preferred directions, there is no need to move any domain walls, provided it is easy to rotate the domain magnetization.

3.7.3 Coercivity

Suppose we have a domain within a ferromagnetic material, oriented along a preferred direction. How can the magnetization become reversed by an external magnetizing field? If we can answer this question then we will be able to propose some desirable features for soft, easily reversed material and others for hard material. There are two ways to reverse a domain's magnetization:

- all atoms simultaneously rotate their magnetism through $180°$;
- a $180°$ domain wall sweeps through the domain, rotating atomic magnets a few at a time in a cascade.

In the first case, the domain will regain its shape exactly but it will have 'flipped' direction by large numbers of atomic magnets going off the preferred axis together, requiring a large energy. In single-domain particles unable to sustain domain boundaries, this may be the only way. In the second case, the domain wall might come from a reverse domain of the same crystal or it might have been nucleated for the specific purpose at the crystal boundary or some other inhomogeneity where a weakness in magnetic order can arise.

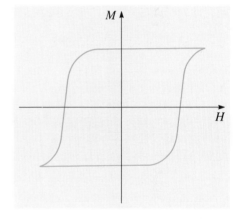

Figure 3.37 M–H loop for single domain particles

Let's look first at an isolated single-domain particle. In order to reverse its magnetization we must at some stage reduce the flux density in the external field direction to zero so that from Equation (3.2)

$$0 = \mu_0(H + M_s)$$

Thus, an estimate of the size of the field required (H_c) is

$$H_c \approx M_s \approx 10^6 \, \text{A m}^{-1}$$

When several single-domain particles are close together, it may not be possible to attain quite such a high coercivity, but the M–H loop will be almost square. (See Figure 3.37.)

Turning now to a saturated but multiple-domain specimen, we must consider first the cost of nucleating a $180°$ wall and second the cost of sweeping it through a domain. Figure 3.38 shows one type of nucleation. Work is done in establishing the wall (surface energy σ_w) and this is supplied as we increase the field up to H_c against a magnetization of M_s.

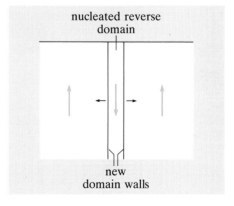

Figure 3.38 Reverse domain nucleation

But, as we have already noted, the wall is established by twisting atomic magnets against anisotropy forces which favour certain alignments. It turns out that for this case

$$H_c \approx \frac{K}{M_s \mu_0}$$

$$\approx 5 \times 10^4 \text{ A m}^{-1} \text{ (for iron)}$$

For typical values we find H_c twenty times lower than in the single domains, because in reversing only a small fraction of the domain at any one time we don't get stung by the magnetostatic energy barrier. The reverse domains can be nucleated in regions of low anisotropy K (such as different phases, impurities, or grain boundaries) and then grow into the bulk, with a correspondingly lower coercivity. On the strength of this we can conjecture that a single impurity type (size, shape and composition of impurity), uniformly and sparsely distributed, could serve to nucleate reverse domains quite efficiently. Since the impurity nucleates the domain boundary, it will be near the centre of a growing domain, where it is literally not in a position to impede wall motion (see 'Moving domain walls'). Further, we can surmise that a range of impurity types (sizes, shapes, compositions) could subsequently pin the seeded walls, which would lead to a harder material. The coercivity of the better soft materials ($\approx 10 \text{ A m}^{-1}$) can in part be accounted for by wall nucleation, in regions of low anisotropy. Of course, the situation is more complicated, for example when magnetoelastic interactions are included: local stresses and magnetostrictive strains may contribute to raise or lower reverse domain thresholds. Material without anisotropy (no easy axes) and no magnetostriction would reverse easily.

The conclusions to be drawn are these:
● Single-domain particles can have high coercivity, especially if elongated in the external field direction. (We saw earlier however that if elongated across the field, a low coercivity, low permeability material can be obtained.)
● Grain boundaries and impurities can assist in nucleating reverse domains (lower K) but can also impede the motion of domain walls.
● Very low coercivity will require an isotropic material or one with easily nucleated reverse domains of very high purity. (See ▼Permanent magnets▲ and ▼A soft touch▲.)

SAQ 3.8 (Objective 3.7)
Contrast the magnetism of a bulk sample of manganese–zinc (cubic) ferrite with that of barium (hexagonal) ferrite in terms of domains. (See Tables 3.2, 3.3, 3.6 and 3.8.)

▼Permanent magnets▲

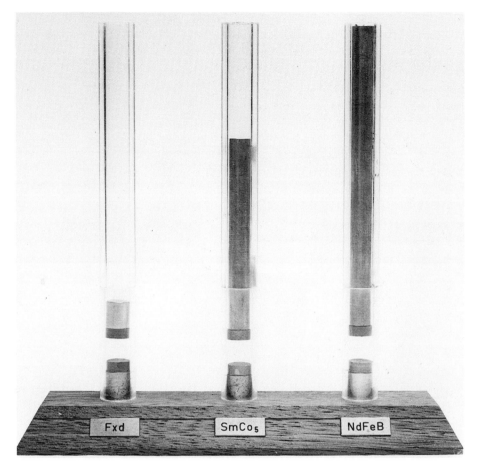

Figure 3.39 Ferrite (Fxd), SmCo₅, NdFeB magnets (courtesy of Mullard Ltd)

Hard magnetic material owes its 'mechanical' description to early high-carbon magnet steels in which impurity-dominated coercivity was almost a by-product of the use of additions for mechanical hardening. However, during the 1930s Al-Ni-Fe and Al-Ni-Co-Fe alloys were developed to provide high-coercivity alloys suitable for permanent magnets. Alnico alloys, as they became known, are dispersion hardened to produce a two-phase microstructure comprising an iron–cobalt rich magnetic phase and a nickel–aluminium rich phase which contributes negligibly to the magnetism. Using all the tricks of microstructure control (composition, heat treatment and also annealing in a magnetic field) magnetic properties can be controlled over a wide range, trading saturation flux density for coercivity. Typical alnicos may contain around 8% Al, 14%Ni, 25%Co, and a few per cent of copper (to slow down atomic diffusion during heat treatment) and titanium (to enhance coercivity). Alnico components are cast or sintered and then heat treated.

Single-domain particles with large uniaxial anisotropy should have high coercivity since magnetization rotation will be strongly opposed. For a similar reason, hexagonal ferrites such as $BaO\,(Fe_2O_3)_6$ were developed for use as 'ceramic' permanent magnets because of their high crystalline anisotropy. The inevitable dilution by oxygen and barium atoms leads to a relatively low saturation flux density. Ceramic magnets are formed by sintering (or bonding in a matrix) the finely powdered barium ferrite, and can be enhanced by field alignment during powder pressing. The chemical stability of ceramics is advantageous and goes some way to mitigate the low Curie temperature.

High resistance to domain nucleation follows from having a large domain-wall energy and this in turn is also favoured by a large magnetocrystalline anisotropy energy K. Sintered powder magnets based on samarium–cobalt (rare earth–transition metal) exploit the high anisotropy of hexagonal $SmCo_5$. The crystals are particularly free from impurities so that only grain boundaries nucleate and impede domain walls. The difficulty of reverse-domain nucleation gives these magnets their high coercivity. Although tricky to process, $SmCo_5$ magnets use the expensive cobalt more effectively than do alnico magnets because of their much higher coercivity, giving a much larger $(BH)_{max}$. Following the success of samarium-cobalt magnets in the 1970s, a further advance was made in the 1980s when neodymium-iron-boron powder magnets were produced. The important phase is tetragonal $Nd_2Fe_{14}B$, which again has a high anisotropy but is even less susceptible to nucleating precipitates within the grains. It is also possible to produce by rapid solidification a neodymium–iron–boron magnet which has single-domain sized grains 20–100 nm across. The coercivity is similar to values obtained by the conventional powder route, but the mechanism of reversal is probably domain rotation owing to the fine grain size.

Figure 3.39 shows ferrite, samarium–cobalt and neodymium–iron–boron disc magnets in opposed pairs supporting weights. The superiority of the more recent material is quite clear. (Consider the effect on the air gap between the discs of interchanging the weights.)

▼A soft touch▲

The description 'soft' for magnetic materials comes from the mechanical properties of pure iron: low impurity content is associated with high ductility and low coercivity; and the characteristics of magnetic softness are low coercivity and high permeability. But in general magnetic and mechanical softness don't go together: permendur (Table 3.8), silicon iron and ferrites are all quite brittle.

EXERCISE 3.19   Comment on the following with reference to Table 3.8.

(a) The maximum saturation is associated with an iron–cobalt alloy.

(b) The maximum permeability is associated with the lowest coercivity.

(c) The decrease in saturation flux density down the table is mirrored by the trend in Curie temperature.

The lower coercivities correspond to the following isotropic materials:

• Purified iron. Reduction of impurities prevents domain wall pinning at non-magnetic inclusions, which are diffused out during high-temperature annealing in hydrogen (followed by slow cooling to avoid thermal stresses).

• Amorphous iron–cobalt. Rapid solidification of about 80% Fe, Co and 20% B, Si, C alloys gives non-crystalline (metastable) microstructures which have correspondingly low anisotropy, which results from internal stresses and magnetostriction only. The low anisotropy allows easy reorientation and growth of domains while the lack of grain boundaries means that reverse magnetization nucleated in one area can easily spread throughout.

• Supermalloy. Low crystal anisotropy

and low magnetostriction together with low impurity content combine to give the low coercivity, high permeability alloys such as supermalloy and mumetal (77% Ni, 5% Cu, 4% Mo). Microscopic magnetic properties result from a combination of material (atoms) and geometry (crystal structure). These alloys place iron atoms in a nicked-dominated FCC structure. The degree of order in the iron-atom placements gives control over magnetic effects.

Table 3.8   Soft magnetic materials and applications

Application	Material	μ_r max	B_s/T	T_c/K	H_c/(A m^{-1})
electromagnets	permendur 2V (49% Co, 2% V)	10^4	2.4	1250	200
	purified iron	10^5	2.2	1043	5
power transformers (low frequency)	silicon iron (3.5% Si)	4×10^4	2.0	1020	20
	amorphous FeCo	$> 10^5$	1.8	630	3.5
audio transformers	permalloy 45 (45% Ni)	2×10^4	1.7	710	30
magnetic shielding	supermalloy (79% Ni, 5% Mo)	10^6	0.7	670	0.2
high-frequency transformers	manganese-zinc ferrite	2×10^3	0.5	420	20

3.7.4 Strategies for efficient power transformer cores

At the end of Section 3.3 I asked you to narrow the field of potential core materials (Exercise 3.11). Let's pick up that story again, armed with an awareness of the role of domains, to see how crystalline and amorphous silicon-iron based alloys have evolved.

Non-magnetic additions

The serious contenders were amorphous iron-boron-silicon and crystalline silicon-iron alloy. Both risk diluting the iron with at least one non-magnetic addition, silicon (B_s is roughly proportional to the iron content). Let's consider first the crystalline alloy. Adding silicon increases resistivity (see Figure 3.40a) more than any other common addition, so eddy current losses can be reduced.

EXERCISE 3.20   How much might eddy losses be reduced by increasing the silicon content from $3\frac{1}{2}$% to 10%?

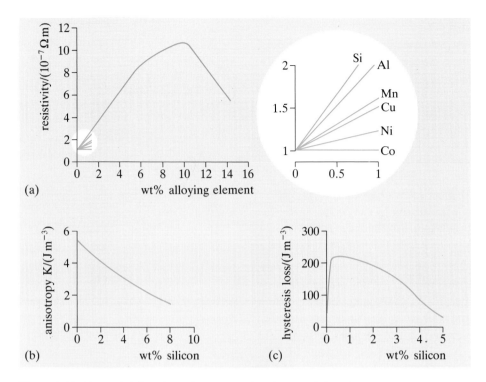

Figure 3.40 (a) Resistivity, (b) anisotropy and (c) hysteresis loss for Si–Fe

But adding silicon to iron does more than this. By lowering the crystalline order (silicon atoms sit at iron sites in the lattice, randomly distributed) the magnetocrystalline interaction is weakened so the anisotropy constant K is reduced, see Figure 3.40(b). Similarly, magnetostriction is affected and actually passes through zero at about 6.3% Si. Low anisotropy keeps the domains' orientations flexible so that the saturation is easily achieved. Reverse domain formation and hence low coercivity are encouraged by low anisotropy. Not surprisingly therefore, hysteresis loss decreases with silicon addition (Figure 3.40c). We'll come to the constraints on how much silicon shortly.

By risking further dilution with a good glass-forming element like boron, and by rapidly quenching the melt, an amorphous alloy can be made. Amorphous metals are inherently more resistive than crystalline ones. This, in addition to the thin-ribbon form of amorphous material produced by the rapid solidification process, gives valuable reduction in the eddy current losses. As we have already seen, amorphous alloys have no crystalline anisotropy, so this material should have low hysteresis losses also.

In transformer applications it is customary to lump the core losses together and quote a watt/kilogram figure, specifying also flux density and operation frequency. A 3% silicon-iron can achieve about 1 W/kg operating at 1.5 T and 50 or 60 Hz, whereas amorphous ribbon at 1.4 T is superior, losing only 0.25 W/kg.

Grains

If amorphous ribbon is so good, why bother with crystalline material? The answer has technological and cultural aspects. First the technology. It is harder to handle several thin strips compared with fewer thicker ones. Also, given sufficient ductility, crystalline sheet can be rolled, stamped and bent and then magnetically restored by annealing (whereas amorphous material is prone to irrecoverable damage from accidental stress). In fact the processing of crystalline material puts limits on how much silicon it can contain. The popular solution for decades has been to use about 3% silicon since above this, an alloy is formed which is too brittle to cold roll and cold form. Grain orientation by controlled cold roll and annealing treatments produces a texture with easy axes along the magnetization direction so that most magnetization need be achieved only by $180°$ wall lateral displacements with minimal rotation. The domain structure is further refined by scribing (see below).

A 6.5% silicon–iron alloy has virtually zero magnetostriction, which makes it an attractive material for separating magnetic effects from mechanical effects. It is too brittle to cold roll, so it has to be produced to final thickness. A rapidly quenched (but not amorphous) casting route can give typically 60 μm ribbon with 1 W/kg total losses, without further refinement. This is already as good as the better low-silicon alloys, but the material is not as handleable.

And so to the cultural aspects. One way to make transformers is to wind coils around a core. Another is to wind a core around some coils. Using thin-strip material, wound cores are very convenient (for either approach) and provided the transformer is a simple type with one primary and one secondary (single-phase), amorphous or 6.5% Si-Fe wound cores are a very attractive proposition. For three-phase work, the greater complexity of three primaries and three secondaries on one core (three phases) is much more amenable to a traditional laminated-sheet core design. Because of the pattern of population (and other non-technical factors beyond reasonable control), in the UK it is convenient to distribute electricity as three parallel phases right up to the final step-down from 11 kV within a few metres of the consumer. In the USA however, the phases are much more frequently split up and separately distributed, often to remote locations, for the final step-down by a single-phase transformer mounted on the distribution pole. Because of these basic differences, there is little market for wound amorphous or 6.5% Si core transformers in the UK (and Europe) but an enormous one in the USA, Canada and other countries with low population densities. The distribution companies have to trade the inherently greater efficiency of a single, large (three phase) transformer for savings in the cost of cabling (single phase) if consumers are widely spread.

Scribing domains

Further refinement of the domain structure of 3% SiFe is achieved by **scribing**. Grain orientation produces large aligned crystals. During magnetizations, 180° domain walls sweep through these grains so that locally there is a changing flux (dB/dt) and hence associated eddy current losses. Bigger grains contain faster moving walls and so produce larger dB/dt, and larger losses, than narrower grains under the same conditions. In the scribing process, lines of surface stress are drawn across the surface by means of the local melting and resolidifying under a focused laser beam or spark discharge. This process promotes narrow domains as the stress lines act as artificial grain boundaries. As we established earlier, effectively smaller grains will house smaller domains. Losses can be reduced by over 7% by this method. Figure 3.41 shows domains before and after scribing.

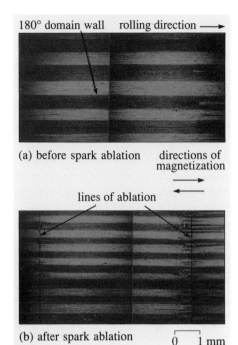

(a) before spark ablation

directions of magnetization

lines of ablation

(b) after spark ablation   0   1 mm

Figure 3.41 Domains in 3% Si–Fe before and after spark scribing. Courtesy of British Steel

SAQ 3.9   (Objective 3.8)
(a) Complete the Table 3.9, which summarizes part of Section 3.7.
(b) Complete Table 3.10 by briefly indicating the processes used to control the particular properties of low-loss transformer core material. (P_e and P_h are eddy-current and hysteresis losses; up and down arrows imply increase and decrease respectively.)

Table 3.9   For SAQ 3.9(a)

Property	Principle	Implications for processing	Remarks
high B_r	many randomly oriented easy axes		isotropic materials — easy axis always on hand
	aligned easy axes		anisotropic materials — no rotation into saturation
high μ_r	aligned easy axes		
high H_c	hard to reverse single domains		uniaxial anisotropy exploited
	impaired reverse domain nucleation		
low H_c	special low anisotropy alloys		non-crystalline: sensitive to stress
			crystalline: low magnetostriction

Table 3.10 For SAQ 3.9(b)

Property	Principle	Process
resistivity, ρ	$P_e\downarrow$ as $\rho\uparrow$	(i) (ii)
sheet thickness, t	$P_e\downarrow$ as $t\downarrow$	(i) (ii)
domain width, d	$P_e\downarrow$ as $d\downarrow$	
coercivity H_c anisotropy constants K and λ_s	$P_h\downarrow$ as $H_c\downarrow$ $H_c\downarrow$ as $K, \lambda_s\downarrow$	(i) (ii) (iii)

Summary of Sections 3.6 and 3.7

● Domain shapes and boundary widths are largely determined by competition between exchange and anisotropy forces.

● Single-domain particles can exist below a certain particle size (about 50 nm).

● Domain structure and environment govern magnetization and hysteresis phenomena.

● Magnetic anisotropies (K and λ_s) and hence domain properties can be controlled chemically (through composition) and physically (through microstructure).

3.8 Magnetic recording

3.8.1 Introduction

Have a look at a tape recorder (audio or video) or a computer disc drive. These are the products we must keep in mind as we now delve into some of the materials science behind their function. Information is represented electronically as a sequence of voltage levels (plus and minus, or larger and smaller). It is frequently necessary to store such information. Storage of information as patterns of magnetization in a magnetic material is a well established principle, dating from the search for audio recording schemes at the end of the nineteenth century. Since then, magnetic recording has developed into a mature technology. Recently it has gained from intensive research prompted by competition from newer technologies like semiconductor and optoelectronic storage.

EXERCISE 3.21   For each of the applications
● audio recording
● archival storage
● computer memory
suggest a favourable and an unfavourable feature of the use of magnetic records.

Before we look at the details of particular recording technologies, it will be helpful to use the ideas developed earlier to assess the potential of magnetic media for data storage. Exercise 3.22 suggests how to do this. In the exercise as in all of this section I will only refer to digital data. If you want to think in terms of analogue signals, remember that the most demanding digital signal is . . . 1010101010 . . . (a square wave of alternating 1 and 0) which is recognizable from its fundamental sinewave (period = two bits). So, for example a 1000 bit per second signal is in a sense equivalent to an analogue signal at 500 Hz (sketch it if you need to).

EXERCISE 3.22 A magnetic 'bit' (binary digit) can be represented by the direction of magnetization in a certain region of matter.
(a) How small (physically) could one isolated bit be? Choose from: subatomic; atomic; 10^3 atoms; μm; mm.
(b) What factors might limit the size of a bit in practice?
(c) Suppose we have a disc coated with single-domain bits ($\approx$ 0.1 μm). Estimate the bit-capacity of a disc with 0.01 m^2 of useful area.
(d) The disc in (c) spins so that the tangential speed is about 36 m s^{-1}. At what rate could data be transferred to or from the disc.
(e) What sort of magnetic materials will be involved?

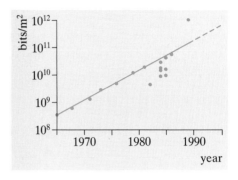

Figure 3.42 Improvements in bit density for various data recording discs

Figure 3.42 shows how the storage density of magnetic recording has advanced. An upper bound of 10^{14} bits m^{-2} was discussed in Exercise 3.22(c), based upon single-domain bits. Three basic technologues will be outlined in the next subsection.

3.8.2 Recording technologies

We have seen that magnetism is a co-operative process between atoms, and that this co-operation favours the formation of regions of aligned atoms, or domains. It is in the disposition of these domains that information is recorded. You will recall that Barkhausen heard them by electromagnetic induction, Kerr saw them by a magneto-optic interaction and Bitter delineated them when small magnetic particles were moved by non-uniform fields at their boundaries. Features of these classic observations are exploited in the recording technologies that we will examine.

We shall look at the materials used to store data in Section 3.12. First we will examine the 'machinery' of three different systems in order to determine the specifications for appropriate recording materials. The three systems are: tape/disc systems, magneto-optic and magnetic bubble devices.

3.9 Tapes and discs

3.9.1 General principles

In this class I want to consider the writing of data by means of an electromagnet moving relative to the recording medium (I will refer to a tape but discs are equally implied). Let's begin by thinking about writing a 'bit' with a very simple electromagnet, a single strand of wire lying across a tape. (See Figure 3.43.) From here we'll go on to invent the conventional recording head. The medium will be magnetized close to the wire wherever the magnetizing field exceeds the coercivity of the tape.

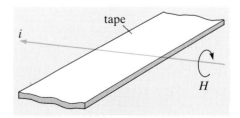

Figure 3.43 Writing with a wire

EXAMPLE Estimate the current required to write on a tape of coercivity 10^4 A m^{-1} to a depth of 1 µm.

ANSWER Ampere's law tells us that at a radius r from a current i in a wire the magnetizing field is

$$H = \frac{i}{2\pi r}$$

If $H = H_c$ at a distance of 1µm from a negligibly thick wire, the current in the wire would have to be

$$i = 2 \times \pi \times 10^6 \times 10^{-4} \, A$$
$$\approx 0.06 \, A$$

Comment. This can't be a copper wire since a rough design rule for insulated copper wire is not to exceed a current density of 10 A mm^{-2} (or 10^7 A m^{-2}). Even if our 'write wire' has a 1 µm radius we have 2×10^{10} A m^{-2}.

EXERCISE 3.23

(a) Restricting the current density in the wire to 10^7 A m^{-2}, at what radius of wire can we magnetize a tape of negligible thickness and what then is the recording current?

(b) Restricting the current density in a 1 µm radius wire to 10^7 A m^{-2}, what is the largest magnetizing field which we could produce?

Exercise 3.23 suggests that writing small magnetic records with a copper wire seems hopeless. If the wire is small enough, magnetizing currents are three orders of magnitude too small. If the current and wire are right, the current density is three orders too large! But of course a few orders of magnitude are just what separate magnetic materials from non-magnetic ones, so we should try to incorporate some magnetic material in the writing system. One way is to introduce magnetic material into the writing unit and this is what we'll consider here, but another is to use the recording medium itself to increase the efficacy of write-currents as in ▼Old fashioned computer core▲. Figure 3.44 shows the principle of the **gapped core** head which is commonly used in writing tapes and discs. This is how it works.

(i) Current in the coil produces a magnetizing field in the core and in the gap. Several turns can be used to alleviate current density constraints.
(ii) The core is made from soft magnetic material (lots of flux for very little magnetizing field) and is several orders of magnitude easier to magnetize (more permeable) than the air in the gap.
(iii) A little bit of the magnetizing field leaks out of the edges of the gap (**fringing field**) and we write with this.

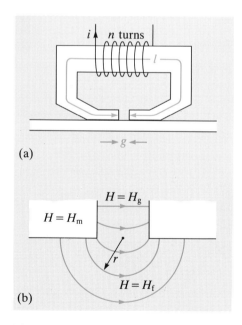

Figure 3.44 (a) Schematic gapped core. (b) Fringing field

EXERCISE 3.24   Complete the following equations. (Refer to Figure 3.44 for symbols.)

(a) Ampere's law around the core and gap gives

$$H_m l + H_g g =$$

(b) For a small gap ($g \ll l$), the flux density in the core and gap are roughly equal (B). Suppose the B–H curve of the core is roughly given by $B = \mu_r \mu_o H_m$ and for the gap, $B = \mu_o H_g$. Using (a) to eliminate H_m,

$$H_g =$$

So if $\mu_r = 1000$, $H_g \approx ni/g$.

(c) For a path across the gap and around the fringing field H_f (approximate it to a semicircle) Ampere's law gives

$$H_f = -$$

(d) So if we want to magnetize to a depth of 1 μm a medium of coercivity 10^4 A m^{-1} we have

$$ni = - \qquad \text{ampere turns}$$

(for example, 10 turns of a wire carrying 3 mA).

▼Old fashioned computer core▲

During the 1950s and 1960s, active computer memory was predominantly 'core store' (Figure 1.21). Bits were stored as the magnetization direction in a small toroid (or doughnut) of ferrite (a core). The write step for a single core involves magnetization by pulses of current on wires which thread it. Figure 3.45 shows the core and its B–H loop. A near rectangular loop is needed. Suppose the core is currently storing a 'zero' and sitting at P in the B–H loop of Figure 3.45. In order to overwrite this with a 'one', we must take it to point R on the loop via Q. If a magnetizing field H_1 takes the magnetization to the brink of change, $2H_1$ should reverse it for a rectangular loop. Now thinking of Ampere's law we can relate H_1 to the required threading current i (which had better not be too large if each element is to be kept small). In practice the $2i$ write current is very conveniently supplied as i in two separate wires each of which thread a line of cores. Only that core threaded by $2i$ is addressed, others on the row or column are only moved by $+H_1$ along the loop and back with no change. There are several ways to read a core but all risk destroying the data and require a

write-back cycle if data are to be kept. For example, a third 'sense' wire threads each core and the read step is to write a zero ($-2H_1$ provided by $-i$, in the row and in the column wires). A one would become a zero and the changing core flux induces a voltage pulse in the sense wire (indicating there had been a one there) but a zero would give no sense pulse (indicating there had been a zero there).

What magnetic materials should be used? Well, computers have always tried to go fast so the material had better reverse quickly. Eddy currents driven by the changing flux will slow its response, so insulating ferrite is used. The rectangular loop is encouraged by going for a polycrystalline material with lots of easy axes so that the remanent state is a small rotation away from saturation — a good candidate is a cubic ferrite with axes along the four body diagonals. The toroidal core is relatively insensitive to stray fields, being a closed loop, so since write currents scale directly with coercivity, a modest 10^3 A m^{-1} or so (almost 'soft') should suffice. Manganese magnesium ferrite is a typical candidate.

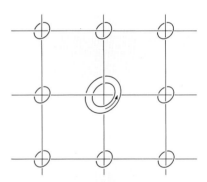

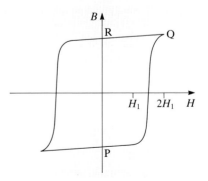

Figure 3.45 (a) Magnetic core.
(b) Corresponding B–H loop

▼Recording heads▲

Recording heads have to fulfil first and foremost an electromagnetic function. Gapped core heads must transduce electric current into magnetic fields for writing, and magnetic field variations into electric voltages for reading. A secondary requirement is that they should not cause excessive wear of, nor be worn by, the recording medium with which they are working.

The electromagnetic function favours high permeability (soft) materials such as NiFe permalloys or MnZn ferrite. For high-density recording it is also important to consider eddy-current losses in the core, so metallic cores must be laminated. The less permeable ceramic is, however, inherently a superior high frequency material, as we have already seen. (Did you notice the similarity in magnetic specification between transformer cores and recording head cores?)

The problem of saturation is mainly associated with the material adjacent to the gap. The corners of the gap material begin to saturate when the gap field exceeds about half the saturation flux density of the material. Saturation of the gap corners is not only inefficient but it also degrades the stored signal quality and the frequency response of the writing process.

EXERCISE 3.25 Why is saturation of the gap corners not usually a problem during reading?

Because gap corners must be of high saturation material, generally we should use metals for them. But these wear much worse than the lower saturation ferrites, so there is scope for compromise.

EXERCISE 3.26 Figure 3.46 shows three recording head core designs. Comment on magnetic and mechanical properties in each case.

Tapes and flexible discs require the recording head and the recording medium to be in contact to ensure reproducible and satisfactory read-write operations. For 'hard' computer discs, the wear problem is greatly reduced by allowing the head to fly on an air cushion a fraction of a micron above the spinning magnetic surface. The wear is then restricted to stop and start operation when the head lands. Incidentally, the 'crash' which occasionally incapacitates computers is a euphemism for a particularly nasty head-disc collision which can occur when the flight control is lost.

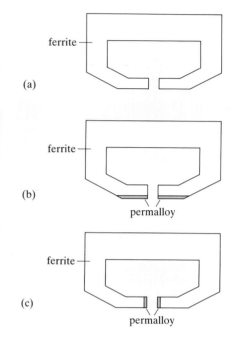

Figure 3.46 (a) Ferrite, (b) metal tipped ferrite and (c) metal-in-gap

The magnetic material is deployed so as to concentrate the magnetizing field where we need it and all we have to do is to make sure it is permeable enough and that it doesn't saturate before we've managed to magnetize the tape. We have here the makings of an interesting materials problem when it is realized that the head and tape may have to be in contact to maintain micron separations between tape and head, so one or other or both may suffer wear (see ▼Recording heads▲).

3.9.2 How big a bit?

In tape and flexible disc recording a write-head, in contact with the medium, produces fields sufficient to magnetize the medium in one or other direction. An ideal material will only switch its magnetization when $H = H_c$, whereupon complete reversal occurs, that is it has a rectangular M–H loop, Figure 3.47(a). In practice, real materials have skewed loops, Figure 3.47(b), and often retain less than saturation flux, so that the reversal is gradual. Suppose that at some instant the write-field is reversed as the tape passes the record head. Figures 3.47(c) and (d) show how the transition is translated onto the tape as a sharp step for the ideal

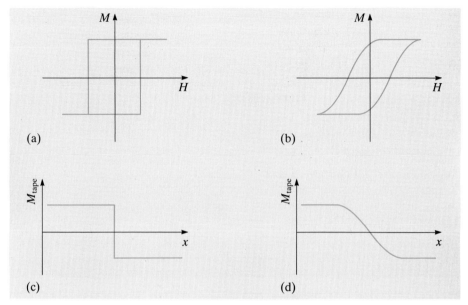

Figure 3.47

▼ S, S* and SFD ▲

Rectangular (M–H) hysteresis loops can be characterized in at least three different ways. We'll look at two of them. The first looks at the degree to which magnetization is retained in remanence. The quantity defined is called the **remanent squareness** S (for unsound geometric reasons!) and is given by

$$S = M_r/M_s.$$

Figure 3.48 shows an M–H loop in which four straight lines are included to approximate the curve by a parellelogram. The closeness of the approximation to a rectangle is reflected by the value of a quantity called **coercive squareness** S^*, which is defined in the figure.

An ideal recording medium will have S near unity to assist the reading process. It will also require S^* to be near unity to prevent wide transition regions between opposite bits; that is, the distribution (or spread) of fields capable of switching some of the magnetization must be narrow. The **switching field distribution** (SFD) must be narrow for two further reasons. First, for digital discs, overwriting of data held on a medium with wide SFD will contain errors where the head-to-switch tail of previously written data fails to be completely overwritten by a worn or under-driven head. Secondly, tapes can be corrupted if the magnetization stored in one layer is sufficient to switch the tail of the SFD in an adjacent layer.

tape and as a slope for the skew material. At this level, the skew material limits the shortest bit we can write but the ideal rectangular material shows no limit. For short bits we need:

- a medium with as near a rectangular loop as possible (see ▼ S, S* and SFD ▲),
- a head design which produces steep gradients in the field (dH/dx) near where the trailing edge field passes the critical level of H_c (the coercivity).

Suppose we have a perfectly rectangular loop material. If we try to make very sharp transitions, magnetostatics steps in. The sharp transition is energetically less favourable than the gradual change (remember domain walls?). It turns out that materal thickness is important and if we want micron bits we will need only micron-thick magnetic media.

3.9.3 Reading

Where the recorded flux changes direction, we can picture microscopic magnets lined up on the tape 'head to head' in opposition. One thing they agree upon, though, is that the space outside the magnetic layer should be magnetized. A moving tape therefore can give rise to varying magnetic flux in nearby stationary loops of wire (or coils). Faraday's law determines the magnitude of the induced voltages. As you probably know, the same head can be used to write and read a tape. In reading, the core of the head directs some of the surface flux through the coil. Figure 3.49 shows schematically the reading operation. As the tape moves at a steady speed u, the flux (Φ) diverted through the coil varies in proportion with the tape magnetization. Sharp spatial variations along

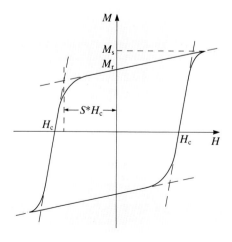

Figure 3.48 M–H loop for S and S^*

the tape make rapid temporal flux changes in the coil. In place of $d\Phi/dt$ in Faraday's law we write $u d\Phi/dx$ so the voltage induced in n turns is

$$\varepsilon = nu \frac{d\Phi}{dx} \qquad (3.10)$$

EXAMPLE A play-back head experiences mean flux densities of 0.01 T up to 1 μm above a disc, linking them through a ten-turn coil. The track is 10^{-5} m wide. What will be the approximate magnitude of voltage registered by a flux reversal over 2 μm when read at 36 m s^{-1}?

ANSWER Use Equation (3.10). For the derivative, use the change of flux divided by the distance over which it changes.

$$\varepsilon = nu \frac{\Delta\Phi}{\Delta x}$$

Inserting values (with $\Phi = BA$),

$$\varepsilon = 10 \times 36 \times \frac{[0.01-(-0.01)] \times 10^{-5} \times 10^{-6}}{2 \times 10^{-6}} \text{V}$$
$$= 3.6 \times 10^{-5} \text{V} = 36 \text{ μV}$$

Notice that in Figure 3.49, the signal which is generated on reading only marks the changes in tape magnetization and not the actual magnetization. In contrast, during recording, the pattern switches whenever the current signal switches. Clearly, regenerating the original electrical signal is a job for our electronic engineering colleagues. (In fact the 'read' signal is the differential of the write signal, so we must integrate to recover the original data.)

Notice too that even if we could write perfect steps, flux changes which switch one way and back again within the length of the read-head gap will not be correctly read.

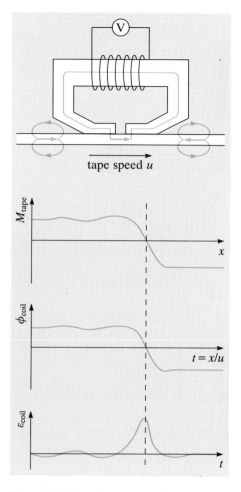

Figure 3.49 Reading process for tapes

3.9.4 Keeping and recycling

An important specification for any storage medium is how long it is be kept and how long before it is to be rewritten. In the case of paper documents this may be a thousand years if well cared for, and clearly archaeologists have been able to exploit the comparable longevity of certain magnetic ceramics in the thermomagnetic dating of pottery. The coercivity of the medium determines its magnetic resilience and we should seek the higher values. For tapes, an important consideration is the effect of adjacent layers of recorded material on the spool. Print-through (magnetization of one layer by another) is to be avoided. The shelf-life is limited by environmental conditions — the magnetic material must be either chemically stable (which suggests using oxides) or else protected.

Tapes and discs are not just used for archives, and in order to compete with other memory systems, multiple rewrite is essential. In audio tapes the previous recording is first erased by randomizing the magnetization pattern, but for digital recording the previous data are directly overwritten by the new 'saturation' state. Erosion of the recording medium owing to contact with the head is a limit to re-use.

3.9.5 Magnetic material requirements

SAQ 3.10   (Objective 3.10)
So far we have established constraints on the following: Remanent magnetization M_r; coercivity H_c; M–H loop shape; thickness of medium δ; head–tape separation d; head gap g; head permeability μ_r; wear properties. For each, say briefly why it has emerged, and what we should look for (high or low values where appropriate).

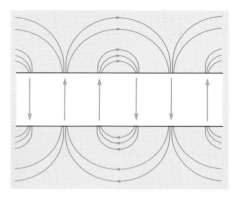

Figure 3.50 Perpendicular recording

One of the requirements on a material arises from the need to be able to read the recorded signal: sufficient flux must be squeezed outside the medium into the range of the read head. So far it has been tacitly assumed that the magnetization lies in the plane of the tape, but a good way to get more field outside is to specify a vertically magnetized medium. Figure 3.50 shows a perpendicular magnetization pattern. There are implications for recording density and material properties. The density can be increased as shorter transitions can be achieved owing to the flux-closing role of an adjacent opposite bit; clearly a strong uniaxial anisotropy will be needed to sustain a magnetization across the short direction.

Another way to tackle the playback problem is to consider other ways of reading the tape. For example, magneto-optic interactions in the tape surface are not reliant on fields external to the medium itself. In fact, as we shall see in the next section, a perpendicular magnetization is favoured for such schemes.

3.10 Magneto-optics

3.10.1 General principles

In this section we will discuss a type of magnetic recording which is characterized by a thermomagnetic writing step and a magneto-optic reading step. Suppose a region of magnetic material is heated above its Curie temperature (T_c) and then allowed to cool in the presence of a magnetizing field (H_1). We could do this quite quickly over a small area with a focused laser. We expect the spot to acquire the magnetization direction of the external field. This could therefore be the write step in a recording system. In fact, since the coercivity generally falls to zero at T_c, it may not be necessary to reach T_c, so long as the coercivity falls below

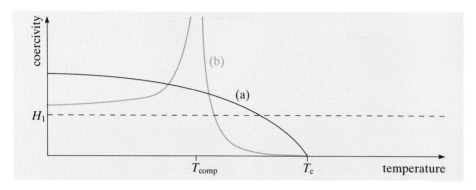

Figure 3.51 Curve (a) shows variation of H_c with temperature for ferromagnetics. Some ferrimagnetics show the behaviour of curve (b)

the strength of our magnetizing field. This field may itself be comfortably below the material's room-temperature coercivity so that surrounding, unheated material is immune. Figure 3.51, curve (a), gives a schematic illustration.

An alternative critical regime exists for certain ferrimagnetic materials which have a compensation point (see ▼Temperature and ferrimagnetism▲). These ferrimagnets have strongly temperature-dependent coercivity. Figure 3.51, curve (b), illustrates the principle of compensation point (T_{comp}) recording. The coercivity goes infinite near T_{comp} because the compensation gives the external magnetizing field virtually nothing to get hold of to reorient the remaining magnetization. But just beyond T_{comp}, the coercivity falls rapidly long before the Curie point so this region is useful for thermomagnetic memory. Materials which show large and temperature-insensitive magneto-optic effects have been developed.

3.10.2 Reading

The domain structure of a thin layer of magnetic material can be observed by using the Kerr or the Faraday magneto-optic effects, using polarized light. By analysing the direction of rotation of the polarization plane, it is easy to distinguish between opposite magnetization directions, even though the angles involved are only a fraction of a degree (see Figure 3.25b). Kerr actually has three separate configurations in his effects, according to the disposition of the surface, the plane of polarization and the magnetization direction. The polar effect occurs when magnetization is perpendicular to the surface. The Faraday effect also requires a magnetization component perpendicular to a transmitted light beam (which suggests thin or transparent material). From now on we'll think only of polar-Kerr.

Magneto-optic reading of the perpendicular-domain magnetization is a characteristic step of this technology, and like the writing step, it is essentially a non contact (no wear) process. Furthermore, in principle the same light source can be used to read and write although obviously the 'read power' should be less than that for writing. Now, the angle through which the polarization is rotated scales with the strength of the atomic magnetization, so that near T_c as thermal energy weakens the magnetic order, reading is difficult. In contrast, the microscopic order near T_{comp} is not much affected by thermal agitation. For this reason reading the compensation-point recording scheme offers superior signal-to-noise-capability over that of Curie-point recording.

▼Temperature and ferrimagnetism▲

Ferrimagnetism is distinguished by having two unmatched but opposing magnetic structures interpenetrating. Each structure is associated with different sources of magnetism (for example Fe^{2+} and Fe^{3+} in magnetite, and Mn^{2+} and Fe^{3+} in manganese ferrite). Furthermore, the two structures are generally associated with specific atom placings (such as tetrahedral A sites and octahedral B sites in ferrites). Therefore, as the temperature of a ferrimagnet is increased, we can't expect each magnetic structure to respond in exactly the same way. For example, the B-site magnetization M_B may start larger than that of the A site, but may then fall off less rapidly with temperature, so that the difference $M_B - M_A$ (which is what we observe externally) at first increases with temperature, before rapidly decreasing to zero as T_c is approached. Figure 3.52(a) illustrates this. Figure 3.52(b) is an intermediate case. The magnetization falls away monotonically, not unlike that in a ferromagnet. The other extreme case, Figure 3.52(c), finds M_B dropping from above M_A to below it so that at one temperature, the **compensation point** T_{comp}, the net magnetization $(M_B - M_A)$ is zero. In practice domain magnetization vanishes at T_{comp}, reappearing with the opposite set of atomic magnets aligned with the external magnetizing fields.

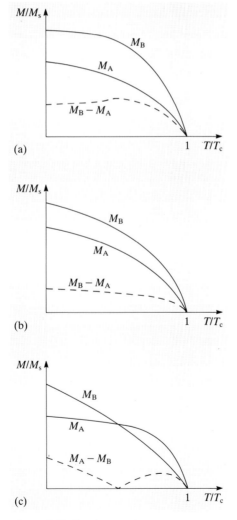

Figure 3.52 Ferrimagnetic temperature dependences

3.10.3 How big a bit?

The size of the remagnetized region in the thermomagnetic process can be controlled by the size of the heated zone in a uniform magnetizing field by using a focused laser beam to do the heating. There is none of the physical contact and associated wear problems of electromagnetic recording schemes.

In order to estimate the bit size, we can compare the scale with that of the photolithographic resolution encountered in semiconductor fabrication. Here, routinely, micron-sized features can be defined optically, so the square micron bit is a distinct possibility.

During formation, of an isolated bit (a 1 between zeros), a boundary region will be established which must be sharply defined across the track. A ragged edge will increase the minimum bit size and may be a source of noise. If the boundary is a classical domain wall we should specify a narrow wall but with a high energy. Large anisotropy energy (K) favours both.

EXERCISE 3.27   A 10 mW solid-state (diode) laser is focused onto a one-micron-thick layer of magnetic material. What temperature rise might we expect from a 100 ns pulse if the focal spot is 10^{-12} m²? (Data: specific heat capacity c is 500 J K^{-1} kg^{-1} and density ρ is 8×10^3 kg m^{-3}. Neglect losses to surrounding matter.)

Although laser beams are very steerable, it is unlikely that we would want an optical system to sweep the whole surface. In fact spinning magneto-optic discs (Figure 3.53) are used to keep the data-space evenly available, but for this, careful steering is needed to keep the beam on target. The optical-tracking guides the beam along a flat (reflective) narrow data track between two equally narrow grooves, by continuously trimming a mirror angle while a second feedback system optimizes focus by finely adjusting the position of a lens. Both control elements are actuated electromagnetically in the same way as a loudspeaker cone, so this part of the magnetic materials technology is well established.

3.10.4 Keeping and recycling

As with conventional tapes and discs, shelf life is predominantly a chemical problem. Although thermomagnetic material is a little more sensitive to heat, the rapid increase in coercivity of compensation point material around T_{comp} enhances magnetic stability up to this point.

The non-contact read and write steps make this technology very attractive for repeated use, provided that thermal cycling is not a limitation on the magnetic material or associated support and protective matter.

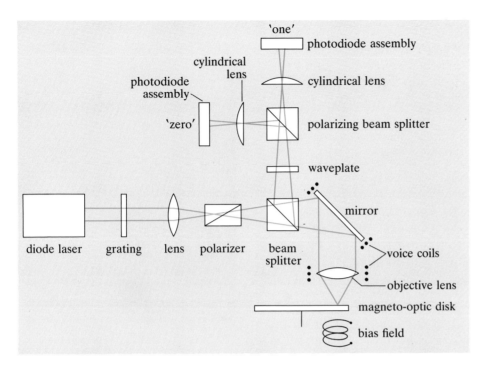

Figure 3.53 A magneto-optic disc system

3.10.5 Magnetic material requirements

SAQ 3.11 (Objective 3.11)
In this section I have indicated constraints on the following:
orientation and strength of magnetization M_r; uniaxial anisotropy K;
compensation or Curie temperatures T_{comp} or T_c; thickness δ;
reflectivity R; Kerr polar angle (θ_k); thermal fatigue. For each, say
briefly why each has emerged and what we should look for (high or
low as appropriate).

Magneto-optic media need to be magnetized perpendicular to the
surface. Figure 3.54 shows a magnetization pattern. Remember that the
track width may be comparable with bit length, so we may have in
Figure 3.54 a 'blob' of reverse magnetization. Remember too that the
blob's boundary will be a simple domain wall and domain walls can be
moved. In the right material such pockets can preserve their shape as
they move, driven by gradients of an external magnetic field (Bitter
moved tiny single-particle magnets in non uniform fields too!). This is the
basis of magnetic bubble systems. No need for spinning discs if the data
can be pushed about!

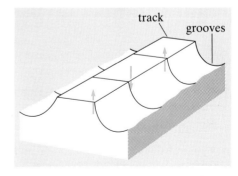

Figure 3.54 Magnetization pattern along a
magneto-optic track

3.11 Bubbles

3.11.1 General principles

Magnetic bubbles are a consequence of a strongly uniaxial material in a thin layer resisting magnetization reversal by an external field. The anisotropy must be large enough to sustain perpendicular magnetization, and in such materials the unmagnetized state is characterized by strip domains wiggling around the layer, as shown in Figure 3.55(b).

Suppose we have a thin, pure, single crystal with perpendicular magnetization and with a small external bias field (H_o) favouring one magnetization direction (Figure 3.56a). We know that we will magnetize the crystal by expanding those domains which are parallel while contracting the antiparallel ones, and this can be done simply by moving 180° domain walls. But in the thin structure as the antiparallel domains contract, the section changes from being short and fat to being long and thin. As we saw in the study of motor magnets, longer thinner shapes are more resilient to demagnetization than shorter fatter ones, so as the bias field is increased, the antiparallel domains shrink just enough to be able to withstand the demagnetizing influence. The domains can still be very long as revealed in Figure 3.56(b).

If the bias field is increased further, the antiparallel-aligned strips become narrower. When these domains are about as wide as the layer thickness, it becomes energetically favourable to break up into blobs (generally known as **magnetic bubbles**). Further increase of the bias field compresses the blobs and as the shrinking bubbles decrease their wall area, the surface energy of the boundary must be brought into consideration. Figure 3.57(a) shows the beginnings of bubble formation and Figure 3.57(b) shows a complete array of bubbles.

At still larger fields (H_o approaching M_s in magnitude) the bubbles collapse, unable to resist further. Stable bubbles exist over a range of bias field between the bubble collapse threshold and the 'strip break up' threshold. ▼Bubble domains▲ considers the bubble development in more detail.

EXERCISE 3.28   How do you think the pictures of Figures 3.55(b), 3.56(b), 3.57(a) and (b) were generated? (What are you seeing? Is the material thick or thin?)

The bubbles will move freely in the pure single crystal. As we have seen, impurities and crystal defects or boundaries impede domain wall motion; but the perpendicular 180° domain walls do not have to move past many such obstacles here. Furthermore, since the bubble is a reverse domain — under pressure from the external field to switch its magnetization — it is not surprising to find bubbles congregating in regions of weaker field. Bubbles also repel each other, rather as two side-by-side bar magnets would.

(a)

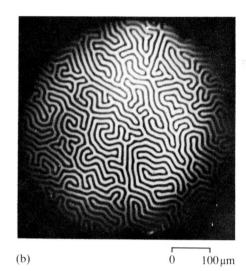
(b)   0   100 µm

Figure 3.55 (a) Section through thin-strip domain material. (b) Plan view of domains in garnet

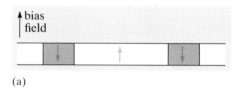
(a)

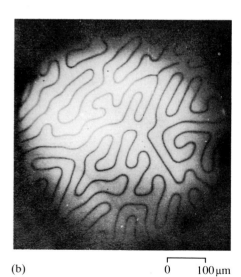
(b)   0   100 µm

Figure 3.56 (a) Section through domains in a bias field. (b) Plan view of garnet

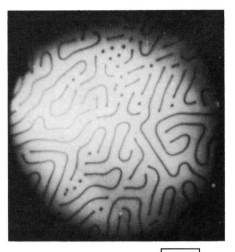

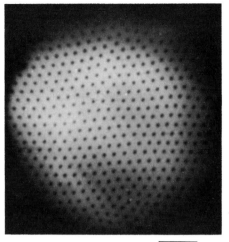

Figure 3.57 (a) Plan view of strip break up. (b) Plan view of bubble array

(a) 0   100 μm   (b)   0   100 μm

▼ Bubble domains ▲

To study bubble domains rigorously involves some elegant but involved physical and analytical arguments for which I regret we have little time. Let's leave the detailed numbers to others and just concentrate on the general ideas.

We begin with the strip domains. Figure 3.58(a) shows how the diameter-to-height ratio D/h varies with the bias field, H_o. As H_o is increased the strip domains are pushed to narrower and narrower equilibrium widths until when H_o equals M_s (the saturation magnetization), the strip vanishes.

But, examination of the energy associated with strips and bubbles shows that at a critical threshold field, the strips can break up to reach lower-energy configurations as bubble domains. Surface energy combines with the external field to favour smaller radii, while the internal field opposes them. As Figure 3.58(b) shows, beyond a second threshold, the domain-wall energy spontaneously collapses the bubble to zero radius. It turns out that for stable bubbles, the ratio D/h is usually around unity.

Two useful characteristic quantities can be used to compare bubble media. The first, Q, compares the anisotropy energy with magnetostatic energy for the material (remember we need high anisotropy for perpendicular magnetization):

$$Q = \frac{K}{\frac{1}{2}\mu_0 M_s^2}$$

The second number, λ, comes from comparing the domain-wall surface energy

(σ_w J m^{-2}) with the magnetostatic energy ($\frac{1}{2}\mu_0 M_s^2$ J m^{-3}) and is a length (m). It appears naturally in the full analysis of bubbles:

$$\lambda = \frac{\sigma_w}{\frac{1}{2}\mu_0 M_s^2}$$

Bubble materials must have $Q > 1$ (so that perpendicular magnetization is preferred) and it turns out that the bubble diameter is usually between 5λ and 15λ. So smaller stable domains need larger magnetization M_s (which may in turn force a need for larger anisotropy to keep $Q > 1$).

EXAMPLE   Using ideas from 'Between domains; width and energy' (Section 3.6) estimate the diameter of bubbles in a garnet with a lattice constant a of 1.2 nm, Curie temperature 600 °C, magnetization 10^5 A m^{-1} and anisotropy energy density K of 10^4 J m^{-3}. Boltzmann's constant = 1.38 × 10^{-23} J K^{-1}.

ANSWER   Let's say diameter $D = 10\lambda$.

$$D = 10\lambda = 10 \times \frac{\sigma_w}{\frac{1}{2}\mu_0 M_s^2}$$

From Equation (3.9)

$$\sigma_w = \pi\sqrt{\frac{2\,AK}{a}}$$

Putting in the values (and note $A = kT_c$)

$$\sigma_w = \pi\sqrt{\left(\frac{2 \times 1.38 \times 873}{1.2 \times 10^{-9} \times 10^{19}}\right)} \text{ J m}^{-2}$$

$$= 14.1 \times 10^{-4} \text{ J m}^{-2}$$

Substituting in our equation for D,

$$D = \frac{10 \times 14.1 \times 10^{-4}}{0.5 \times 4\pi \times 10^{-7} \times (10^5)^2} \text{ m}$$

$$= 2.25 \text{ μm}$$

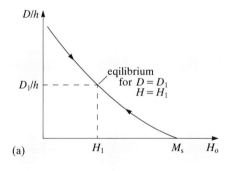

(a)

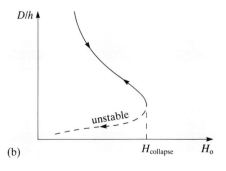

(b)

Figure 3.58 (a) D/h against H_o for strips. (b) D/h against H_o for bubbles

3.11.2 Shifting bubbles

Regions of slightly weakened bias field can be created by using a second, surface-skimming field and a high-permeability island on the surface of the bubble medium; this is illustrated in Figure 3.59. The surface field is concentrated in the high permeability (permalloy) island so that just outside the island there are two regions where the skimming field has a vertical component. The vertical bias field is modulated by this vertical component so that a dip exists on the trailing edge of the island where a bubble could be trapped. If you prefer, you can think of the island as an induced bar magnet with north and south poles, and the bubble domain as a mobile, vertical bar magnet, with its south pole attracted to the island's north.

A pattern of T- and I-shaped islands can be used with two surface skimming fields to produce a series of propagating traps. The two skimming fields are oscillated together but 90° out of phase so that the combined field appears to rotate and traps are handed along the TI structure. Figure 3.60 shows the traps at different instants.

It turns out that smaller devices can be made using more sophisticated patterns. Bubbles can even be generated locally by pulling a lump off a sort of reservoir domain trapped under a permalloy island. But to shrink devices further, it is useful to use an alternative shifting scheme. In the scheme we have been considering, the role of the permalloy is to manage a surface field, and optimizing its geometry (width, length, thickness, distance from surface) is an involved task. Furthermore, its interaction with the bubbles needs to be considered. Suppose that instead we could adjust the surface properties of the bubble material directly on a similar scale, to cause local regions where parallel rather than perpendicular orientation is preferred. We might then dispense with some of the permalloy and the limitations associated with its use. We'll return to this in considering bubble materials.

3.11.3 Magnetic materials requirements

This time it's your turn to pick out any special constraints which have emerged.

> SAQ 3.12 (Objective 3.12)
> Suggest what materials properties and qualities have been constrained by the preceding discussion.

In a practical system, the permalloy islands or other surface features are laid down in tracks so that the bubbles can be shunted along and made to pass under a read head. As a final exercise in this section on technologies, you might like to try to invent a way of reading bubble information for yourself. Exercise 3.29 asks you to try this.

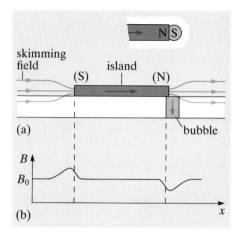

Figure 3.59 Modulating the bias field

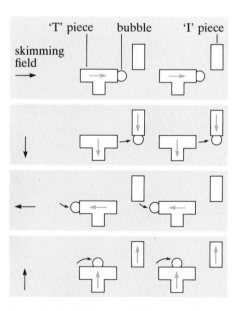

Figure 3.60 A TI track shifting a bubble

EXERCISE 3.29 Suggest one magnetic effect which could be used to detect the presence of a one-micron diameter bubble with a magnetization of 0.05 T.

SAQ 3.13 (Objective 3.9)
Distinguish between electromagnetic and thermomagnetic writing and between electromagnetic and magneto-optic reading of magnetic information.

Summary of Sections 3.9 to 3.11

● Magnetic records can be held on bits of around cubic microns in size.

● Tapes/discs use electromagnetic read and write steps with a contact- or near contact-head, and a moving medium.

● Magneto-optic systems use a thermomagnetic (Curie or compensation point) process for the write step, and the Kerr effect for the read step. Highly reflective surfaces with perpendicular magnetization are required.

● Bubble systems move domains past a reading head. Bubbles are guided around by magnetic surface tracks. The medium needs to have low impurity and defect concentrations and to support perpendicular magnetization.

3.12 Specifying the medium

From the preceding discussion we can assemble a general product design specification (PDS) for magnetic tape or disc or bubble media.

3.12.1 PDS for magnetic recording media

1 They must be uniquely switchable (in a critical field) by some form of 'write' process.

2 They must be stable and resilient to ambient (and self) demagnetization.

3 They must provide sufficient flux for the reading process, with adequate signal-to-noise-ratio.

4 They must be tolerant of wear, or else compatible with non-contact schemes.

5 They must be economically viable for production as tapes, etc.

6 They should offer high storage density with capability for reasonable transfer rates.

EXERCISE 3.30 For each entry in the PDS, list any constrained magnetic properties.

This PDS can only be further refined by specifying the technology for which it is intended. In the following paragraphs therefore we will look at media for each of the three schemes outlined earlier.

3.12.2 Particles for tapes and discs

The idea behind particulate media is that we can prepare particles with carefully tailored magnetic properties, and then deploy them in a passive binder on a flexible polymeric sheet (such as PET), or a rigid aluminium disc. The magnetic and mechanical functions are largely separated. By opting for single-domain particles we can exercise control over remanence and coercivity, as we shall see shortly. With this size of particle we also reduce the 'graininess' to substantially below one micron. There is however a limit of smallness (refer back to 'Small domains and superparamagnetism').

The coercivity of single-domain particles is governed largely by the ease with which the magnetization rotates within them. To promote high coercivity, we need to make certain magnetization directions preferable to others ('easy axes'). That is, we need some form of anisotropy. Useful sources of anisotropy are composition (magnetocrystalline energy) and shape (magnetostatic energy). Magnetostrictive (stress induced) anisotropy is to be avoided as it introduces magneto-mechanical coupling which can aggravate noise problems in contact-recording systems.

The remanence M_r of the medium depends upon the saturation magnetization of the particles and on the degree to which domain magnetization has to rotate away from an easy axis as the medium saturates (M_s). Of course we want remanence to be as large as possible to keep signal strength up on reading, and a large saturation must be a good start. To keep remanence high we must either provide several easy axes so that saturation is never far from one of them, or else we should go for just one easy axis but align it along the desired saturation direction. I've already indicated the kind of choices we have in promoting easy axes.

The sharpness of transition between a zero-bit and a one-bit could be limited by the $M-H$ loop of the medium if it were far from rectangular. To encourage a narrow switching field distribution (SFD), we could restrict the particle size- and shape-distributions and align easy axes during coating of the substrate. (Look back at Figure 3.2 — recognize anything?)

The coating thickness and uniformity must be tightly controlled. Perhaps the worst problem is associated with variations of thickness when these

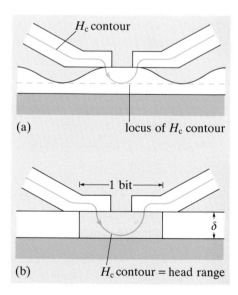

Figure 3.61 (a) Rough tapes missing H_c contour. (b) Bit width and depth

▼γ iron oxide▲

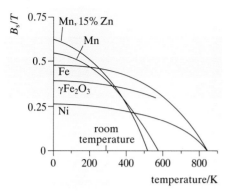

Figure 3.62 Saturation flux density vs temperature for some ferrites

Figure 3.62 shows the variation of saturation flux density with temperature for a few ferrites. On the basis of this we might be tempted to investigate the iron ferrite (magnetite or iron II oxide) as a potential particulate recording material. Not only is it stronger than the others, but it is also least temperature sensitive at room temperature. Unfortunately, magnetite (the original lodestone magnet) is not chemically stable, particularly when powdered to form single-domain particles. All that surface area lays it wide open to further oxidation from Fe_3O_4 to Fe_2O_3. In other words the Fe^{2+} ions can hardly wait to give up another electron and join the rest of the Fe^{3+} ions. The resulting iron III (ferric) oxide is essentially non-magnetic (α iron oxide). The Fe^{3+} ions in αFe_2O_3 pair up so that there is no net magnetic moment. Larger lumps of iron II oxide (ferrite) are passivated by the coating of iron III oxide which first forms.

However chemists have long been aware of a naturally occurring magnetic form of Fe_2O_3, now called γ iron oxide. The γ oxide is metastable, reverting to the α form on heating a little above 250 °C in air, so it's nominal Curie temperature of 590 °C is never reached. Synthetic γ iron oxide is available, and a preparation route has been developed to produce needle-shaped particles (required for good shape anisotropy) like those in Figure 3.2. An iron II hydroxide slurry is oxidized to particles of $\alpha Fe_2O_3.H_2O$ under conditions which are critical to the final particle shape. Dehydration to γFe_2O_3 (reddish), is followed by reduction to the inverse spinel structured Fe_3O_4 (blackish) iron ferrite.

Finally, the ferrite particles are oxidized at about 200 °C (and in water vapour) to γFe_2O_3 (brown), which essentially retains the inverse spinel structure, but the extra oxygen leaves even more interstitial sites vacant.

Most of the coercivity of γFe_2O_3 is due to the shape anisotropy and it is hard to improve upon the 5:1 length-to-diameter ratios already achieved. Again the chemists can help. Remarkable improvements can be achieved by coating the tiny particles with cobalt hydroxide, $Co(OH)_2$, before final oxidation. This is done to try to capture some of the increased coercivity which cobalt is known to confer on magnetite, through modification of the crystalline anisotropy. By restricting cobalt to the surface, the undesirable temperature dependence of the coercivity (always a risk when exploiting crystalline anisotropy) is found to be suppressed, and yet the cobalt still impedes reversal of magnetization.

The materials γFe_2O_3 and the cobalt modified $Co\gamma Fe_2O_3$ (one per cent or so Co) are very popular particles for magnetic recording.

Table 3.11   Particles for tapes and discs

	γFe_2O_3†	$Co\gamma Fe_2O_3$†	CrO_2	$BaO(Fe_2O_3)_6$	Fe
Coercivity H_c/(kA m⁻¹)	20	50	60	200	80
Temp. coeff. of coercivity/K⁻¹	-10^{-3}	-2×10^{-3}	-5×10^{-3}	-3×10^{-3}	-6×10^{-4}
Saturation flux density $\mu_0 M_s$/T	0.5	0.5	0.6	0.3	1.3
Shape of particles and	needles	needles	needles	hexagonal	needles
size/μm	0.3 × 0.06	0.3 × 0.06	0.5 × 0.05	plates 0.025 × 0.1	0.5 × 0.05
Curie temperature/K	(860)	860	390	590	1050
Magnetostriction λ_s	-5×10^{-6}	-10×10^{-6}	$+1 \times 10^{-6}$	—	4×10^{-6}
S*	0.8	0.8	0.8	0.9	0.8

† See ▼γ Iron oxide ▲

are of the same order as the size of record head — or less. Figure 3.61(a) shows how rough tapes can miss the strongest recording fields. The thicker the coating, the more flux will be available for reading but there is little point in using coatings thicker than the desired bit size. To understand why, see Figure 3.61(b). When the head is sensitive to a single bit of approximate length $2y$, say, it is also responding in the read process to a magnetic contour of depth y into the coating. So $\delta \leqslant y$ gives an upper limit to the useful coating depth.

The coating's magnetic-particle volume fraction is in practice between 20% (rigid discs) and 40% (flexible media). As the coercivity of an isolated single domain is typically halved by interparticle effects at 40% volume fraction, there seems little scope for using a higher fraction. Yet continuous-film media which are virtually 100% magnetic material are a viable alternative to particulate coatings. We shall come to these next.

EXERCISE 3.31   Table 3.11 lists the properties of some single domain particle media.
(a) Eliminate unsuitable particles for a data-recording medium on board a spacecraft. The operating conditions include temperatures approaching 420 K and stray fields of up to 15 kA m^{-1} (consider Curie temperature and coercivity).
(b) Suggest a material suitable for the magnetic strip of electronic banking cards. (Think about what it must do to fulfil its function.)

3.12.3 Thin films for tapes and discs

One of the major drawbacks of single-domain media is the worsening coercivity as the packing density increases. The saturation flux density of the tape, which increases with particle volume fraction, is thus limited. We might as well give up the isolated single-domain ideal and go back to polycrystalline materials, with domain boundaries, if we want more signal or thinner layers. Why thinner layers? To permit smaller bits and higher bit densities. Also, thinner layers are on average closer to the head so they are easier to read.

In practice though, thin-film recording media are more distinguished by their preparation (▼Thin films▲) than by their magnetic credentials. Microscopically it is as if we had a fully dense particulate medium. The domains are polycrystalline, that is they may extend over several crystal grains, each one essentially a single domain but strongly coupled to its neighbours and not preformed (previously we encountered multi-domain crystals). Herein is the trick of the thin film: to keep the coercivity high — in spite of intergranular co-operation — domain walls are pinned in the classical way with impurities and defects. Since here domains contain grain boundaries, and since grain boundaries are preferred resting places for impurities and defects, we've got the impurities and defects just where we need them: inside the domains. In practice good results are obtained when domain walls, film thickness and crystals are of comparable size ($\leqslant 1$ μm). As Table 3.12 shows, thin-film coercivity is as good as that of the particulate medium (compare with Table 3.11).

Table 3.12   Thin films for tapes and discs

	Co	Fe	Co–Fe	CoγFe$_2$O$_3$
Coercivity/(kA m^{-1})	90	70	90	70
Saturation flux density/T	1.3	1.9	1.8	0.25

Near fully dense metal film allows the highest possible saturations to be achieved. The price paid for this is the need to cover the film with a wear- and corrosion-resistant overcoat, which means we need much of the extra saturation to cope with the increased separation of head and film.

You might be thinking that a non-metal such as $Co\gamma Fe_2O_3$ would be more wear resistant and give a good saturation. Oxides wear 'well', but they are inherently low-magnetization materials and $Co\gamma Fe_2O_3$ is very hard to prepare pure in film form. It looks as if the magnetic contribution to tape and disc media is doing its best.

▼ Thin films ▲

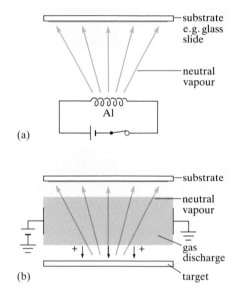

(a)

(b)

Figure 3.63 (a) Evaporation.
(b) Sputtering

Thin-film technology is the backbone of manufacturing with modern electronic materials. Most semiconductor devices are fabricated in very thin layers of material specially deposited or prepared, on a scale of microns. That is just the size of operation required for magnetic-recording media, so it is not surprising to find the same machinery in use. We will be looking at micro fabrication in more detail in connection with semiconductor memory, but it is worth noting here some of the basic techniques for laying down thin films.

There are two broad classes of deposition:

physical and chemical. In the physical approach, material is deposited from a vapour or from a liquid containing atoms or ions of the material. In chemical processes there is a chemical reaction at or very near the deposition site, in which the deposited species is produced. We'll concentrate on physical vapour processes here.

Again we can divide the group according to how the vapour is created: thermal and non-thermal. So, for example evaporating aluminium wire in a vacuum onto a glass plate to make a mirror is a thermal process. Figure 3.63(a) illustrates this process, which is clearly restricted to readily melted materials such as non-refractory metals. Figure 3.63(b) illustrates the non-thermal approach known as **sputtering** which can be used for just about everything. Ions from a low pressure electrical gas discharge or other source are accelerated (in a vacuum) onto a target (or targets) made up of the material which is to be deposited on the substrate. Physical erosion of the surface (bond breaking) by the energetic ion beam results in ejected target material arriving at the substrate. The deposition rate and substrate conditions can be controlled to encourage or discourage expitaxial growth on appropriate surfaces.

EXERCISE 3.32 A thin film of polycrystalline γ iron oxide is required on a polymeric substrate. Suggest starting materials and subsequent treatments.

A particularly useful feature of physical

vapour depositions is that stress in the film can be controlled to some extent. So for example, it is found that by evaporating at oblique incidence some amorphous material can be grown with strong uniaxial anisotropy perpendicular to the film. Amorphous films can exploit neither shape nor crystalline anisotropy so this is a useful process. By contrast evaporation at normal incidence produces a more isotropic layer.

Figure 3.64 shows a thin-film recording head, prepared using the same methods. It consists of a sputtered ferrite core with evaporated and etched coil windings. Microfabrication is not just for integrated circuits.

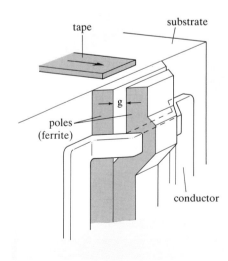

Figure 3.64 A thin film head

3.12.4 Magneto-optic materials

In the quest for high bit density, I suggested that a perpendicularly magnetized film or coating would be better able to cope with the demands of . . . 0101010 . . . . Necessarily therefore such a medium is more stretched by broader regions such as . . . 00 . . . 0011 . . . 1100 . . . 00 . . . , where it has to hold regions of 'short fat' magnetization. Remember that a short fat magnet tries to demagnetize itself, because it has a high magnetostatic energy cost. We must counter this tendency by making the perpendicular direction energetically so much more preferable than the in-plane directions. We must seek highly anisotropic preparations and, in contrast to the stuff of tapes and discs, we don't have to be too fussy about where the anisotropy comes from. Magneto-optic systems are a totally non-contact technology so magnetostriction need not be feared half so much, especially if it helps the perpendicular preference.

Some oxides such as barium ferrite and the garnets have very strong crystalline anisotropy, but they are not very useful for a Kerr (reflection) effect. In fact amorphous metals have proved to be a good reflecting environment for Kerr effects.

The larger the Kerr rotation angle, the better the signal-to-noise ratio. Lanthanide and actinide ions are magnetic through 4f-shell effects with 5s, 5p and 6s closed shells around. Transition elements are d-shell magnets with only one shrouding s shell. Thus it is necessary for the alloy to incorporate a transition metal to ensure that the light can get at the magnetism.

Whichever critical temperature is used (T_{comp} or T_c) it need not be more than 100 K or so above ambient to make magneto-optic systems thermally superior to most non-magnetic technologies (paper, semiconductors etc.), the upper limit being constrained by considerations of laser heating power and thermally induced microstructural changes. The strong compositional variation of T_c in magnetic alloys is a distinct advantage here, and if we're looking for ferrimagnetic alloys (for compensation-point behaviour) we must expect the use of some exotic components.

Alloys of rare earths with transition metals, prepared as thin amorphous films, provide all that we have specified. In these ferrimagnetic alloys, the rôles of A and B sites are played respectively by rare-earth and transition-metal atoms. The rare-earth atoms have full 5s, 5p and 6s shells outside the magnetic 4f shell, so their magneto-optic interaction is weak, leaving the transition metal predominant. The lack of crystalline structure denies us one form of anisotropy but to exploit it fully would have called for some directional growth. In fact thin-film amorphous material makes use of more than one consequence of the production route. For a start, thin films frequently incorporate stress owing to the bonding interactions with the substrate material and entrainment of impurity atoms. Magneto-elastic coupling (magnetostriction) then permits this to induce preferred magnetization directions. Also, short-range magnetic interactions, which are the stuff of ferrimagnetism anyway, can give rise to larger scale anisotropies if the conditions of film growth are slow enough to allow ordering during deposition.

The final structure also needs mechanical support and physical protection; Figure 3.65 shows a section through a magneto-optic disc.

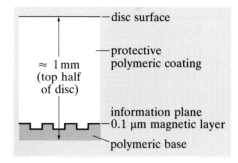

Figure 3.65 Section through a magneto-optic disc

EXERCISE 3.33 Table 3.13 lists five potential magneto-optic recording materials. All are amorphous alloys. Suggest two further areas where data would be required to enable the suitability of these materials to be assessed.

Table 3.13   Amorphous alloys for magneto-optic media

	Type of recording	T_c/K	θ_k/degrees	H_c/(kA m^{-1})
TbFe	Curie	≈ 400	0.25	500
GdCo	compensation	≈ 700	0.33	100
GdFe	Curie	≈ 450	0.5	30
GdTbFe	Curie	≈ 430	0.35	100
GdTbCo	compensation	≈ 650	0.33	200

3.12.5 Bubble materials

The primary requirement for a bubble-domain material is that it will support bubbles — it must favour perpendicular magnetization in thin films. Again we must specify a large uniaxial anisotropy. A second requirement is high bubble mobility. Domain boundaries around the bubble edges must be free to move across the film. We know from our studies of soft magnet material that this implies a need for low impurity and defect concentrations — grain boundaries are not desirable. It is also likely that poor conductors, with their low eddy currents, will be preferred.

Add to this the ability to control magnetic properties, and chemical stability, and the specification is complete. Any ideas? Well, we've already seen exotic alloys fitting a similar specification for magneto-optics and we've also noted the uniaxial prowess of hexagonal ferrites. These together with garnets and orthoferrites are possibilities. In fact orthoferrites were first used in the study of bubbles, but useful devices were developed in garnets, which support much smaller bubbles. ▼Inside a garnet▲ looks into these versatile ferrimagnets. Ferrites in general are very amenable to magnetic tweaking. They are made by melting oxides in controlled proportions. Their magnetic characteristics are intermediate between those of their constituents and although precise prediction of magnetic characteristics is not easy, recipes have been developed which give reproducible results.

▼Inside a garnet▲

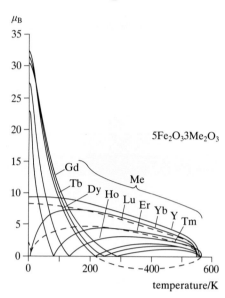

Figure 3.66 Magnetization (per formula unit) vs temperature for some garnets

We saw in Section 3.4 that cubic ferrites based on the spinel structure could be alloyed to produce a range of properties. An even wider variety of properties is achieved by the chemists who manipulate cubic ferrites with a garnet structure.

Garnets are more complicated than spinels, but can be summed up in a comparable manner. The natural mineral garnet is Mn_3 $Al_2 Si_3 O_{12}$, with the aluminium and silicon octahedrally and tetrahedrally co-ordinated respectively. However, magnetic interest is in related crystals which have three classes of trivalent ion, for example

$$\{Y^{3+}_3\} [Fe^{3+}_2] (Fe^{3+}_3) O^{2-}_{12},$$

where the brackets distinguish {dodecahedral}, [octahedral] and (tetrahedral) sites. The yttrium ions can be replaced by rare-earth ions (single and mixed), revealing a range of magnetic properties. Figure 3.66 illustrates the varied behaviour of a few single garnet ferrites.

The magnetic structure is due to superexchange coupling between Fe^{3+} ions on octahedral and tetrahedral sites, aligning $[Fe^{3+}_2]$ antiparallel to (Fe^{3+}_3), so that one Fe^{3+} ion in five is unpaired. In addition, the rare-earth (or yttrium) ions are also superexchange coupled, although somewhat weakly, to the tetrahedral ions. At low temperatures, magnetism is predominantly due to the rare-earth ions. At such temperatures, they have more atomic magnetism than, and are opposed to, the unmatched Fe^{3+} ions. But at higher temperature the weaker coupling diminishes and after passing through a zero net magnetization, the Fe^{3+} ion dominates. Thus these garnets show a compensation point.

EXAMPLE   It is found that non-magnetic scandium ions (Sc^{3+}) preferentially occupy octahedral sites in garnets. Show that if small amounts of scandium do not significantly reduce superexchange coupling, then the magnetization of samarium garnet ($Sm_3 Fe_5 O_{12}$) is increased by scandium substitution ($Sm_3 Fe_{5-x}$ Sc_xO_{12}, where $0 < x \leqslant 1$). Sm^{3+} contributes 1.5 units of magnetism, and Fe^{3+} contributes 5.

ANSWER   First express the formula in terms of the ionic sites:

$$\{Sm^{3+}_3\} [Fe^{3+}_{2-x}] [Sc^{3+}_x]$$
$$(Fe^{3+}_3) O^{2-}_{12}$$

with { } and [ ] aligned against ( ). The magnetization is

$$-3\{1.5\} - (2-x)[5] - x[0] + 3(5)$$
$$= 0.5 + 5x$$

Therefore as x increases, the magnetization of the formula unit also increases. (For x up to 0.1 we might expect little change in coupling so we could almost double the magnetization.)

Single and mixed garnet crystals can be grown from the melt using liquid-phase epitaxy in a manner similar to that by which semiconductor-grade silicon boules are produced. A prepared seed crystal of a non-magnetic garnet (such as $Gd_3Ga_5O_{12}$) is dipped into and slowly withdrawn from a melt of appropriate composition. As it is withdrawn, it is rotated. During growth, stress induced by the slight lattice mismatch, and preferential ion alignment, combine to give the grown crystals a large uniaxial anisotropy in the growth direction. Ordinarily the garnet structure would not promote anisotropy.

The garnet structure is very tight. All the large cages between oxygen ions are occupied by positive ions. As a result, these materials are very stable. (Compare this with the situation in inverse spinel ferrites.)

EXERCISE 3.34   Why is $\{Gd^{3+}_3\} [Ga^{3+}_2]$ $(Ga^{3+}_3) O^{2-}_{12}$ non magnetic?

The quest for higher bit-density requires smaller bits. Smaller bits in general need a larger magnetization (to hold one direction against demagnetizing influences) and stronger anisotropy (to keep the magnetized regions narrow), which is the reason why preliminary work in orthoferrites gave way to a development of devices in garnets. It is interesting to notice that for some time the scale of the technology was limited not by the bubble material but by the properties of the tracks (and their associated fields) used to shift the bubbles around. Just as you beat one limit another pops up elsewhere!

One way to overcome a technological limitation is to rework the design, eliminating the offending component. In bubble devices, this approach can be used to increase capacity at least by a factor of four. Suppose that we can make permalloy tracks to a width of just 1 μm. It turns out that bubbles must have diameters of at least 2 μm to operate with these tracks. Now, an alternative shifting scheme, which I will describe in a moment, can be used with bubble diameters as small as 1 μm. The bubble area in the new scheme is therefore up to a factor of four smaller than for the permalloy overlay. So what is this scheme?

First, you should note that garnets are strongly anisotropic as a result of the production route (liquid-phase epitaxy), resulting in preferred perpendicular magnetization. Secondly, in-plane preference can be created in regions penetrated by shallow ion bombardment at about 50 kV (H^+ or He^{2+} ions can be used). The third factor is that bubbles are attracted to the boundaries between ion-implanted and unimplanted regions. They can be pushed along the boundaries by the same surface-skimming fields as before, but now the permalloy is replaced by the modified surface layer. We don't need to get involved in the detailed mechanism here. The point is this: direct manipulation of the surface of the bubble material on a submicron scale allows a denser memory system to be developed. This time a development of the production process extracts enhanced performance from an established material.

SAQ 3.14 (Objectives 3.10 and 3.12)
Why can mechanically hard materials like garnets be used in bubble devices without any worry about recording-head wear, whereas ferrite-particle tapes have a deleterious effect on permalloy heads?

Summary of Sections 3.12 and 3.13

● Particles (such as γ iron oxide) and thin films (such as Co–Fe) can be prepared to give suitable media for tapes and discs.
● Amorphous alloys of rare-earth and transition metals are used for Curie-point and compensation-point magneto-optic media.
● Specially grown single-crystal mixed garnets are used for bubble media. Surface modification by ion beams allows small steering features to be drawn on the crystal.

3.13 Epilogue

The source of intrinsic magnetism is within atoms, but magnetic strength comes when these atoms are able to co-operate through exchange and superexchange interactions to produce large numbers of atomic magnets acting in unison. We have looked at the atoms and the domains into which they assemble themselves, and we have seen how preparation and processing interact with the magnetic order, at times enhancing it, at times opposing it. Through the manipulation of this magnetic microstructure, technology has found many valuable commodities with applications throughout electrical and electronic engineering.

I've chosen to discuss a few examples: motors, transformers, magnetic data storage; but don't suppose that the story stops there. There are other applications of what we have studied (ferrites in microwave devices, for example) and there is much more to study about atoms and magnetism, such as the science behind nuclear magnetic resonance (NMR — see Chapter 7). What we have studied in this chapter is indeed 'something about magnets' (and it is enough for now), but it is really only the beginning.

Objectives

You should now be able to do the following.

3.1 Describe hard and soft magnetic characteristics in terms of the B–H loop and associated quantities. (SAQ 3.1)

3.2 Discuss the property requirements of magnetic materials in motors, transformers and related devices such as transducers. (SAQ 3.3)

3.3 Suggest magnetic materials for specific applications in motors, transformers and related devices such as transducers. (SAQ 3.4)

3.4 Account for the magnetism of ions and atoms, in particular those of the transition metals. (SAQ 3.5)

3.5 Contrast magnetism in certain elements, alloys and oxides at an atomic level. (SAQ 3.6)

3.6 Describe the factors which contribute to the sizes and shapes of domains in bulk solid and particulate matter. (SAQ 3.7)

3.7 Explain hard and soft magnetic characteristics at the domain level. (SAQ 3.8)

3.8 Discuss the microstructural control of magnetic properties. (SAQ 3.9)

3.9 Describe the principles of various schemes for magnetic data recording. (SAQ 3.13)

3.10 Discuss the property requirements of magnetic materials for conventional tape/disc memories. (SAQ 3.10)

3.11 Discuss the property requirements of magnetic materials for magneto-optic memories. (SAQ 3.11)

3.12 Discuss the property requirements of magnetic materials for bubble memories. (SAQ 3.12)

3.13 Define or distinguish between:

B–H loop, M–H loop

Coercivity, remanence, permeability

Crystal anisotropy, magnetostriction

Eddy current loss

Exchange, superexchange

Ferromagnetic, ferrimagnetic

Hysteresis loss

Magnetization (M), magnetic flux density (B), magnetizing field strength (H)

Answers to exercises

EXERCISE 3.1

(a) A fat loop is characteristic of hard magnetic material; a slender one is characteristic of soft magnetic material. (But notice the scales.)

(b) Hysteresis loss is the energy spent in taking a specimen around its B–H loop. In fact the area enclosed by a B–H loop is equal to the energy loss per unit volume (J m^{-3}) in going once round the loop.

(c) Within a magnetic material, many differently directed regions of magnetism (domains) contribute to the magnetism of the bulk. By changing the size and orientation of these regions, the bulk magnetism can be altered.

(d) In soft materials domains grow, shrink and rotate their orientation with little opposition. In hard materials, adjustments to the domain structure are difficult; that is, there is a large energy cost.

(e) At the Curie temperature T_c the magnetic co-operation between atoms is overwhelmed by thermal vibrations. As T_c is approached, the material becomes less magnetic, losing all its strong magnetism at T_c.

EXERCISE 3.2 From Equation (3.1)
$H_1 = - H_2 l_2/l_1$. The negative value means that this is a demagnetizing field.

EXERCISE 3.3

(a) $2\pi R H = Ni$.

(b) $B = kH = kNi/2\pi R$.

(c) $\dfrac{d\Phi}{dt} = \dfrac{r^2 kN}{2R} \dfrac{di}{dt}$.

(d) $V = \dfrac{r^2 kN^2}{2R} \dfrac{di}{dt}$.

Notice how the inductance ($r^2 kN^2/2R$) depends on shape (r^2/R), on material (k) and on the number of turns (N^2).

EXERCISE 3.4 Using Equation (3.2)
$B = \mu_0(H + M)$, so if M is proportional to H (say $M = \chi H$) we would write $B = \mu_0 (1 + \chi)H$. The relative permeability is therefore $(1 + \chi)$, and is constant.

EXERCISE 3.5

(See Table 3.14.)

Table 3.14

Quantity	Symbol	Remarks	Typical figure
Saturation flux density or induction	B_s	The maximum contribution to the magnetic flux density from the material — all atomic moments maximally aligned.	0.25–2 T (soft materials)
Remanence	B_r	The magnetic flux density in the material when the net magnetizing field strength H (due to external currents and the material) is zero.	0.2–1 T (hard materials)
Coercivity (and intrinsic coercivity)	H_c (and H_{ic})	The magnetizing field strength required to reduce the net flux density B (or else the flux density due to the material alone, B_i) to zero.	1–10 A m^{-1} (soft) 50–900 kA m^{-1} (hard)
Relative permeability	μ_r	$\mu_r = B/\mu_0 H$. The ratio of flux density to net magnetizing field strength.	$(\mu_r)_{max}$: 10^3–10^6 (soft)
Differential permeability	μ	$\mu = \dfrac{1}{\mu_0} \dfrac{dB}{dH}$. Note: this is not the same quantity as $\mu_0\mu_r$.	
Saturation magnetization	M_s	$B_s = \mu_0 M_s$	$\approx 10^6$ A m^{-1}
Hysteresis loss or energy loss per cycle	W_h	A frequently used figure of merit for soft materials: the energy cost of one cycle of magnetization.	30 J m^{-3} cycle^{-1} (soft)
Energy product	$(BH)_{max}$	A frequently used figure of merit for permanent magnet materials indicating capacity to support external fields.	10–200 kJ m^{-3}

EXERCISE 3.6
Simply substituting material Y for X would result in trying to operate with the soft components saturated (P'). The material change would require a design change to allow operation near P''. This is a good example of product–property interaction.

EXERCISE 3.7
You should have got some of the following points.

Alnico. Relatively poor coercivity so design must allow for the high demagnetization risk. Furthermore, the magnetization curve is not very linear so even a small demagnetizing field will cause operation not on the major part of the loop, but on a minor one with a consequent loss of flux density. (Alnico is relatively insensitive to temperature variations.)

Ferrite. A relatively low remanence (which worsens with increasing temperature) is in part compensated by abundance and ease of fabrication (by sintering or bonding) into almost any shape.

Samarium–cobalt. There is one major disadvantage only: samarium is expensive being a particularly rare, rare earth element, and cobalt is not so plentiful.

Neodymium–iron–boron. At first sight this material is very attractive for p.m. motors. Its only weaknesses are a low Curie temperature and a high temperature sensitivity for the remanence and coercivity. Otherwise it is superior to samarium–cobalt.

EXERCISE 3.8 Current in the primary builds up and changes in such a way as to generate an e.m.f. in the primary which just matches the applied voltage.

EXERCISE 3.9 Higher permeability gives lots of B for not too much H, and H is here caused by the currents. So a high permeability core needs a lower magnetizing current. Since the primary coil will in practice have some resistance, the magnetizing current is a source of power loss.

EXERCISE 3.10 There are two sources of heat in an off-load transformer core: hysteresis loss generated by the pinning and release of domain walls; and eddy current loss. You might also have mentioned resistive losses from the magnetizing current, but strictly these are in the primary coil, not the core.

EXERCISE 3.11 Elimination is often a safe approach. Air looks good on eddy current and hysteresis fronts but is hopelessly impermeable, so magnetizing currents will be very large. Coupling the flux between the coils may be difficult, so this material is rejected. The ferrite is also promising on core loss, but its low saturation and low permeability are unfavourable. Being ceramic it would be difficult to cool. That leaves the very permeable amorphous metal with its low eddy loss and the higher saturation silicon–iron. Lamination will greatly compensate for the silicon–iron's low resistivity; both materials could be seriously considered.

Notice that both are based on iron. We need $d\Phi/dt$ to induce voltages in primary and secondary coils. For a given operating voltage, operating at larger B allows a smaller cross-section of core. Iron is an obvious choice.

EXERCISE 3.12

(a) Suppose there are N octahedral holes and so $2N$ tetrahedral holes in a lump of ferrite. $2N/8$ tetrahedra provide A sites with Fe^{3+} ions. $N/2$ octahedra provide B sites and half contain Fe^{3+}. Thus $N/4$ B sites have Fe^{3+} ions, which is the same number as A sites with Fe^{3+}.

(b) Mn^{2+} and Fe^{3+} have the same electronic structure which is $1s^2 \ldots 3p^6 3d^5$. The five d shell electrons are unpaired and

provide a unit each to the magnetism, giving 5 μ_B. In the sequence $Mn^{2+} \to Zn^{2+}$, an extra d shell electron is added, pairing up with one already there, successively reducing the magnetism by steps of μ_B to zero at Zn^{2+}.

EXERCISE 3.13 Possible applications of magnetostriction include:

• Transducers. Sound waves may be launched or received by a magnetoelastic coupling.
• Thermal compensation. Magnetoelastic coupling in certain alloys is very temperature sensitive so that thermal expansion or elastic modulus may be rendered almost temperature independent over certain ranges (for example, in the alloys invar and elinvar respectively).

EXERCISE 3.14 If freshly smelted iron had a net magnetization, there would be a substantial magnetic field outside the material (see Figure 3.30a). This costs energy. Figures 3.30 (b) and (c) have no net magnetization but still have field outside the magnet. Figures 3.30 (d) and (e) are more likely since there is no external field. The freshly smelted iron's domain structure would probably likewise try to keep the fields internal and hence have no net magnetization.

EXERCISE 3.15 We could end up with a material comprising many single-domain grains, closely coupled by exchange forces across grain boundaries. Although grains may be below the superparamagnetic threshold, the interactions will keep ferromagnetic behaviour dominant — after all the solid may be near 100% density. We will almost certainly have domains extending over several crystal grains, so the material could be quite hard (because impurities and defects congregate at grain boundaries, where domain walls are therefore pinned).

EXERCISE 3.16 The minor loop is traced out when, prior to saturation, the magnetizing field is reversed, taken to some point again short of saturation, and then returned to the starting point. Just as in initial magnetization, favourable domains are expanded and rotated.

The demagnetizing path falls progressively

shorter of saturation at each reversal. In so doing, each time the orientation of fewer domains is changed so that a sort of randomizing of orientation and size is encouraged.

EXERCISE 3.17 Stresses built up around the edges during stamping will restrict domain wall motion, giving rise to losses. Transformer cores must be 'soft' and have free moving domain walls. Annealing will remove these stresses.

EXERCISE 3.18 Referring back to 'Between domains; width and energy' and 'Within domains; thickness', you can see that the width of walls and the width of domains are seen to scale roughly with $1/\sqrt{K}$ and $1/\lambda_s$ respectively, so small K and λ_s imply broad walls and large domains.

EXERCISE 3.19

(a) Iron has on average about two electrons per atom contributing to its magnetism; cobalt has fewer than two. In the alloy there is evidently some slight increase in the time spent unpaired by some of the 3d electrons as the saturation flux density is higher.

(b) Easy unopposed enhancement of external fields (high permeability) requires readily rearranged domains (low coercivity).

(c) A large saturation will be associated with strong atomic magnets, which in turn will have a strong magnetic (exchange) interaction, which is retained in the face of a large amount of thermal agitation and hence a high Curie temperature.

EXERCISE 3.20 From Figure 3.40(a), the resistivity is doubled between $3\frac{1}{2}$% and 10% Si, so eddy losses could be halved if all else remains unchanged.

EXERCISE 3.21 You may have chosen from among the following.

Audio. Magnetic tape readily stores analogue or digital information; can be erased and rewritten; multiple replay with little loss of quality. But: susceptible to corruption by strong magnetic fields.

Archival storage. As for audio, Also, magnetic recording not corrupted by ionizing radiation.

Computer memory. As for archival storage. In addition, magnetic memory is not 'volatile' (needs no power supply to sustain data).

(Note the Achilles' heel of magnetic records: strong magnetic fields.)

EXERCISE 3.22

(a) A single atom is like a tiny magnet which could be directed one way or another to record a bit. Stern and Gerlach worked with atomic 'bits'; but let's stick to solid matter and say 'a few atoms' would record a bit in principle.

(b) If it's too small (too few atoms) it may be: hard to find; insufficiently strong to give acceptable signals; randomized by thermal energy.

(c) A single-domain bit would have an area of 10^{-14} m^2. 10^{12} would therefore fit on 10^{-2} m of disc. Let's revise this figure to 10^{11} (0.3 μm per bit) to allow space to prevent exchange coupling between the single-domain particles. 10^{11} bits is about 10 giga-bytes. (Note: these figures for capacity can be treated as an upper bound.)

(d) Data passes the head at 36 m s^{-1} and there are 0.1×10^{-6} m bit^{-1}, so the data rate (bits per second) is $36/10^{-7} = 3.6 \times 10^8$.

(e) Probably hard material for the recording medium but hard and soft in associated apparatus.

EXERCISE 3.23

(a) Amperes law gives

$$H = \frac{i}{2\pi r}$$

But the current is related to current density j by

$$i = j\pi r^2$$

and j is restricted. Putting $H = H_c$, to magnetize at radius r a tape of negligible thickness $H_c = jr/2$. With $H_c = 10^4$ A m^{-1} and $j = 10^7$ A m^{-2}, we must have a wire of radius $r = 2 \times 10^{-3}$ m and so the current is

$$i = 10^7 \times \pi \times 4 \times 10^{-6} \text{ A}$$
$$\approx 120 \text{ A}!$$

It looks as though we have invented a way of measuring large currents rather than recording information.

(b) Using the relation obtained above, $H_c = jr/2$, the largest field we could produce with 10^7 A m^{-2} at 1 μm would match a coercivity of

$$H_c = \frac{10^7 \times 10^{-6}}{2} \text{ A m}^{-1}$$
$$= 5 \text{ A m}^{-1}$$

Such a value typifies the softer magnetic materials, which are useless for permanent records!

EXERCISE 3.24

(a) $H_m l + H_g g = ni$

(b) $H_g = \frac{ni}{g} \frac{1}{(1 + l/g\mu_r)}$

(c) $H_f = -\frac{H_g g}{\pi r} = -\frac{ni}{\pi r}$

(d) $ni = -10^4 \times \pi \times 10^{-6}$ ampere turns
$$\approx 3 \times 10^{-2} \text{ ampere turns}$$

EXERCISE 3.25
When the stored magnetization is being read, the flux linked by the read head is only a fraction f_1 of that which is remanent, which is in turn only a fraction f_2 of the saturation, which is in turn only a fraction f_3 of the original flux in the write head. The fraction of write flux density which might return on reading is $f_1 f_2 f_3$, which is likely to be significantly below unity.

EXERCISE 3.26

(a) The ferrite head will have a relatively low saturation but very good wear resistance.
(b) The metal-tipped head-gap corners will not saturate as readily as the ferrite but will wear badly.
(c) The metal-in-gap head combines the favourable attributes of both (a) and (b).

EXERCISE 3.27
The volume of heated material v is

$$v = 10^{-6} \times 10^{-12} \text{ m}^3 = 10^{-18} \text{ m}^3$$

The mass of heated material m is

$$m = \rho v$$
$$= 8 \times 10^3 \times 10^{-18} \text{ kg}$$
$$= 8 \times 10^{-15} \text{ kg}$$

The energy input u is

$$u = 10^{-2} \text{ W} \times 10^{-7} \text{ s} = 10^{-9} \text{ J}$$

The temperature rise ΔT is

$$\Delta T = \frac{U}{mc}$$
$$= \frac{10^{-9}}{5 \times 10^2 \times 8 \times 10^{-15}} \text{ K}$$
$$= 250 \text{ K}$$

The reflectivity of the surface and neglected losses will reduce this figure, but clearly it is of the right order.

EXERCISE 3.28
We are seeing domains in 'thin' material. Magneto-optic effects could have been used to get these pictures, for example polar Kerr (reflection) or Faraday (transmission) (or possibly electron microscopy). In fact they are Faraday-effect pictures.

EXERCISE 3.29
Possibilities include:

(a) Electromagnetic induction, as in a tape head.
(b) Hall effect. (You may not have come across this yet.)
(c) Magnetoresistance. Electrical resistivity is affected by magnetic fields, especially in a material like permalloy.
(d) Faraday/Kerr magneto-optic effects.

Comment: Methods (a)–(c) would need a tape-head sort of construction with a gapped core or else a small active element comparable in size to the bubble. Method (c) is commonly used for bubble detection. Method (d) would need a magneto-optic type focused beam.

EXERCISE 3.30
The numbering below relates to the numbering of the items in the PDS before Exercise 3.30.

1 SFD (or S^*); maximum value of H_c (e.g. to prevent head saturation); T_c; T_{comp}.
2 T_c; minimum value of H_c; λ_s; magnetic (and chemical) stability.
3 M_r — its average value and its uniformity in regions of 'constant' value.
4 (Wear characteristics are not a magnetic property.)
5 (Need more information — scale of operation, tape or disc or what?)
6 Perpendicular/longitudinal orientation, SFD, switching time, layer thickness (together with system-dependent parameters).

EXERCISE 3.31

(a) Curie temperature T_c. The CrO_2 would get uncomfortably close to its limit; reject. All others all right.

Stray fields H_{stray}. The γFe_2O_3 is threatened. Stray fields can of course be screened by supermalloy or similar, but this adds weight; reject.

Coercivity H_c. The $BaO(Fe_2O_3)_6$ looks a little hard to magnetize. Power and weight limits may militate against it; reject.

(b) I should like mine to be barium ferrite (just look at that high coercivity, which will guard against accidental erasure!), but it is rather weak in terms of saturation and hence readability. A card I analysed in 1989 used γFe_2O_3.

EXERCISE 3.32   We could try to sputter magnetite onto a polymeric base capable of withstanding up to 250 °C. After forming the inverse-spinel polycrystalline layer, exposure to oxygen in a moist atmosphere at 200 °C (see 'γ iron oxide') should convert the layer to the γ oxide. (The sputtering of magnetite may not be stoichiometric. Therefore it may be necessary to adjust the oxygen content of the starting material, say by mixing iron powder or else some γ oxide with the magnetite to ensure that the deposited material is magnetite, Fe_3O_4.)

EXERCISE 3.33   You may have chosen from the following.
• Thermal cycling stability. Under operating conditions is the amorphous phase stable or does crystallization begin?
• Anisotropy: is the material capable of supporting perpendicular magnetization?
• Chemical stability. Can the alloy be suitably stabilized against oxidation and other corrosion?
• Production. Is the production route viable?
• Reflectivity. Does the alloy reflect sufficient power to give adequate read signals?

EXERCISE 3.34 Ga^{3+} is non magnetic (its ionic structure is $1s^2 \ldots 3p^6\, 3d^{10}$). There is therefore no magnetic ion for the Gd^{3+} ion ($1s^2 \ldots 4d^{10}, 4f^5, 3s^2\, 5p^6$) to superexchange with and so no co-ordination of the magnetic ions.

Answers to self-assessment questions

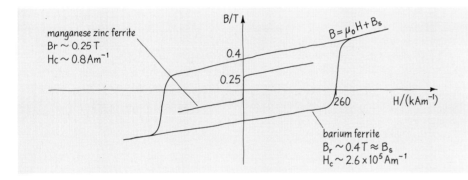

Figure 3.67 Answer to SAQ 3.1

SAQ 3.1   See Figure 3.67.

SAQ 3.2   The magnitude of the force will be Bi (newtons if i is in amperes and B in tesla). If B is along a z-axis and i is along a y-axis, F will be along an x-axis.

SAQ 3.3   Saturation flux density. The material should not saturate a long way below the design capabilities of the source of magnetic field (coil or permanent magnets).

Permeability. A low permeability is needed. We don't want to expend a lot of effort just magnetizing this material; field in the gap is what is required.

Resistivity. Eddy currents may be excited in the magnetic material, with attendant losses and heating. High resistivity laminated structures may be required.

Hysteresis loss. If the material is cycled round its B–H loop we will probably want to specify a low-loss material.

Energy product — this is only a relevant quantity with reference to the efficient use of permanent magnet material.

In general there must be cost specification as well since high performance material commands a high price.

SAQ 3.4   The magnet has a short fat section, enabling a low-remanence material to almost saturate the much narrower section 'soft' components which guide the flux into the soft, shaped lug and into the core of the coil. A mild steel or low-silicon iron should suffice here for most of the soft path. The shaped lug could be made from a pressed and sintered manganese–zinc ferrite, although the ends of the lug may saturate. The magnet itself could be

barium ferrite unless space or weight are at a premium.

SAQ 3.5   The electronic structure of the ions groups them as follows

Sc^{3+}, Ti^{4+}	$3d^0$
no magnetism	
Ti^{3+}, V^{4+}	$3d^1$
one unpaired spin	
V^{3+}	$3d^2$
V^{2+}, Cr^{3+}	$3d^3$
Cr^{2+}, Mn^{3+}	$3d^4$
Mn^{2+}, Fe^{3+}	$3d^5$
all unpaired spins, so strongest	
Fe^{2+}	$3d^6$
one paired, so four unpaired	
Co^{2+}	$3d^7$
Ni^{2+}	$3d^8$
Cu^{2+}	$3d^9$
Zn^{2+}	$3d^{10}$
full d shell so no magnetism	

SAQ 3.6   Pure iron is ferromagnetic. Exchange interactions between spins on neighbouring atoms promote massive co-operation between atoms with co-aligned magnetic moments. The transition-metal ferrites have inverse spinel crystal structure. Superexchange interactions (through the special action of oxygen ions) are able to cancel the iron-ion magnetism (Fe^{3+} on tetrahedral sites are antiferromagnetically coupled to Fe^{3+} on octahedral sites). The observed ferrimagnetism is then a consequence of

the magnetism of the divalent transition metal ion. Zinc ferrite has a non magnetic ion (Zn^{2+}) in the tetrahedral sites; there is no superexchange and the magnetic Fe^{3+} ions are not even strongly coupled. In fact they act entirely independently.

SAQ 3.7 Large domains will appear in large chunks of material. If the material is polycrystalline, it will have large domains if the crystals are large. Equally important however is a small magnetostriction so that large closure domains are easily accommodated, and high-surface-energy walls so that we don't get too many. Now, the wall width is kept down by a high anisotropy energy K, penalizing atoms which don't align along crystal easy axes, and high anisotropy will help to keep the surface energy of the wall high.

SAQ 3.8 The cubic ferrite is a 'soft' material. It magnetizes fairly easily having several easy axes and low anisotropy. The domain within a crystal would have relatively wide walls. Since the material is soft, these walls must move fairly easily. (Broad walls are hard to pin.) The hexagonal ferrite has a high uniaxial anisotropy. The narrow-walled domains will be predominantly aligned along the easy axis. Reverse domains will be hard to nucleate owing to the high surface energy associated with the high anisotropy.

SAQ 3.9
(a) See Table 3.15.
(b) See Table 3.16.

Table 3.16 Answer to SAQ 3.9(b)

Property	Principle	Process
resistivity, ρ	$P_e\downarrow$ as $\rho\uparrow$	(i) Alloy with Si.
		(ii) Vitrify with B and Si.
sheet thickness, t	$P_e\downarrow$ as $t\downarrow$	(i) Roll alloys to 0.25 mm strip (limits Si addition).
		(ii) Rapid quench alloy or glass to $\approx$ 30 µm ribbon.
domain width, d	$P_e\downarrow$ as $d\downarrow$	Laser/spark scribes to define narrow domain structure.
coercivity H_c	$P_h\downarrow$ as $H_c\downarrow$	(i) 'Pure' defect-free (annealed) alloy grains, aligned to minimize domain rotation.
anisotropy constants K and λ_s	$H_c\downarrow$ as K, $\lambda_s\downarrow$	(ii) Alloy with Si to lower K and λ_s.
		(iii) Eliminate K_z by amorphous material.

Table 3.15 Answer to SAQ 3.9(a)

Property	Principle	Implications for processing	Remarks
high B_r	many randomly oriented easy axes	Keep texture random. Avoid accidental stress. Anneal after mechanical work.	isotropic materials — easy axis always on hand
	aligned easy axes	Roll/field anneal to promote reoriented crystal texture or align single-domain particles.	anisotropic materials — no rotation into saturation
high μ_r	aligned easy axes	(As for high B_r.)	
high H_c	hard to reverse single domains	Sintered powders.	uniaxial anisotropy exploited
	impaired reverse domain nucleation	Avoid large-scale low anisotropy or non-magnetic inclusions. Disperse impurity (around grain boundary for example). Sintered powders.	
low H_c	special low anisotropy alloys	Rapidly solidified amorphous alloys.	non-crystalline: sensitive to stress
		Needs to be chemically and structurally 'pure'.	crystalline: low magnetostriction

SAQ 3.10 Magnetization M_r. Read signal is proportional to M_r. High value required.

Coercivity H_c. Relates to permanence of record. High value required.

M–H loop. Rectangular loops keep transitions narrow.

Thickness of medium δ. Thinner tapes favour shorter transitions. Low value required.

Head-tape separation d. Read signal falls rapidly with distance from head. Low value required.

Head gap g. Fixes maximum bit length, which is comparable with g. Low value required.

Head permeability μ_r. Write currents create gap field and fringe (which writes). High value required.

Wear. Arises from contact between head and medium. Resistance to wear must be adequate.

SAQ 3.11 Magnetization M_r. Perpendicular for polar Kerr effect. Microscopically high value required.

Uniaxial anisotropy K. Domains are perpendicular (to surface) and should have smooth walls. High value required.

Compensation temperature T_{comp}. H_c rises below T_{comp} and falls rapidly above T_{comp}.

Thermomagnetic recording possible at above 400 K.

Curie temperature T_c. H_c falls to zero as temperature approaches T_c.

Thickness δ. This must be sufficient for perpendicular magnetization. (Not critical.)

Reflectivity R. The process exploits Kerr effect in the surface reflection. High value required.

Kerr polar angle θ_k. Bits 0 and 1 are separated by $2\theta_k$. High value required.

Thermal fatigue. Adequate resistance required.

SAQ 3.12 We are seeking thin layers with the following attributes:

Magnetization M_r. We have specified a need for perpendicular magnetization which must in turn require strong uniaxial anisotropy.

Uniaxial anisotropy K. See above.

Thickness δ. The bubble domain size is comparable with layer thickness, so δ of the order of μm is expected.

Mobility/purity. High purity material is favoured to permit easily moved domains. High speeds and oscillating drive fields also suggest high resistivity.

μ_{island}. High-permeability islands, again of μm dimensions, must be patternable on the layer surface.

SAQ 3.13 Electromagnetic writing. The electromagnetic process uses a magnetic field bursting out of a gapped core which is wound by a coil. Currents in the coil produce the 'write field'.

Thermomagnetic writing. The thermomagnetic process raises a local region to a temperature at which the coercivity is substantially below its room-temperature value and lower too than an imposed writing field.

Electromagnetic reading. Changes in surface-leaking flux are coupled into the core of a coil in the read-head. Voltages are induced in proportion to the rate of change of flux linked.

Magneto-optic reading. Polarized light is shone at normal incidence into the surface. The polar Kerr effect rotates the plane polarization of the reflected light according to the local magnetization.

SAQ 3.14 In bubble systems, neither the head nor the medium moves: only the bubbles move. The wear properties of head and medium are irrelevant. In tape recording, good strong signals require intimate head-tape contact so abrasion can be a problem.

Chapter 4 Transducers

4.1 Introduction

Chapter 1 showed that transducers are our way into and out of electronic circuits. Chapter 3 on magnets gave examples of transduction in which signal exchange was essentially via an inductive route (Figure 4.1a). That sort of transduction involved Faraday's law, $V = d\Phi/dt$. Here we shall pick up ideas from Chapter 1 on conductive properties, and from Chapters 2 and 3 on dielectric materials and domains, in order to look at transducers which employ either $V = Q/C$ or $V = (dQ/dt)R$ (Figures 4.1b and c). As always, we want to understand the scientific background, the materials choice, the performance and the manufacture of these devices. Transducers of this sort can translate a great variety of environmental information into electrical signals. Alternatively, they can convert electronic signals into other forms of energy. Here they may be used in either control or power mode.

The starting point in making transducers is always to find a substance which shows a relevant 'effect'. Then, as we saw in Chapter 3 for magnetic transducers, it will usually be necessary to modify composition and microstructure to enhance the effect or suppress unwanted behaviour. This is where the science of the materials has to be applied. When an adequate suite of material properties has been arranged, the device geometry has to be designed to balance sensitivity, response time, signal-to-noise ratio and so on, so that the transducer meets a performance specification.

We shall concentrate on transducers made of metal oxides, particularly oxides in ceramic (polycrystalline) form. These materials have enjoyed a wide extension to their range of application in recent years. Few of the applications to be described here *demand* an electroceramic device, however, and single-crystal oxides as well as metals, semiconductors and polymers are other widely used materials.

The potential advantages of ceramics are that they can be:
- cheap to produce in any shape or size, offering versatility in device design,
- easy to control in composition and microstructure, offering versatility of function.

These advantages often give ceramics the edge in the competition for marketable transducers. ▼Process routes for ceramics▲ outlines how they are made.

We shall look at the following as 'case studies':
- Two ways of making 'electrochemical' transducers for monitoring the oxygen content of a gas. This is an example of a general need for electronic chemical analysis.

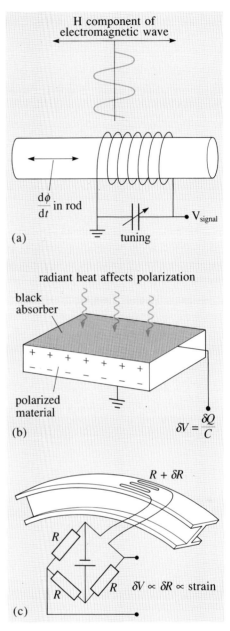

Figure 4.1 (a) Ferrite rod in aerial. The flux change in the rod, produced by the magnetic component of the transmitted wave, enhances the strength of the received signal according to Faraday's law $V = d\Phi/dt$

Figure 4.1 (b) Polarized ferroelectric is heat sensitive. Capacitor gives signal by changing charge

Figure 4.1 (c) Change of resistance under strain puts the Wheatstone bridge off balance and a voltage signal $i\delta R$ appears

- A pyroelectric transducer (that is, one that converts radiant heat energy to an electric signal) used in an intruder alarm system.

- An optical modulator in which one of the many 'opto-electronic' effects controls the transmission of sudden flashes through goggles. It was designed originally to protect the eyes of military aircrews, but it might sell to welders.

- Two piezoelectric devices (interconverting mechanical and electrical energy) used in sonar submarine detection systems. The ultrasonic power output device and the hydrophone deploy a common material but with different modifications to meet the different roles.

The last two items require anisotropy, since the stimuli and responses are *vectorially* related. For example, in the piezoelectric effect a stress vector generates a polarization vector. The optical modulator uses the interaction of polarized light with an axis of anisotropy in the material. There are many ways of making a polycrystalline material anisotropic. Steel for transformers is made anisotropically responsive to magnetic field by rolling it. The crystals become preferentially aligned. A permanent magnet is also anisotropic in that the direction of the flux within it defines a unique axis. Similarly ferroelectric ceramics can be 'poled' to have an anisotropic charge distribution by the influence of a strong electric field. The crystals are more likely to polarize with an axis favourably oriented with the field to give a net polarization.

Our first case study further explores mechanisms of conduction in ionic solids showing how to control and use them.

▼Process routes for ceramics▲

'Traditional' ceramics are based on natural materials, and their compositions are based on what is in those raw materials blends. Modern ceramic alloys are in no sense 'natural' materials, and their composition is not restricted by what nature provides. Much of the research game has therefore been adding pinches of this or that in order to enhance or suppress various aspects of a suite of properties.

The standard process route can easily accommodate complex recipes. The raw materials are oxides at required purity from specialist suppliers. Chosen proportions are ball-milled together to produce an intimate mixture at very fine particle size. The correct chemical substance is then formed by **calcining** (fiercely heating) this mixture. This material is then ground up and artefacts made by compacting and sintering. **Damp compacting**, using pressure and an organic binder (such as sugar solution), is a cheap

and versatile method. It calls for a little care in burning out the binder slowly so that gas generation doesn't burst the object or leave it too porous. When really high quality ceramic is demanded, **hot pressing** may be used. This is a more costly process in which powder with no binder is sintered, in a press, under a pressure of several thousand pounds per square inch.

Alternative routes in which oxides are precipitated from solution are gaining ground. They are more expensive than the methods above. This is the **sol-gel** method. Here, to get oxides into solution, organic derivatives are used. Compounds such as **ethoxides**, $C_2H_5O.MO$, where M is virtually any metal, will dissolve in alcohol–water mixtures. By adjusting the acidity, the metal hydroxide can be precipitated. This is filtered, dried and roasted to become oxide. If the details of solubilities are known, recipes can be devised where the precipitate has the various metal oxides

mixed in chosen ratios at a nanometre-scale of intimacy. Use of this technique yields these advantages:

- more precise and homogeneous compositions,
- finer powders which sinter to full density rapidly, and at lower temperature.

A common problem in getting accurate final compositions is that the constituent oxides may be differently volatile during the heat treatments. Two ways of controlling this are to adjust the initial mix to allow for loss and to control the furnace atmospheres to prevent evaporation. Often the basic recipe has additives which reduce the sensitivity of some property to such irregular process variables. We shall see examples when we look at the control of electrical conductivity in oxides.

4.2 Materials for oxygen sensors

Electrochemical transducers offer the possibility of continuously monitored chemical analysis. In Chapter 1, a transducer sensitive to the oxygen concentration in the exhaust gas of a petrol engine was mentioned as a way of deciding whether the fuel–air mixture was too rich or too lean. Two sorts of transducer are possible:

• a 'chemo-voltaic' transducer, which interprets the difference between the oxygen partial pressures in the exhaust pipe and that in the air as a voltage;
• a 'chemo-resistive' device, whose resistance varies with the oxygen partial pressure in the exhaust.

Both use oxide ceramics in which conductivity depends on local oxygen concentration. However their mechanisms are different.

4.2.1 The voltaic approach

Chapter 1 revised the fundamental ideas of batteries as sources of an electromotive force (e.m.f.), which can drive an electric current. The e.m.f. arises because the reactions between the metal electrodes and the electrolyte come to equilibrium with different amounts of charge on the metal. The voltaic oxygen sensor is based on the same principle, but with some modifications (Figure 4.2). The electrolyte is an oxide ceramic solid able to conduct by oxygen-ion diffusion, and the electrode reactions are between the electrolyte and *the oxygen gas in the atmospheres around each electrode*. The thin platinum metal electrodes have a dual function: to provide the electron-conducting part of the battery, and to catalyse the reaction between the atmosphere and the electrolyte. Unlike the metal electrodes of an ordinary battery, they are not sources or sinks of ions.

To approach an explanation of how the device works first look at Figure 4.3(a). Here two vessels contain oxygen at different pressures and the tap in the linking tube is shut. If the tap is opened it's easy to see what happens: oxygen molecules move from the high-pressure side to the low-pressure side until the pressures become equal. The sensor in Figure 4.3(b) is a battery essentially trying to achieve the same thermodynamic equilibrium. The reaction of gaseous oxygen with the electrolyte is

$$\tfrac{1}{2}O_2 \text{ (atmosphere)} + 2e^- \text{ (metal in electrode)} \rightleftharpoons O^{2-} \text{ (electrolyte)}$$

From right to left in the equation, oxygen ions in the solid are giving electrons to the metal and leaving the solid to become gas molecules in the atmosphere. In Figure 4.3(b) you can see that if the cell were short circuited, the flow of electrons through the wire and the flow of oxygen ions through the electrolyte would effectively transfer oxygen *molecules* from the high-pressure side to the low-pressure side. Obviously the final equilibrium would be when there is no pressure difference. As long as there are different concentrations of oxygen in the two atmospheres on either side of the electrolyte, these currents could flow. For a current to

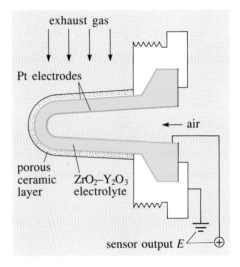

Figure 4.2 Voltaic oxygen sensor

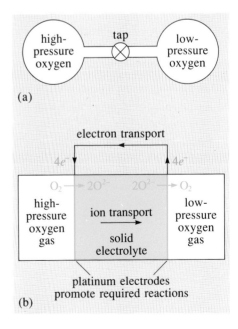

Figure 4.3 (a) Vessels containing oxygen at different pressures. (b) Operation of voltaic sensor

flow, there must be an e.m.f. between the electrodes, which we could measure with the cell open-circuited. As SAQ 1.1 showed, the e.m.f. of a battery depends on the equilibrium concentrations of the reactants at the electrodes. Here it is only the concentrations of oxygen in the two atmospheres which are significantly different. The oxygen-ion concentrations in the electrolyte scarcely vary. So the oxygen pressure difference is what decides the e.m.f. If one side is a standard oxygen pressure (partial pressure of oxygen in air is 0.2 atmosphere) the e.m.f. becomes a measure of the other oxygen partial pressure.

To get nearer to a practical device we must first know how to make a solid electrolyte in which conduction is entirely by oxygen-ion transport, since only that effect is relevant to the thermodynamic drive to balance the oxygen gas pressures. ▼Designing an oxygen-ion electrolyte▲ takes up this matter.

Then there are a couple of other important questions to investigate:

● How does the e.m.f. vary with the equilibrium positions of the reactions at each electrode?
● What controls the rates of the electrode reactions, and hence response time of the device?

Let's consider these in turn.

▼Designing an oxygen-ion electrolyte▲

An effective electrolyte must be a good conductor. In Chapter 2 we were concerned with ionic solids (ceramics) as insulators and tried to find ways of stifling any conductivity. What judicious changes of composition in an ionic solid might enhance the mechanism of oxygen-ion conductivity?

Since diffusion depends upon vacancies, adjusting the vacancy population enables conductivity to be controlled. When discussing ceramic dielectrics in Chapter 2, we hinted that one role for the addition of niobia, Nb_2O_5, to barium titanate was to *fill* oxygen vacancies by substituting for TiO_2. Niobia brings more oxygen per cation than the crystal structure requires. Also it was mentioned that a possible source of vacancies in the oxygen lattice was pollution of the raw ceramic materials by alumina, Al_2O_3. Alumina does not provide enough oxygen to build the perovskite structure. This is the clue we need. To make an oxide into a conductor it must be doped with significant quantities of another oxide which leaves the crystal deficient of oxygen. For example, into oxides of formula MO_2 (two oxygen ions per metal ion) could be added M'_2O_3 (one and a half oxygens per cation), where M' just represents another metal. The important thing about the substitution is that it should be a material which is soluble in the host.

EXERCISE 4.1   What are the implications for crystal structure in the assertion that the additive should form a solid solution with the host? How would you choose the additive so as to achieve substantial solubility?

Such doping will increase the population of oxygen vacancies to an extent far exceeding that of the thermally generated vacancies. Thus there is the possibility of a highly conductive ceramic. But which oxides should be choose as the host and dopant? We want the electrode reactions to have an equilibrium position which is *strongly* dependent on oxygen pressure.

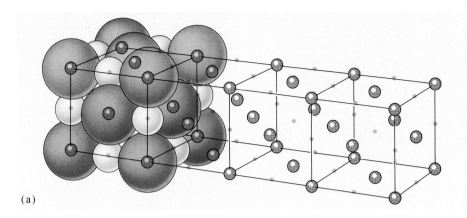

(a)

Figure 4.4 (a) MgO structure: edge-sharing octahedra; all cages hold an Mg^{2+} ion

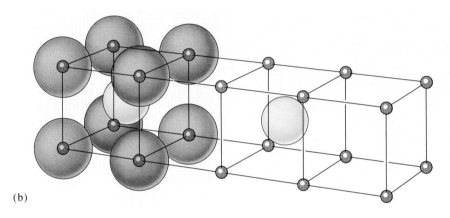

(b)

Figure 4.4 (b) Zirconia (ZrO_2) structure: face touching cube cages, composed of O^{2-} ions; half the cages hold Zr^{4+} ions, the other half are empty

What are the chemical qualities of the electrolyte that will ensure this? How would these qualities be achieved crystallographically? These are questions which a little chemical intuition can penetrate.

Obviously the oxide electrolyte must be able to react with the atmosphere. In other words, it should be relatively easy to move oxygen ions in and out of the solid. A very strongly bonded oxide will be too chemically stable; in such a material, movement of oxygen ions in and out would be difficult. So we should choose a relatively weakly bonded oxide, characterized probably by a rather open

crystal structure. This is in addition to providing many oxygen vacancies to encourage diffusion.

EXERCISE 4.2 Figure 4.4 shows two more oxide crystals which you can compare with the perovskite structure of Figure 2.15. Which of these three best meets the specification of the previous paragraph? How would you dope your chosen structure to provoke oxygen vacancies?

Pure zirconia actually crystallizes in distorted versions of the structure in Figure

4.4(b), with a phase change at 1000°C. So as a ceramic being taken up and down in temperature in the exhaust pipe of a car, it may prove to be mechanically unstable. However the gods are with us for once! The additives needed to make oxygen vacancies also stabilize the cubic fluorite crystal structure over the whole working temperature range. For zirconia used in mechanical applications the customary crystal stabilizers are yttria, Y_2O_3, or lime, CaO. In spite of its much higher cost, yttria is favoured for this electrical application as it gives a much higher ionic conductivity (Figure 4.5).

SAQ 4.1 (Revision)
The straight line relations of Figure 4.5 show the conductive mechanism to be thermally activated (obeying Arrhenius's law). What is being thermally activated; the breaking of oxygen ions from their sites, or the generation of vacancies for them to move into? Why are the lines in Figure 4.5 nearly parallel? Evaluate the activation energy for the conduction mechanism. (Take Boltzmann's constant to be $k = 86 \ \mu eV \ K^{-1}$.)

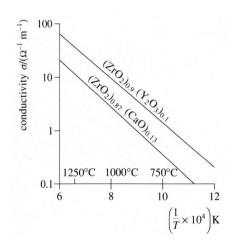

Figure 4.5 Arrhenius plot. Conductivities of lime- and yttria-doped zirconia.

EMF and equilibria

When the electrodes are in chemical equilibrium with their oxygen atmospheres, then we say, thermodynamically, there is a condition of minimum free energy. We can express this free energy in terms of equilibrium concentrations of reactants. This is because the numbers of particles of each species in the reaction mixture is related both to the entropy and the enthalpy of the system. (The enthalpy encompasses the chemical and electrical potential energies of the system, which are determined by chemical bonding and charge separation. Both of these are related to the equilibrium concentrations of reactants.) I will not trace this argument from the beginning; to do so would involve a lot of chemical thermodynamics. The outcome is the **Nernst equation**, which relates the e.m.f. E of a cell to the equilibrium conditions at its electrodes. For the oxygen sensor, with only the different oxygen pressures P_{low} and P_{high} as variables, Nernst's equation becomes simply

$$\frac{P_{low}}{P_{high}} = K \exp\left(\frac{-4eE}{kT}\right)$$

The left-hand side compares the concentrations of the reactants at each electrode. The right-hand side states the familiar battle between an orderly energy form, $4eE$, and thermal rattle, kT. The term K is a constant, which need not concern us. Taking logarithms of both sides of the equation lets us extract the e.m.f. as

$$E = \text{Constant} - \frac{kT}{4e} \ln \frac{P_{low}}{P_{high}}$$

This can be made yet simpler for the situation under discussion. First, if the high pressure side is electrically earthed, the constant becomes our local zero of potential. Secondly, since P_{high} is the *fixed* partial pressure of atmospheric oxygen, the only variable is the lower oxygen partial pressure. Thus,

$$E = A \ln P_{low}$$

expresses how the e.m.f. varies with oxygen partial pressure. The A term is a calibration constant of the instrument. It is largely determined by the chemistry of the chosen electrolyte, but is also influenced by temperature according to Nernst's equation. Doped zirconia (ZrO_2) gives a sensitivity at 700 K of about 50 mV per factor of 10 change in oxygen pressure. Since, in a car exhaust, P_{low} goes from perhaps 10^{-2} atmosphere for a slightly lean air/fuel mixture to say 10^{-20} atmosphere for a mix of the rich side of stoichiometry, a signal voltage change of nearly a volt over the range of interest is to be expected. The size of this voltage step will be a bit higher or lower depending on whether the sensor is hotter or cooler but, since the control circuitry only has to determine whether the mixture is lean or rich, this variability is not a problem.

Rate and response time

Now we must think about what controls the rates of electrode reactions, since the rate is what will govern the response time of the transducer. How quickly can the low-pressure electrode move from one state of chemical equilibrium to another? Reaction rates obey the rules of thermal activation, so a high-temperature environment will be very helpful for achieving a quick response. The exhaust gas heats the surface at which the reactions occur to about 1000 K, so the environment is certainly hot. However, one of the reactants is a solid, and diffusion in that phase may be a limiting factor. By choosing an open structure for the electrolyte (Figure 4.4b) and loading it with oxygen-ion vacancies we have done our best to hasten diffusion there. But above all, it is necessary to reduce the activation energy of the reaction by providing a catalyst. Platinum is the chosen electrode material. It successfully catalyses the redox reaction between the oxide and the atmospheric oxygen and suppresses any competing electrode reactions. The resulting instrument can respond fast enough to follow changes of stoichiometry within a few revolutions of the engine.

SAQ 4.2 (Objective 4.3)
Figure 4.6 shows a set up for fool-proof welding of stainless steel pipes. The problem is that at the high temperatures attained by the metal during welding, much of the alloyed chromium may be oxidized, spoiling the corrosion resistance of the steel. The outside of the pipe is protected during welding by inert gas being pumped through the welding torch. The inside is to be protected by inert gas contained in a temporary chamber formed by the two plugs (see Figure 4.6). Safe welding requires the oxygen concentration to be less than 30 p.p.m. The chamber is flushed with argon. The gas is sampled and presented to a voltaic sensor in a furnace controlled at 700°C. The welder can only switch on the torch when the sensor discerns that the oxygen concentration is low enough.

Contrast this application of the oxygen sensor with the car-exhaust gas monitor in respect of response time and accuracy. Why is the furnace necessary?

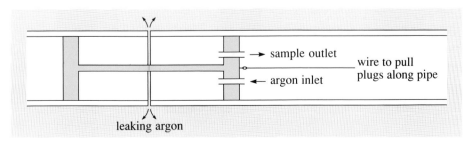

leaking argon

Figure 4.6 Welding system for SAQ 4.2

4.2.2 The resistive approach

The zirconia voltaic oxygen sensor has had commercial success for exhaust gas analysis. Its main snag is a slightly too slow response at high engine speeds. The Ford Motor Company has researched an alternative; a tiny oxygen-sensitive resistor which lies in the gas flow and has a potentially faster response. Rather than monitoring a voltage, with this sensor we monitor its resistance. The material is titanium dioxide, TiO_2 (also called titania).

Although from their adjacent places in the periodic table, one above the other, you might expect titanium and zirconium oxides to behave similarly, nature has a surprise for us. The smaller size of the Ti^{4+} ion compared to Zr^{4+} makes TiO_2 crystallize into a tighter structure (Figure 4.7). In this structure oxygen-vacancy diffusion is more difficult and a new mechanism of conduction comes into play. See ▼Extrinsic conduction in oxides▲

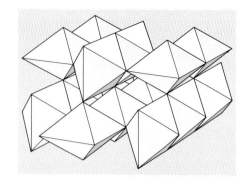

Figure 4.7 Structure of TiO_2

▼Extrinsic conduction in oxides▲

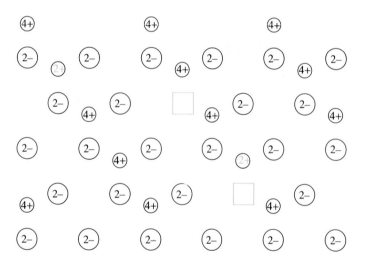

Figure 4.8 In this section of crystal there are 26 anions (O^{2-}) and two oxygen vacancies. A total of 52 electrons have therefore been transferred from the 14 cations. The electrons come from 12 cations of 4+ charge and 2 cations of 2+ charge:
$(4 \times 12) + (2 \times 2) = 52$

In ionic compounds, bonds form by complete transfer of electrons from the metal atom to the non-metal atom. Each atom becomes an ion with an inert-gas electron structure, and these ions are so stable that it is very difficult to disrupt them to remove electrons or make holes. Hence as we saw in SAQ 1.9, intrinsic conduction in ionic crystals is negligible: the band gap is too big to allow significant thermal excitation of electrons from the valence band to the conduction band.

As you know, in silicon we can have *extrinsic* conduction by doping the pure element with other elements which do not have the right number of electrons to fit the bonding regime exactly. Phosphorus doping, for example, gives surplus electrons, which are mobile. Atoms with insufficient electrons (for example boron) produce incomplete bonds, which can be thought of as holes, and which also readily move from atom to atom. Both modes of conduction are extrinsic.

Imperfect bonding in ionic crystals can lead to extrinsic conduction too. Oxides prepared as ceramics are 'non-stoichiometric', that is, of inexact composition, and this non-stoichiometry is the source of the imperfect bonding. The degree of non-stoichiometry depends on the chemistry of the elements concerned. We have just seen how hot zirconia can either take oxygen into the crystal or reject it to the atmosphere, depending on the pressure of oxygen in the surroundings. Such exchanges take place during the high-temperature processes (calcining and sintering) of all ceramics and is a principle cause of non-stoichiometry.

Thinking of oxygen as the variable element, there are two cases to consider:

- oxygen-deficient oxide (or metal-rich, which is the same thing),
- oxygen-rich oxide (or its equivalent, metal-deficient).

Each has different implications for extrinsic conductivity. It will be easiest to see what they are if we use a specific example. Consider titanium oxide. It ought to be TiO_2, with all the titanium atoms ionized to Ti^{4+}. Depending on how it is made, it may end up with a bit less or a bit more oxygen than the required ratio of 2 oxygens to every titanium. We'll take oxygen deficiency first.

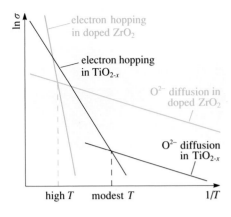

Figure 4.9 Conductivity variations and conduction mechanisms of oxygen-deficient titania and doped zirconia

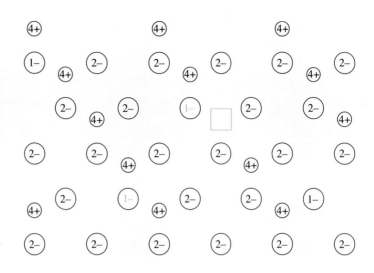

Figure 4.10 In this section of crystal there are 28 anions but only 13 cations. The cations supply $4 \times 13 = 52$ electrons. There are 24 O^{2-} anions and 4 O^- ions to achieve charge balance: $(24 \times 2) + (4 \times 1) = 52$ electrons as supplied

Oxygen deficient oxide

In this case, as the titanium atoms give electrons to the oxygen atoms, a point will be reached where each oxygen has collected two electrons, but not all the titaniums have shed four electrons. Some of the titanium atoms are therefore left in their alternative valency state of Ti^{2+}.

Now the situation is as Figure 4.8: a few Ti^{2+} ions in a sea of Ti^{4+} ions, and a few oxygen vacancies in a lattice of O^{2-} ions.

There are now two possibilities for conduction:

• O^{2-} ions may diffuse using the vacancies,
• electrons on Ti^{2+} ions may hop to Ti^{4+} ions.

If the electron-hopping dominates, the oxide will be an n-type extrinsic conductor.

You will notice in this case that there are equal numbers of Ti^{2+} ions and O^{2-} vacancies so in the conductivity equation

$$\sigma_c = \sum_i n_i e_i \mu_i$$

it is the *mobilities* of the different carriers which determines the conductivity. But what decides mobility?

Each conduction process has its characteristic activation energy (since both rely on diffusion). At some temperature, the mechanism with the higher activation energy will become dominant. Oxygen-deficient titania becomes an n-type conductor at a modest temperature (Figure 4.9), in contrast to the behaviour of doped zirconia.

Surplus oxygen

The opposite possibility, oxygen-rich (or metal deficient) oxide, leads to p-type conduction. Staying with titania, if the crystal is lean of metal, when all the titanium atoms have given up four electrons not every oxygen atom will have gained the two it needs to become O^{2-}. There will be some O^- ions in the crystal as Figure 4.10 shows, together with some cation vacancies.

The possible mechanisms for conduction are now:

• diffusion of cations using the vacancies,
• electrons hopping from O^{2-} into the holes which O^- provide.

Activation energies for cation diffusion in oxide crystals are always high. Thus the dominant mechanism is 'hole' conduction as the deficit of electrons among the oxygen ions moves from ion to ion. The material is then a p-type conductor.

Both n-type and p-type electronic conduction can afflict the yttria doped zirconia electrolyte used in the voltaic oxygen sensor. In spite of the overwhelming population of oxygen vacancies deliberately introduced by the doping, and the low activation energy for their diffusion in the cubic crystal, at high enough temperatures the oxygen-deficient side of the cell may become an n-type conductor and the oxygen-rich side a p-type. Figure 4.11 shows the boundaries for the onset of these electronic effects. It is precisely because the oxygen vacancy conduction persists to very high temperatures that this material is chosen for these voltaic sensors.

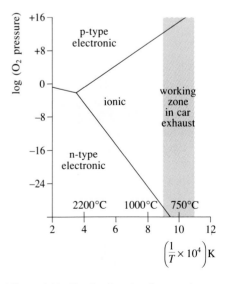

Figure 4.11 Conductive dominances in yttria-doped ZrO_2 at various temperatures and oxygen pressures

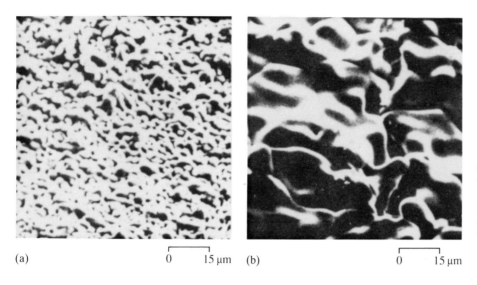

(a)   0   15 μm   (b)   0   15 μm

Figure 4.12 (a) TiO_2 with porous structure. Resistivity is sensitive to both temperature and oxygen partial pressure

Figure 4.12 (b) TiO_2 with impervious structure. Resistivity is sensitive to temperature only

The TiO_2 in the resistive sensor is oxygen deficient, and is therefore n-type electronic conductor. At high temperature, and with a catalytic surface film of platinum, TiO_2 reacts with surrounding oxygen:

$$O^{2-} + Ti^{4+} \rightleftarrows \tfrac{1}{2} O_2 \text{ (gas)} + Ti^{2+} + O_{\text{vacancy}}$$

Notice that the reaction can go either way, so the number of Ti^{2+} ions — and hence the resistivity — depends on the surrounding oxygen partial pressure.

For the piece of titania to change its resistivity quickly, oxygen must be able to diffuse into and out of the ceramic and to react at the surface. Therefore, to make a responsive detector working on this principle, a very porous ceramic of small grains is provided. The atmosphere reacts with a large surface area and diffusion distances are short. Figure 4.12(a) shows this microstructure.

There is a snag. The activation energy for electrons hopping from Ti^{2+} to Ti^{4+} is 1.6 eV, which gives rise to an enormous temperature coefficient of resistivity (4% K^{-1}) for this electronic conduction at the working temperature in a car exhaust pipe. Resistance changes may then have little to do with changing oxygen concentrations, and everything to do with changing temperatures. The solution is typical of transducer practice when two stimuli produce the same response: compensation. The patentees of the resistive transducer propose putting *two* titania resistors in the probe in the exhaust pipe. The two resistors are connected in series into a Wheatstone bridge (Figure 4.13). The second resistor is made of oxygen deficient titania of dense microstructure (Figure 4.12b), so though it suffers the same temperature variation, it is not susceptible to variation with oxygen pressure. The bridge then cancels the temperature change, but registers resistance changes in the porous element due to oxygen concentration effects. You'll frequently come across this technique of compensation of unwanted responses by transducers. It turns up again in our next study.

SAQ 4.3   (Objective 4.1)

Oxygen-deficient titania could conduct electricity both by diffusion of oxygen ions via oxygen vacancies or by electrons hopping from Ti^{2+} ions to Ti^{4+} ions. So $\sigma_e = \Sigma nq\mu$ has two terms, one for each mechanism. Write out the equation in full and comment upon the population and mobilities of each charged entity so as to explain why the electron hopping mechanism predominates.

SAQ 4.4   (Objective 4.2)

Assess the veracity of each of the following statements.

(a) Zirconia doped with calcium oxide (CaO) will be an n-type semiconductor because electrons can jump from Ca^{2+} ions to Zr^{4+} ions.

(b) Transition metal oxides are more likely to exhibit n-type semiconduction than oxides of non-transition elements because transition metals can readily form ions of different valency.

(c) Doping zirconia, ZrO_2, with niobia, Nb_2O_5, would make it conduct by oxygen-vacancy diffusion.

(d) Oxygen-rich titania is a p-type conductor because it contains O^- ions.

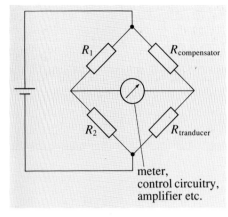

Figure 4.13 Wheatstone bridge, used for temperature compensation

Summary of Section 4.2

A solid oxide material can conduct electricity by oxygen-ion diffusion under an electric field. Such diffusion requires that there be oxygen vacancies for ions to diffuse into. Oxygen vacancies may be thermally generated, or deliberately introduced by controlled doping. A dopant carrying fewer oxygen ions per cation than its host will introduce vacancies and enhance conductivity — provided the dopant is soluble in the host.

A doped (conducting) oxide can be used as the electrolyte in a voltaic oxygen sensor, provided oxygen can easily diffuse into and out of the electrolyte at both the anode and cathode sites. Zirconia doped with yttria is a suitable electrolyte. The voltage generated is an indication of the relative partial pressures of oxygen at each electrode. The Nernst equation relates voltage to partial pressures.

Oxide conduction may also happen by electron (or hole) drift under an electric field. Electron drift is possible in oxygen-deficient material. Here, to maintain overall charge neutrality, some cations carry more electrons than they would in stoichiometric oxide. If these 'extra' electrons are mobile, n-type conduction is possible. Conversely, surplus oxygen causes some oxygen ions to be electron-deficient. These 'holes' may render the material p-type.

N-type titania can be used as a resistive oxygen sensor. In practice, compensation is needed to mask the thermal response of the material's conductivity.

4.3 Ferroelectric materials in transducers

4.3.1 Parallels with ferromagnetics

The rest of the case studies all use ferroelectric ceramics. You have already met this class of materials as capacitor dielectrics in Chapter 2, but a greater knowledge of them is needed to understand their use in transducers and to appreciate materials selection and processing for these applications.

Just as a ferromagnetic material spontaneously magnetically polarizes below a critical temperature known as its 'Curie temperature', so ferroelectric materials exhibit spontaneous electric polarization below their **Curie temperature** T_C. A strong enough applied field will 'saturate' the polarization (that is, the polarization will reach a maximum) and, as with ferromagnetics, the magnitude of that maximum grows swiftly as temperature falls from T_C. Compare Figure 4.14, which shows the saturation polarization versus temperature curves for some ferroelectrics, with Figure 3.62 for some ferromagnetics. For the moment just note the variety of saturation polarizations and Curie temperatures which are available. Not all the materials on the graph have the perovskite crystal structure of barium titanate, the ferroelectric archetype, but all become ferroelectric by virtue of crystallographic phase changes which give non-uniform charge distribution. Can you remember what happens in barium titanate crystals as they cool below 125°C?

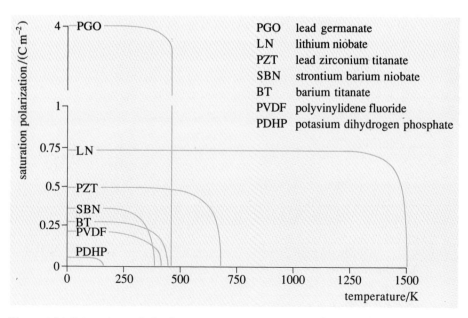

PGO   lead germanate
LN   lithium niobate
PZT   lead zirconium titanate
SBN   strontium barium niobate
BT   barium titanate
PVDF polyvinylidene fluoride
PDHP potasium dihydrogen phosphate

Figure 4.14 Saturation polarization versus temperature curves for some ferroelectrics

SAQ 4.5   (Objective 4.4)
Compare what happens in iron and barium titanate as they cool through their Curie temperatures. What is the significance of the Curie temperature in terms of 'sticking' and 'rattling' ideas?

When spontaneous magnetization or electrical polarization develop, at or just below T_C, the effect is cooperative. That is to say, a little bit of local progress towards polarization helps the surroundings to polarize. You can imagine that two aligned iron atoms would influence a wider zone than any single atom would; these two atoms orient their neighbours, so the urge increases. The process is further aided by the magnetostrictive stress induced in the region (that is, the small dimensional change accompanying the magnetization). Similarly, and in this case largely driven by the stress factor, a local nucleation of tetragonal phase in the barium titanate cubic structure must encourage neighbouring cells of the crystal to make the change. For these reasons, spontaneous magnetic or electric polarizations increase rapidly as the temperature falls below the Curie temperature.

Parallels with ferromagnetics continue: the idea of domains is used to explain the absence of bulk polarization in a specimen and the onset of polarization when it is subjected to an electric field. By changing the applied field the polarization of a specimen can be driven round a hysteresis loop. Figure 4.15 shows the hysteresis loop for polycrystalline barium titanate (ceramic); compare this diagram with Figure 3.7, and note the parallel ideas of saturation, remanence and coercive field.

The hysteresis loop informs us that a piece of ferroelectric in zero field may be either polarized or unpolarized, points 0,R and 0,0 on Figure 4.15. Both conditions are useful, just as in magnetics there are uses for both permanent and electromagnets.

EXERCISE 4.3   Remind yourself of ferromagnetic equivalents of the 0,0 and 0,R states of material.

Piezoelectric devices use the 'poled' condition with a piece of material carrying positive and negative electric charges on opposite faces (more about this later). Ferroelectric material used as the dielectric of a capacitor is in the 0,0 condition. From here it can respond to small changes of applied field (changing voltage on the capacitor) with a large and rapid change of polarization; that is, it has a high relative permittivity. As with magnetic materials, the distinction between the permanent and temporary polarized states is to do with the ease or difficulty of movement of domain walls. Relations between domain and polycrystalline microstructures are complicated. ▼Domains in ferroelectrics▲ outlines the principles from which practical applications can be understood.

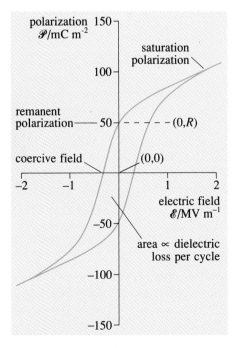

Figure 4.15 Hysteresis loop for ceramic BaTiO₃

In tetragonal ferroelectrics:

● The high relative permittivity of capacitor dielectrics is achieved by the easy motion of 180° domain walls.

● To pole a specimen, it must be brought to saturation by a strong field and then allowed to relax back to the remanent condition. To create saturation, 90° walls must be moved. That means grains changing shape and generating internal stresses. Poling is done as hot as the Curie threshold allows, and the conditions are held for some hours. This allows some diffusion at grain boundaries to help accommodate shapes, but Curie temperatures are much lower than the ceramic sintering temperatures at which decent rates of transport between grains can occur, so it is inevitable that poled ceramics carry residual internal stress.

▼Domains in ferroelectrics▲

When a phase transformation occurs in a crystal, the new phase is likely to be nucleated at several places within the original crystal. The nuclei of the new phase grow until the whole volume has been transformed. In a polycrystalline specimen, each *grain* will contain crystals of the new phase, variously oriented because of the multiple nucleations. Remember that the phase transformations we are concerned with here happen in the *solid* state, so the new phases have a definite relationship with the axes of the original grain. The boundaries between the phases within a grain are defined by abrupt changes of crystal structure, without the zone of disorder which is characteristic of a grain boundary.

When a ferroelectric cools through its Curie temperature, the sort of transformation described above occurs. Different distortion directions nucleate within the grain, and since these fix the direction of electric polarization, a grain will contain regions with different directions of polarization. These regions are **ferroelectric domains**, and are closely analogous with ferromagnetic domains. The phase boundaries are domain walls. Perovskite ferroelectrics transform to a tetragonal, a rhombohedral or an orthorhombic phase, Figure 2.19. Barium titanate undergoes all three transformations in the solid state at different temperatures.

In principle, the overall polarization of a

grain can be changed by domain walls moving, that is by certain domains growing and others shrinking. The stability or mobility of domain walls is determined by energy considerations. Let us think about these effects in more detail for the particular case of a tetragonal crystal. The new, tetragonal crystals are slightly taller (≈ 1%) and narrower than the symmetrical cubic structure. (They are also, of course, electrically polarized in the direction of their elongation — which is why we are interested in them.) The elongation and shrinkage are parallel to the original cube axes. Thus when the new phases are created, their axes are in mutually perpendicular directions (Figure 4.16a).

Where one phase meets another (at a domain wall), the difference between polarization directions on either side of the wall can only be either 90° or 180°. Clearly, at 180° walls the adjacent crystal structures fit, because their longer tetragonal axes are parallel. However, at 90° walls the crystals do not fit: the longer lattice spacing on one side abuts with the shorter spacing on the other (Figure 4.17). The result is internal stress within the wall, relieved to some extent by the transition taking place over several crystal cells, much as the spin directions in magnetic domain walls change gradually (Figure 3.21). Similar but geometrically more complicated arguments apply to the rhombohedral or orthorhombic

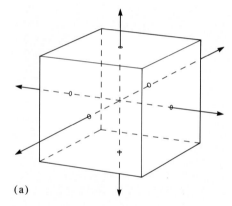

(a)

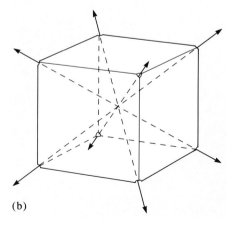

(b)

Figure 4.16 (a) Tetragonal phase change has 6 easy directions deriving from cube. (b) Rhombohedral transformation has 8 easy directions

● Ageing, the slow decay of remanent polarization as diffusion relaxes internal stresses, is common — Figure 4.19, curve A. If a steady remanence is required it is best to get this over and done with quickly (curve B) by allowing the domain walls to move in response to the residual stress. Domain walls move most freely in crystals which do not have a lot of oxygen vacancies. On the other hand if walls can move easily the material will have low coercivity. To avoid this walls have to be 'pinned', and then ageing persists for a long time (curve A).

The remanent polarization of a poled specimen is, like the saturation polarization, temperature dependent. As the Curie temperature is approached, it falls to zero. This is the 'pyroelectric' effect which enables

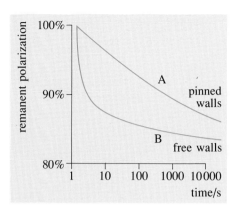

Figure 4.19 Pinning domain walls slows rate of ageing

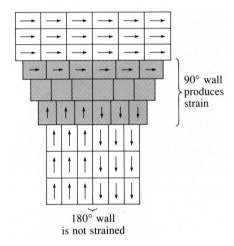

Figure 4.17 Crystal lattice distortions at 180° and 90° walls

transformations. Exercise 4.4 asks you to work out the angles between adjacent domains for the rhombohedral distortion. Notice that this transformation offers *eight* easy directions of polarization (Figure 4.16b).

EXERCISE 4.4   The domain polarizations for the rhombohedral crystal form are along the body diagonals of the cube. Figure 4.18 shows the relevant dimensions (deduced by Pythagoras' theorem) which are needed to calculate the angles $A\hat{X}B$ and $A\hat{X}C$ between pairs of these diagonals. Calculate these angles.

Strain energy is the dominant term in the ferroelectric domain wall energy budget, and 90° and 180° walls are thus distinctly different in behaviour. It is easy for 180° walls to move under the influence of an applied electric field. It merely requires the cations to move cooperatively to the other ends of their cages without any shape changes. Walls at 90° are relatively difficult to move. Successive layers of crystal have to change shape (imagine it as a wave passing through the structure pictured in Figure 4.17). In a polycrystal such shape changes will be powerfully inhibited by the constraint of neighbouring grains. Evidently these stresses between grains must be influential both when the domains are forming as the material cools through its Curie temperature and during any subsequent attempt to change the state of polarization.

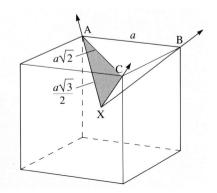

Figure 4.18 Dimensions for Exercise 4.4

transduction of radiant heat energy into electric signals. ▼Designing a pyroelectric burglar detector▲ describes one of the simplest deployments and also leads us to the family of ferroelectrics which has met the best commercial success: PZT.

Summary of Section 4.3

Ferroelectric materials spontaneously polarize below their Curie temperature T_C. This polarization arises from asymmetric ionic displacements, consequent upon a change of phase.

Spontaneous polarization happens at many sites throughout a crystal, in many different directions. This leads to the development of domains of polarization.

Domain walls move by atomic displacement, transforming one orientation of crystal distortion into another.

Poled material carries remanent polarization.

Remanent polarization is temperature dependent. Thus poled ceramic (such as lead zirconate, PZ) can be used, capacitively, as a radiant-heat sensor and transducer.

Detailed design of a radiant-heat sensor must take into account a range of materials properties, for which merit indices may be devised.

▼Designing a pyroelectric burglar detector▲

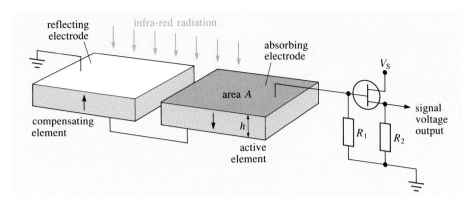

Figure 4.20 Principles of pyroelectric detector. The compensating element does not respond to infra-red radiation (because of its reflecting surface), but it does respond to changes in ambient temperature. Its ambient temperature response cancels that of the active element

Figure 4.20 shows the principle. A tiny piece of polarized material is the dielectric of a capacitor. The polarization dictates the charge held by the capacitor and hence the voltage across it. The idea is that the radiant energy from an intruder warms the material and the consequent change in polarization alters the voltage and transduction has been achieved. A low noise, high impedance amplifier conveys the signal into the system and a split second later the peace of the neighbourhood is shattered by an excruciating din.

The pyroelectric element is a very thin sheet so that it heats up quickly and uniformly and so that, with metallization evaporated onto its faces, it is indeed a capacitor. It is also thermally well insulated from its surroundings so that it actually does change temperature, though perhaps by as little as 10^{-6} K. To analyse

its performance, suppose it has area A, thickness h (volume Ah) and, absorbing a burst of radiant heat energy ΔH, its temperature rises by ΔT.

$$\Delta T = \frac{\Delta H}{Ah\rho S}$$

with ρ and S representing respectively the density and the specific heat of the material. Considered as a capacitor the device has capacitance

$$C = \varepsilon_0\varepsilon_r\frac{A}{h}$$

Remembering (from Chapter 2) that polarization $\mathscr{P}$ is the surface charge density on the dielectric, the charge released when the temperature changes by ΔT is

$$\Delta q = A\Delta T\frac{d\mathscr{P}}{dT} = \Pi A\Delta T$$

if we use the symbol Π to denote the

'pyroelectric coefficient' $d\mathscr{P}/dT$.

Now bring all these together into $q = CV$ and we find a voltage step

$$\Delta V = \frac{\Pi\Delta H}{\varepsilon_0\varepsilon_r A\rho S}$$

That's all very well if there are sudden bursts of heat. More realistically, there is some radiant power flux (energy per second per square metre) and the capacitor presents a changing voltage:

$$\frac{dV}{dt} = \frac{1}{A}\frac{dH}{dt}\left(\frac{\Pi}{\varepsilon_0\varepsilon_r\rho S}\right)$$

This is an interesting expression because all the terms in the bracket are materials properties so together they make a 'merit index'. Table 4.1 gives some values.

SAQ 4.6
Burglars radiate about 60 W. Assuming all the power incident on an alarm is absorbed, what rates of voltage change would a burglar induce in a pyroelectric detector made of the most and least meritorious materials at 10 m range? (Model the burglar as a point source.)

In SAQ 4.6 all you have calculated is the initial rate of rise of the voltage of a transient pulse. Can you see what will stop the voltage rising continuously, even when the radiant input is steady?

There are two things. Firstly, as soon as the temperature of the device exceeds that of its surroundings, it will start to lose heat. At some excess temperature the rate of heat loss will match the radiant power

Table 4.1 Room temperature data for potential pyroelectric element materials

Material (and condition)	$\dfrac{T_C}{K}$	$\dfrac{\Pi}{\mu Cm^{-2}K^{-1}}$	ε_r	$\dfrac{\rho S}{MJm^{-3}K^{-1}}$	$\dfrac{\Pi/\varepsilon_0\varepsilon_r\rho S}{mVm^2J^{-1}}$	$\dfrac{\rho_e}{\Omega m}$	$\dfrac{\varepsilon_0\varepsilon_r\rho_e}{s}$	$\dfrac{\alpha}{\mu m^2 s^{-1}}$
PZ (ceramic)	470	350	250	2.6	60	10^8–10^{11}	0.2–220	0.38
PVDF (polymer film)	> 373	30	10	2.4	140	10^{14}	9000	0.05
LT (single crystal)	900	180	54	3.3	110	10^{13}	4800	1.3
SBN (single crystal)	390	650	380	2.3	80	10^{10}	34	≈ 0.3
TGS (single crystal)	320	280	38	2.3	360	10^{13}	3400	0.26

Key: T_C; Curie temperature; Π, pyroelectric coefficient $d\mathscr{P}/dT$; ε_r, relative permittivity; ρ, density; S, specific heat capacity; ρ_e resistivity; α, thermal diffusivity.

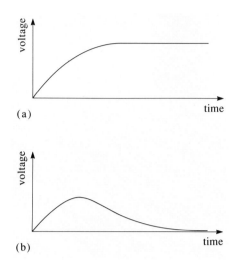

Figure 4.21 (a) Voltage reaching a plateau. (b) Voltage falling back to null as charge leaks away

input. The temperature stops rising, and the voltage stops increasing ($dV/dt = 0$, see Figure 4.21a). Secondly, the new voltage on the detector arises from the production of a certain amount of 'new' charge Δq (this is caused by the new polarization of the dielectric). This 'new' charge leaks away, either through the dielectric or through the input impedance of the amplifier, and eventually the voltage on the detector falls back to its null value (Figure 4.21b).

The net effect is this: the detector does not respond to continuous radiation. It responds to a *change* in received radiation by giving a voltage 'blip' when the radiation changes. The blip will certainly be contaminated by noise, so for safety the blip should be as large and long-lasting as possible. That means minimizing the leakage of charge once it has appeared by, for instance, making sure the dielectric is highly resistive and using a high-impedance amplifier. But even with those refinements, a system that registers single blips is hardly convenient or reliable. We would prefer a system that gives a continuous indication of continuous radiation.

A couple of techniques have proved useful for this. One approach is to have a rotating shutter in front of the detector. Each fresh exposure of the detector produces a new blip, so that over time there is an average, non-null voltage as long as there is radiation falling on the system. (This technique requires that the shutter speed and the thermal response of the device be so matched that the detector 'cools' sufficiently during the period when the shutter is closed. Clearly we want a big dV/dt if the detector is only going to get brief bursts of radiation.)

Another approach is to have radiation reflected onto the detector by a segmented mirror made up of parabolic sections (see Figure 4.22). Each section reflects radiation from a different zone. Provided the burglar is obliging enough to keep moving, the variation in radiation falling on the detector as the intruder moves from zone to zone is registered as fluctuating irradiation, and a sequence of voltage pulses is produced. (A segmented Fresnel lens in front of the detector can give the same effect as the mirror.)

EXERCISE 4.5 Advanced analysis shows that to minimize the effect of leakage through the dielectric, the product RC for the device should be large. A thin pyroelectric element will help increase the capacitance but will decrease the resistance. Does the thickness matter as a determinant of RC? Do these conflict with the desire for a large dV/dT?

Exercise 4.5 has thrown up another merit index, $\varepsilon_0 \varepsilon_r \rho_e$, which governs RC. Table 4.1 has a column for this and you will notice

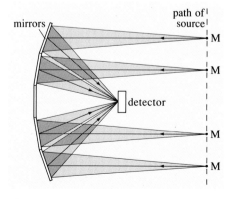

Figure 4.22 Use of segmented mirror to scan movement

that the entries are mostly large numbers. Except for the material PZ in its lowest resistivity form, charge leakage *through the dielectric*, is not a problem. In fact, it's the bias resistor R_1 in the amplifier that is the dominant resistance in the product RC.

So which material should be used? Remember this is a 'consumer durable'; the watchwords are simplicity, reliability and cheapness. Let's look again at the options in Table 4.1 and I think the choice will make itself!

Single-crystal materials are intrinsically expensive, so you need a special reason to choose one.

- TGS (tri-glycine sulphate) is the most sensitive performer but its Curie temperature is only 47°C and under strong sunlight might even get hot enough to 'depole'. It is also mechanically weak and hygroscopic. Forget it.

- LT (lithium tantalate) is famous for its high T_C. Its special field of application is as a sensor in liquid-sodium cooling systems; hardly a big market so there's no benefit from its being in bulk supply.

- SBN (strontium barium niobate) has no perceptible virtues over ceramic materials; why bother?

- PVDF (polyvinylidenedifluoride), interestingly, is both a ferroelectric and a polymer, and so is open to processing routes forbidden to the others. In particular it could come into its own for a large-area thermal imaging system where its low thermal diffusivity and ability to be made very thin would avoid thermal crosstalk (!) between tiny elements on a single sheet. None of that helps here though.

That leaves PZ, or lead zirconate. This is a member of a well used family of ferroelectric ceramics (we shall meet further applications in Section 4.5). As such, it is readily available. Furthermore, as we shall see in Section 4.5, its resistivity can be altered by small modifications to its composition. Thus by deliberately making the dielectric slightly conductive, we can incorporate the amplifier's bias resistor R_1 into the detector itself. This saves the manufacturer the cost and trouble of fitting a resistor. Trivial as this may seem, in multi-element detector arrays such as those used in infra-red viewing and surveillance devices it is a considerable boon. PZ is the material to go for.

4.4 PZT; Composition, structure, properties

The PZT ceramics are oxide alloys of lead, zirconium and titanium; that is, alloys of PbO, ZrO_2 and TiO_2. These ceramics are to ferroelectrics what iron is to ferromagnetics. By manipulating the composition and microstructure of PZT alloys, it is possible to control their properties to suit nearly all technical needs. The familiar perovskite structure of $BaTiO_3$ appears in the alloy system of PZT when there are equal numbers of divalent (Pb^{2+}) and tetravalent (Zr^{4+}, Ti^{4+}) cations. In other words, the perovskites are alloys of the binary system $PbZrO_3-PbTiO_3$. (This emphasizes the similarity with $BaTiO_3$.) Property variations of PZT are mostly a question of varying the proportions of zirconium and titanium. The general formula is then $Pb(Zr_zTi_{1-z})O_3$ with z a fraction running from 0 to 1. Figure 4.23 is the equilibrium phase diagram for the system. To understand the origins of these phases, and the possibilities of modifying them, we need to go down to the atomic scale, where the relative sizes and charges of the ions become important. Table 4.2 gives a range of radii for the ions we shall be concerned with.

Let us review what Figure 4.23 shows. At the top is the undistorted cubic phase. This high-temperature phase is a solid solution of the two components covering all compositions. In this phase there is no

Table 4.2   Radii of various ions

Charge/e	Ion species	Radius/nm
−2	O^{2-}	0.132
+1	K^+	0.133
+2	Ba^{2+}	0.143
	Pb^{2+}	0.132
	Ca^{2+}	0.106
	Mn^{2+}	0.090
+3	La^{3+}	0.106
	Bi^{3+}	0.114
	Mn^{3+}	0.070
+4	Ti^{4+}	0.064
	Pb^{4+}	0.084
	Zr^{4+}	0.087
+5	Nb^{5+}	0.070

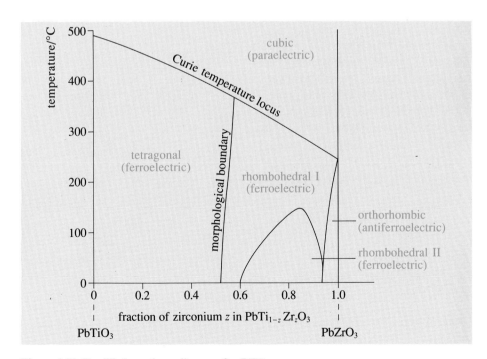

Figure 4.23 Equilibrium phase diagram for PZT

discrimination between octahedra containing Ti^{4+} (edge length 0.28 nm) and those holding Zr^{4+} (edge length 0.31 nm). This 10% variation is accommodated by the rather soggy cuboctahedral cages being somewhat oddly shaped. This phase is *paraelectric*, meaning that an applied field can distort it into a polarized condition, but when the field is removed thermal agitation will quickly relax the crystal back to its unpolarized state. Except when the temperature is close to T_C, the polarization is very weak; but later, in our optoelectric case study, we shall find how it may be used.

As we follow the temperature down on Figure 4.23 we reach the Curie transition. The temperature for the transition varies with composition, but for all compositions it is higher than for $BaTiO_3$ (in which T_C is 125°C). This can be explained in terms of atomic bonding. Ba^{2+} ions have a full outer shell of electrons (inert-gas structure), so can interact with their O^{2-} neighbours by electrostatic bonding only. But Pb^{2+} ions still have two outer electrons beyond their last full shell, and these can participate in covalent bonding with the oxygen neighbours. With this extra bonding, a lot more rattle can be sustained before the low-temperature, polarized, ordered crystals collapse into their higher entropy state. This directional covalent bonding also influences the ferroelectric polarization and is the underlying cause of PZT being a so much more versatile material in transducer contexts than $BaTiO_3$.

In the low-temperature, polarized crystals (below T_C) the size difference of ZrO_6 and TiO_6 octahedra *is* significant because the ability of the octahedra to distort must depend on their oxygen-ion spacing. For compositions at the titanate end of the phase diagram, the polarized form is tetragonal (that is, like a cube stretched parallel to an edge). At the zirconate end, zirconium-rich alloys distort to rhombohedral crystals (like a cube stretched along a body diagonal). These rhombohedral crystals undergo further changes at lower temperatures. Much technical attention is focused upon ceramic compositions near the 'morphotropic boundary' separating the tetragonal and rhombohedral behaviours since optimum piezoelectric properties are found here (more on this in Section 4.6). Compositions within 10% of pure $PbZrO_3$ become orthorhombic (tilted cube) and are actually 'antiferroelectric'. This means that adjacent crystal cells are oppositely polarized, so there is electrical ordering but no net polarization.

PZTs are never used without extensive tuning of properties by judicious composition modification. What tricks can be easily explained?

SAQ 4.7   (Objective 4.5)
What composition of PZT would you recommend for a pyroelectric burglar detector. The phase diagram of Figure 4.23 will help you to choose.

SAQ 4.8   (Objective 4.5)

(a) What is the Curie temperature of $PbTi_{0.4}Zr_{0.6}O_3$?

(b) What angles are to be found between the directions of polarization in adjacent domains of $PbTi_{0.4}Zr_{0.6}O_3$?

(c) What shape is the crystal cell of $PbTi_{0.6}Zr_{0.4}O_3$ at room temperature?

(d) What conditions of polarization would result in a specimen of $PbTi_{0.6}Zr_{0.4}O_3$ after the treatments below?

(i) Strong electric field applied at 480°C, held for some hours, and switched off. Specimen cooled to 20°C.

(ii) Strong electric field applied at 350°C, held for some hours, specimen cooled to 20°C, then field switched off.

(e) Why is it difficult to pole $PbTi_{0.97}Zr_{0.03}O_3$?

(f) Why is it impossible to pole $PbTi_{0.03}Zr_{0.97}O_3$?

(g) What crystal structure would $PbTi_{0.47}Zr_{0.53}O_3$ have after being cooled quite quickly from 500°C and crossing the morphological boundary at 50°C?

4.4.1 Additives to PZT

Small quantities of other metal oxides brought into PZT can modify many properties, such as conductivity, Curie temperature, coercivity, compliance. Even properties not beginning with letter C can be manipulated! Essentially there are four substitutional options for cations within the corner-shared oxide-ion tetrahedra in the perovskite crystal:

• Isovalent cations — that is, other $2+$ ions replacing Pb^{2+}; or other $4+$ ions replacing Zr^{4+} and Ti^{4+}.

• Higher valent cations — that is, $3+$ ions replacing Pb^{2+}; or $5+$ ions replacing Zr^{4+} and Ti^{4+}.

• Lower valent cations — that is, $1+$ ions replacing Pb^{2+}; or $3+$ ions replacing Zr^{4+} and Ti^{4+}.

• Variable valency cations on either site.

Rather than looking at these substitutional options in isolation, we shall look at them in the context of their use for controlling the sorts of properties mentioned above. Even the experts consider property control by these substitutions to be something of a black art, but we should at least aim for a plausible account of some successful techniques.

Substitutions to control conductivity

PZT alloys prepared without additives are p-type extrinsic conductors. This property is conferred by O^- holes, the origin of which is subtle; see ▼p-Type conductivity of PZT▲ for a description. Unless this conductivity can be suppressed, it is difficult to sustain a strong enough electric field in the material to pole it.

▼p-Type conductivity in PZT▲

Firstly, a reminder not to confuse holes with oxygen-ion vacancies. In this context, 'a hole' is an O^- ion which would prefer to be in its O^{2-} ionization state (because that would give it an inert-gas electron structure). The O^- is looking for another electron, so we say it has a hole. Electrons hopping into holes constitute a p-type conduction mechanism. An oxygen vacancy, on the other hand, is the complete absence of an oxygen ion. An oxygen ion diffusing into the vacancy is carrying charge, so this is another (but not a p-type) conduction mechanism.

To follow the development of O^- ions in PZT we should start at the calcining stage of alloy production. Appropriate amounts of PbO, ZrO_2 and TiO_2 powders are intimately mixed and fired to produce the new compound with perovskite crystal structure. Ideally all the lead ions stay as Pb^{2+} and the other cations as Zr^{4+} and Ti^{4+}. But the free energy of the system will be lower if all opportunities for increased entropy are taken up. Significant among these is the possibility of lead ions being Pb^{4+} and fitting into the octahedral sites.

After calcining the material is therefore

$$Pb^{2+}(Zr, Ti, Pb)^{4+}O_3$$

with perhaps only a few per million octahedral cages holding Pb^{4+} ions. The next hot-processing stage is sintering. Some decomposition of the perovskite occurs with loss of PbO, which is rather volatile. The first effect of this loss is to generate both lead and oxygen vacancies in the lattice. More importantly, *some* of the lead lost from the crystal is from those few Pb^{4+} ions. To leave as PbO, these ions must pick up two electrons from the oxygen ions, thus

$$Pb^{4+} + 3O^{2-} \rightarrow Pb^{2+}O^{2-} + 2O^-$$

You can see that this generates O^- ions, which are the holes giving p-type conductivity. Their concentration is of the order of parts per million, whereas that of the oxygen vacancies is *much* higher — perhaps several parts per thousand. But, at ordinary temperatures, hole mobility is so much greater than oxygen-vacancy mobility that p-type conduction dominates in the perovskite structure.

As 'p-Type conductivity of PZT' explained, the loss of PbO during sintering results in non-stoichiometric crystal with many lead and oxygen vacancies, and some O^- ions which act as holes. Adding lanthanum oxide La_2O_3 fills the oxygen vacancies *and* supplies the O^- ions with the extra electron they need to become O^{2-}. Lanthanum is a rare earth element. All the rare earths form large trivalent ions, and La^{3+} is the biggest at 0.106 nm.

EXERCISE 4.6   Would you expect La^{3+} to substitute for Ti^{4+} and Zr^{4+} on octahedral sites, or for Pb^{2+} on cuboctahedral sites (see Table 4.2)? What difference would it make to the population of oxygen-ion vacancies if the lanthanum ion went onto one site rather than the other?

SAQ 4.9   (Objective 4.7)
Which of these oxides Nb_2O_5, Mn_2O_3 and K_2O might also be used to supply oxygen to PZT? Take the ionic sizes from Table 4.2.

So lanthanum oxide fills oxygen vacancies and suppresses any conductivity arising from them. But the p-type holes still need to be attacked if we are to increase resistivity.

Because La_2O_3 brings extra oxygen to the crystal there will be a tendency for oxygen to be lost to the atmosphere during sintering. It enters the system as *ions* and leaves as *molecules*:

$$2O^{2-} \rightarrow O_2 + 4e^-$$

Here is a source of electrons for the O^- ions:

$$O^- + e^- \rightarrow O^{2-}$$

The holes are filled and p-type conductivity suppressed.

So we have a way of reducing the excessive conductivity of naturally oxygen-deficient PZT. But there are circumstances where a controlled amount of conductivity is desirable. This can be achieved by gently encouraging a little p-type conduction. Introducing uranium as U^{6+} will do the trick. Such a strongly positive ion substituted onto the octahedral sites would attract the surrounding oxygen ions very strongly and produce local distortion. The strain energy of the distortion could be reduced by an energy trade-off between chemical energy and strain energy in which some electrons transfer to the uranium ions from the oxygen ions:

$$U^{6+} + 2O^{2-} \rightarrow U^{4+} + 2O^-$$

This transfer of electrons produces holes (on the O^-), which restores conductivity.

Figure 4.24 shows the delicate control which is possible. One application is in the pyroelectric burglar trap where, by introducing a small amount of leakage into the PZT dielectric, the necessary bias resistance of the amplifier can be 'built in' to the detector.

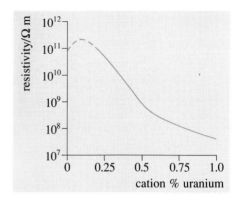

Figure 4.24 Resistivity versus atomic % uranium in PZ

Substitutions to control domains

At the titanate end of the PZT phase diagram (Figure 4.23), towards pure $PbTiO_3$, the tetragonal long axis is 6% greater than the transverse axis. So great is this change that the stresses produced during polarization usually shatter the ceramic. Substituting Zr for Ti, then, can be thought of as a way of controlling the domains. Near the middle of the phase diagram, the crystals can cope with the strains caused by the developing domain structure.

In connection with Figure 4.19 we saw that to get a stable domain configuration in a poled ceramic you had to let it age quickly to relieve the internal stresses set up during poling. This was to be done by having the oxygen sublattice of the crystal free of imperfections, but had the consequence of reducing the coercivity. We have already seen how higher valent substitutions fill oxygen vacancies. So if lanthanum has been added to control conductivity, it will also have had the effect of increasing domain wall mobility. When the engineer wants a material with high coercivity, domain walls have to be pinned by some further addition.

The trick is to introduce *lower* valent cations into either cuboctahedral or octahedral cages. Take K^+ substituting in place of Pb^{2+} as an example. Arriving as K_2O this cation brings less oxygen than the crystal requires so some oxygen vacancies reappear. Now an oxygen vacancy is a point in the crystal deficient in negative charge — there should be an O^{2-} ion there. The cation K^+ compared with the Pb^{2+} is a point deficient in positive charge. It is energetically favourable to preserve local charge balance in the crystal, so it is best if the deficiencies of negative and positive charge stay close together.

As a pair, K^+ and O_{vac} (representing the oxygen vacancy), form an electric dipole so within its domain there will be a lowest-energy alignment for the pair with the dipole lined up parallel with the internal field of the domain. This alignment will be accomplished by diffusion of oxygen ions around the K^+ site until the best site for the vacancy is found. We may have perhaps 1% of the cuboctahedral cages carrying these dipoles and uniformly dispersed through every domain with appropriate alignment. Now consider the energetics of changing the orientation of the domain polarization. All the K^+O_{vac} dipoles in the new domain would be the wrong way round! A significant energy cost for switching the domain has been built in and the existing domain structure is locked in place. An equivalent mechanism uses Co^{3+} substituted onto octahedral $(4+)$ sites. These domain-pinning additives reduce the ageing rate of poled ceramics and increase their coercive fields. You will remember that making hard permanent magnets also depended on techniques for pinning domain walls.

Substitution for Curie temperature control

The high Curie temperatures of PZT alloys have already been explained as a consequence of the Pb^{2+} ion having covalent possibilities because it does not have a full electron shell at its outer skin. The route to lower T_C is therefore to replace lead ions by some which do have a full shell, and Group II elements such as Ba or Sr are the obvious possibilities. These work — but so does our old favourite lanthanum, which we are likely to want in the material anyway because of its other useful effects. The rare earths are transition elements where the deep-lying 4f electron shell is filling while $5s^25p^65d^16s^2$ buzz round outside. In lanthanum the 4f shell has no electrons, so when it has ionized by losing its 5d and 6s electrons the outside world sees the 'inert gas' octet $5s^25p^6$, like xenon. Because it is smaller than Pb^{2+} and generates lead vacancies because it is trivalent — so further weakening the lead bonding — its effectiveness in pulling down T_C is enhanced. It reduces T_C by 25 to 40 K per added percent, so with some 10% lanthanum T_C is below room temperature.

Since lanthanum appears as an additive to PZT in so many applications, many texts refer to PLZT as the basic 'family' of ferroelectrics. The general formula for the material then becomes $Pb_{1-1.5y}La_yTi_{1-z}Zr_zO_3$. The composition map in Figure 4.25 shows the phases corresponding to

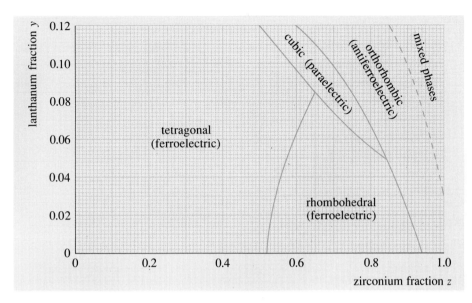

Figure 4.25 PLZT phases stable at room temperature. The values of y and z specify the composition in the general formula $Pb_{1-1.5y}La_yTi_{1-z}Zr_zO_3$

different values of y and z at room temperature. A composition with, for example, the formula $Pb_{0.865}La_{0.09}Ti_{0.35}Zr_{0.65}O_3$ is sometimes specified as 9/65/35, the numbers representing y, z and $1-z$ expressed as percentages.

The zone where the cubic phase abuts the ferroelectric phases represents strongly paraelectric materials whose polarization can be rapidly switched on and off by an applied field. That is the basis of action of the electro-optic device in the ▼Flash goggles▲ case study.

SAQ 4.10   (Objective 4.7)
Which of the additive ions listed in the key would produce which of each of the following effects in PZT?

(a) Reduce Curie temperature.
(b) Increase the ageing rate of a poled ceramic.
(c) Increase coercivity.
(d) Reduce p-type conduction.
(e) Increase p-type conduction.

Key

(i) La^{3+} substituting for Pb^{2+}.
(ii) K^+ substituting for Pb^{2+}.
(iii) Ba^{2+} substituting for Pb^{2+}.
(iv) Cr^{6+} substituting for Ti^{4+}/Zr^{4+}.
(v) Co^{3+} substituting for Ti^{4+}/Zr^{4+}.
(vi) Nb^{5+} substituting for Ti^{4+}/Zr^{4+}.

▼Flash goggles▲

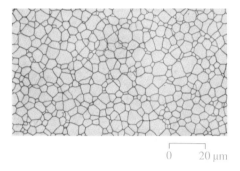

Figure 4.26 Microstructure of hot pressed PLZT (with composition $Pb_{0.865}La_{0.09}Ti_{0.35}Zr_{0.65}O_3$)

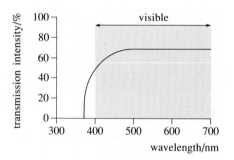

Figure 4.27 Transmission versus wavelength curve for PZT (with composition $Pb_{0.865}La_{0.09}Ti_{0.35}Zr_{0.65}O_3$)

After it had been discovered that PLZT could be made strongly paraelectric at room temperature the search was joined to make it transparent as well, for it was realised that the electric polarization could interact with the electric field vector of light transmitted through it. The secret of transparency is to avoid microstructural features such as pores and grain boundaries which can scatter light.

What size structural features would scatter light?

The size to avoid is anything around the wavelength of the light since this would 'diffract' the light. Visible light has wavelengths in the range 0.4 to 0.6 µm. Hot pressing, that is sintering under pressure, can produce thin sheets with

microstructure shown in Figure 4.26. Grain sizes range from 2 to 10 µm, there is no porosity and, by careful control of purity of ingredients, fine precipitates of other phases at grain boundaries have been avoided. Figure 4.27 shows a light-transmission versus wavelength curve for one particular composition of PLZT (actually $Pb_{0.865}La_{0.09}Ti_{0.35}Zr_{0.65}O_3$). Notice how lucky we are that the cut off is right at the extreme blue end of the visible spectrum. With anti-reflection coatings, the loss due to surface reflection can be cut to as little as 2%.

Figure 4.28(a) shows the intention of the design. The electro-optic plate has a polaroid sheet on each side (P₁ and P₂) and these are crossed. In what follows you have to be careful to distinguish 'optical

polarization' of the light wave by the polaroids, meaning the alignment of the electric vectors of the electromagnetic light wave, from 'electrical polarization' of the PLZT plate, which is its paraelectric response to an applied electric field. When the PLZT plate is not electrically polarized, light passes through it without effect. The first polaroid has aligned the electric vector of the light waves so that they cannot now get through the second polaroid. Transmission can be as low as 0.0015% of incident intensity with efficient polaroids. Now let's put a voltage on the PLZT plate so that it is electrically polarized in its plane. The *wave* electric vector can be thought of as having components parallel and perpendicular to an electrical polarization in the PLZT plate. This electrical polarization slows one

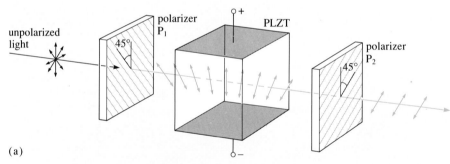

(a)

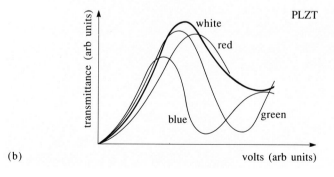

(b)

Figure 4.28 (a) Principle of operation of flash goggles. (b) Transmission versus voltage curve for white light

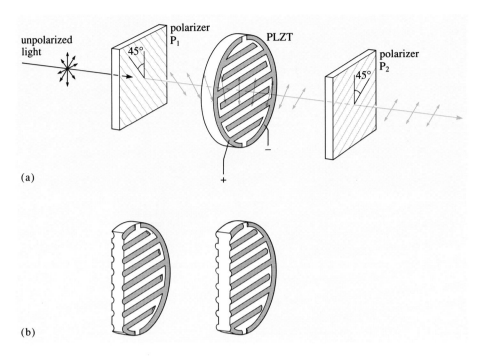

(a)

(b)

Figure 4.29 Electrode arrangement for in-plane polarization

component of the wave relative to the other. Recombining the components on the far side of the PLZT plate, we find the plane of optical polarization has been rotated. The wave can now pass through the second polaroid, at an intensity depending on the applied voltage (Figure 4.28b). So to make flash goggles work, the 'open' (transparent) condition is set up with a polarizing field on the PLZT. Typically the transmission will be as low as 20% because the polaroid sheets absorb rather a lot of light. Since the eye has a logarithmic response to brightness, that is easily accommodated. A photodetector is mounted on the goggles to sense any intense flash and signal via the electronics to switch off the field. The paraelectric polarization relaxes within $\approx 150\ \mu s$ and the goggles go opaque. As soon as the hazard has gone, the field switches back on and opens transmission — again with a delay of the order of $\approx 150\ \mu s$. The very fast response, light weight and deep contrast of some 5000 to 1 between open and shut states makes this the best anti-flash-blindness system yet devised. Figure 4.29 sketches the arrangement of the electrodes which apply the field to the PLZT. The grey electrodes are actually fine lines, spaced at about 1 mm. They are too close to the eye to be focused so do not interfere with vision, and they only block a small fraction of the aperture. Electrode patterns, plate thickness and voltage characteristics are the detailed design matters which make the system work well.

SAQ 4.11   (Objective 4.9)
The general formula for PLZT is $Pb_{1-1.5y}La_yTi_{1-z}Zr_zO_3$. The composition of PLZT used for flash goggles has $y = 0.09$ and $z = 0.65$. Locate this composition on the phase diagram of Figure 4.25. Why would the following compositions be unsuitable for this application?

(a) $y = 0.06$, $z = 0.65$;

(b) $y = 0.12$, $z = 0.65$.

Summary of Section 4.4

PZT is a widely used ferroelectric. Additives give control of a range of properties.

An important additive to PZT is lanthanum oxide. PLZT is PZT doped with lanthanum oxide.

Lanthanum oxide has these effects on PZT:

• It fills oxygen vacancies in PZT, making domain walls mobile so that the polarization of poled ceramics can quickly stabilize.

• It provides spare electrons to fill 'holes' and suppress p-type conductivity.

• It reduces T_C to an extent that can enable paraelectric effects at room temperature. Interaction with the electric vector of light enables PLZT to be used in opto-electric modulators.

Potassium oxide in PZT fixes domain walls and increases coercivity.

4.5 Piezoelectric effects

The last phenomenon I want to discuss in this chapter is the **piezoelectric effect**. This is the direct interconversion of mechanical to electrical energy via a material which changes its state of polarization when stressed (the 'generator' effect) or which changes its shape when its ambient electric field changes (the 'motor' effect). The greatest use of PZT is in piezoelectric devices, for a host of applications ranging from gas igniters, gramophone pickups and hydrophones using the generator effect, to ultrasonic drives, sound sources and micropositioners using the motor effect.

A polarized ferroelectric ceramic specimen will be a bit longer in the poling direction than it was when unpolarized and the field of all the domains summed to zero. This is because poling biases the direction of the elongated axes of the crystal grains. (The transverse dimensions will be shorter since there are fewer grains with their long axes this way.) Figure 4.30(b), the **butterfly curve**, traces the strain changes along the applied field direction as a specimen is taken round its hysteresis cycle (Figure 4.30a). Its slope at any point, the rate of change of strain s with field $\mathscr{E}$, is a piezoelectric 'motor' coefficient m,

$$m = \left(\frac{\partial s}{\partial \mathscr{E}}\right)_\sigma$$

Notice that m is zero for a ceramic in zero field in the virgin, unpolarized condition so the technically useful state is a poled piece in zero field (corresponding to point $\mathscr{P}_r$ in Figure 4.30a). From Figure 4.30(b) you can see that, around this point, the strain is proportional to applied field over quite a range. Piezoelectric devices are operated in this range, care being taken that drive fields do not exceed the coercive field which would

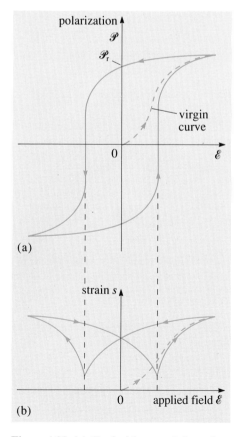

Figure 4.30 (a) Typical hysteresis loop for a high coercive field. (b) Associated strain ('butterfly' curve)

switch the domains and 'depole' the piece. Part of our quest in selecting materials for specific application in our case studies will be to explore how to control this piezoelectric coefficient.

The opposite piezoelectric effect is when an applied stress causes a mechanical strain s, which alters the polarization $\mathscr{P}$. We can define a coefficient for this 'generator' effect;

$$g = \left(\frac{\partial \mathscr{P}}{\partial s} \right)_{\mathscr{E}}$$

Since stress and strain are connected via the Young's modulus E of the material, $s = \sigma / E$,

$$g = E \left(\frac{\partial \mathscr{P}}{\partial \sigma} \right)_{\mathscr{E}}$$

The stimuli, either a mechanical stress or an electric field, are vectors. But the material gives both parallel and perpendicular responses. To be rigorous (and of practical use), we need three-dimensional expressions of these piezoelectric coefficients. My more carefree equations can only be used for the parallel stimulus and response.

The most important piezoelectric characteristic is its **coupling coefficient** k, expressing the efficiency of energy converted between mechanical and electrical forms. It is defined by

$$k^2 = \frac{\text{mechanical energy output}}{\text{electrical energy input}} = \frac{\text{electrical energy output}}{\text{mechanical energy input}}$$

which for a poled ceramic can be shown to be

$$k^2 = \frac{E m^2}{\varepsilon_0 \varepsilon_r}$$

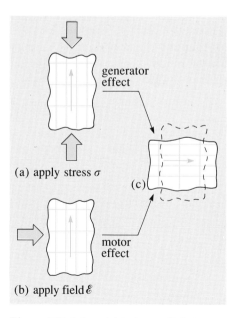

To explain the piezoelectric effects think first about a single grain of the material in isolation. Let's suppose it's a material which polarizes by a tetragonal distortion of its crystal structure. Figure 4.31(a) shows a single grain polarized in a certain direction and subjected to an external stress. The rectangular grid on the diagram symbolizes the tetragonal crystal cells. How can this grain respond to the stress? It needs to get shorter along the direction of the stress and can do this by a crystallographic transformation setting the long axes of the cells at 90° to their original direction (Figure 4.31(c)). Now, of course, since the ion displacements are directly related to the tetragonal long axis, the polarization of the grain has changed. This is the origin of the generator effect. The motor effect is the exact mirror. When an electric field is applied transverse to the direction of the polarization of the grain it will be energetically favourable for the polarization to realign with the new field. Again to do this the crystal cells have to change their long axes and the grain changes its shape; a mechanical response to the electrical stimulus (Figure 4.31(b) and (c)).

(a) apply stress σ

(c)

generator effect

motor effect

(b) apply field $\mathscr{E}$

Figure 4.31 (a) and (c) An applied stress induces a strain which causes a change in polarization (generator effect). (b) and (c) An applied electric field induces a change of polarization which causes a change of shape (motor effect)

The piezoelectric responses are very fast — transducers can follow signals at many megahertz frequency. The mechanism within the material cannot therefore rely on slow thermal diffusion from one structure to the other. Rather we are looking at a shear transformation in which the applied stress drives dislocations through the crystal. You will have met at least one other example of such a transformation.

EXERCISE 4.7   What happens when steel is rapidly cooled from a high temperature at which its structure is FCC (face-centred cubic)? How is this different from the consequence of slow cooling?

The straightforward argument for an isolated grain is seriously modified for grains within a polycrystalline ceramic body. Attempts by each grain to change shape are strongly impeded by its neighbours. Low energy conversion efficiency (k^2) is the result. Such small changes of shape which can be accomplished involve plastic deformation at grain boundaries so that the new grain shapes can conform to one another. Alternatively the internal stresses generated as the grains try to change shape can induce cracking. Remembering that some transducers are vibrators, fatigue fracture becomes a hazard. Ceramics for these applications are carefully processed to avoid pores which can act as stress concentrators and initiate fracture.

The advantage of PZT as a piezoelectric material is that the energy conversion efficiency depends strongly on the crystal morphology of the ceramic. In PZT formulated close to the rhombohedral–tetragonal morphologic transition the coupling coefficient climbs to a very usable maximum (Figure 4.32). The suggested reason for this peak is that here,

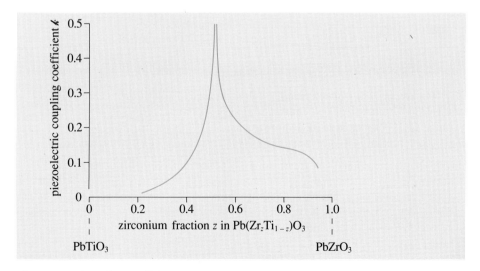

Figure 4.32 Coupling coefficient versus composition for PZT

in response to electric or mechanical stimuli, the grains can find shear transformations between the different cell types as well as between the easy directions of either type. This provides much easier mutual conformation of grain shapes because there are so many more 'easy directions' of polarization for adjacent grains to choose from (Figure 4.16).

The ▼Sonar▲ case study demonstrates piezoelectricity in action.

SAQ 4.12   (Objective 4.9)

A spark igniter consists of a block of piezoelectric ceramic measuring 2 mm square by 1 mm thick, with electrodes on its square faces. A load of 20 N can be applied to the square faces by thumb pressure via a lever. Use

$$g = E\frac{\Delta \mathcal{P}}{\Delta \sigma}$$

and

$$C = \frac{\varepsilon_r \varepsilon_0 A}{t}$$

to assess whether the block can produce enough voltage to make a 1 mm spark in air (breakdown voltage 30 kV cm^{-1}). The characteristics of the ceramic are:

- Young's modulus E $= 70$ GN m^{-2}.
- Generator coefficient $g = 30$ C m^{-2}.
- $\varepsilon_r = 70$ (appropriate for the capacitor using a poled ceramic dielectric).

Take ε_0 to be 9×10^{-12} F m^{-1}.

Summary of Section 4.5

Poled PZT is commonly used for its piezoelectric effect.

Applied mechanical stresses or electric fields can shift the directions of crystal distortion onto alternative axes. Where this happens, domain directions change, causing a net change in polarization of the material. This is the basis of the piezoelectric effect.

The piezoelectric effect may be either a 'motor' or a 'generator' effect, depending on the direction of transduction. The coupling coefficient expresses the efficiency of transduction.

Design of a practical piezoelectric device must take account of materials properties such as strength, fatigue resistance, thermal properties, and of product features such as electrical and mechanical impedance. Composite construction can improve product performance.

▼ Sonar ▲

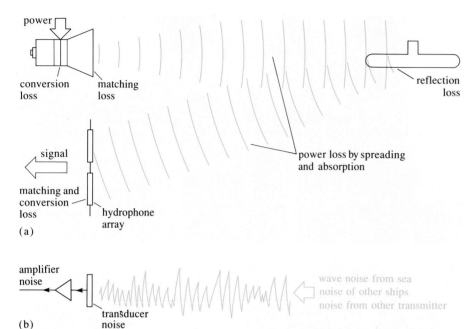

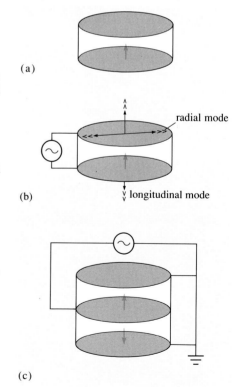

Figure 4.33 (a) Power losses for transmitted signal and echo. (b) Noise sources

In this case study I am going to consider the problems of sonar underwater detection at a sufficiently general level to show the product–property connections for its two transducers. The principle is simple: a piezoelectric 'motor' transducer produces an underwater beam of sound, and a 'generator' transducer is used to detect the echo and transfer it into the electronics. For a close-range system, such as a depth sounder for inshore navigation, it's easy to combine the functions into a single device; but long-range surveillance, for example the detection of submarines, is much more difficult. The performance of secret military installations can only be conjectured at, but we can readily deduce that a range of 20 km isn't going to be easy to achieve. Covering a strategically important seaway, such as that between Scotland and Iceland (a distance of 800 km), would therefore require a string of detectors — at least 40 if a range of 20 km per detector could be achieved, and more if a level of redundancy is to be built in.

The basic technical problem is to achieve adequate signal-to-noise ratio. The echo from a distant target is a very small fraction of the signal originally sent out. Figure 4.33(a) reviews the losses. If 10 kW,

say, of sound power can be transmitted, the phone must be sensitive to perhaps no more than 0.1 µW of echo power. As Figure 4.33(b) suggests, some noise sources might swamp the signal. A technique called pulse code modulation greatly enhances the possibility for sorting signal power from noise (because the detection system 'knows' what pattern of pulses constitutes a true echo). Nevertheless, among the options always to be explored are increased transmitted power and more sensitive phones. Parts of both strategies depend on the property suites of available piezoelectric materials.

Both transducers are (or can be) a slab of poled PZT ceramic carrying metal electrodes on the pole faces. In the 'motor', alternating electric fields presented to the electrodes induce mechanical vibration in the ceramic which have to be coupled out to the surroundings. In the 'generator', a pressure wave arriving at the ceramic sets it in oscillation and a voltage appears across it. Clearly both require efficient energy transfer (the coupling coefficient k must be as large as possible) so compositions near the morphotropic boundary are attractive. But the power output transducer needs $ds/d\mathscr{E}$ to be large,

so that feasible drive voltages produce strong vibrations; and in the hydrophone, a large $d\mathscr{P}/d\sigma$ will give a respectable signal. These distinctions call for different materials. And there are other problems too. Let's develop a design for the sound wave generator first.

Consider a disc of piezoelectric ceramic poled through its thickness and electroded on its circular faces (Figure 4.34a). When an alternating field is applied, the disc responds mechanically with longitudinal and radial vibrations (Figure 4.34b). We need to concentrate the power into a single vibration mode and we'll choose a 'piston-like' longitudinal one. Geometry determines the dominant mode and there is a strong and isolated longitudinal vibration when the thickness of the disc is about 0.4 of its diameter. Next let's make electrical insulation easy by putting two oppositely poled discs on the two sides of the live electrode and earth the outer faces (Figure 4.34c). Also, that trick has neatly doubled

Figure 4.34 (a) Direction of poling. (b) Longitudinal and radial modes of vibration. (c) Sandwich configuration the volume of ceramic available to carry the power. So is that the design? Dip it in the sea and switch on. I'm afraid it won't do. Exercise 4.8 indicates a list of problems that have to be designed around.

EXERCISE 4.8   This exercise should get you to see the problems all together. Although you won't be able to offer a design which solves them all simultaneously, you can think of each in turn and consider whether the solution is through 'materials', 'geometry' or 'electronics'.

(a) The vibrations alternately stretch and compress the ceramic. How well do ceramics sustain tensile stress?

(b) Vibration induces fatigue fracture. What features will encourage such failure?

(c) The piezoelectric energy conversion efficiency is at best 50% and we are contemplating a high-power transfer device. Where does the rest of the energy go? With what consequences?

(d) What controls the transfer of power from the electrical circuit into the ceramic?

(e) What controls the transfer of power from the ceramic into the sea water?

(f) What limits the amplitude of the driving electric field?

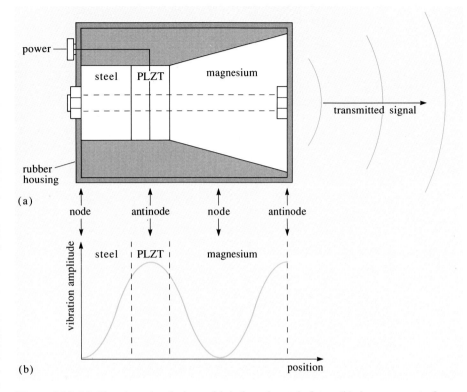

(a)

(b)

Figure 4.35 (a) Signal sender design, with bolt and metal plates. (b) Arrangement of nodes and antinodes in (a)

With the answers to Exercise 4.8 in mind we can follow the features of the design in Figure 4.35. In the middle of the assembly are the back-to-back ceramic slabs, with the electrical connection going to the central electrode. The most obvious extra features are the two slabs of metal (steel and magnesium) and the bolt through the middle of the whole stack.

The bolt is to put the ceramic under compression so that the piezoelectric oscillations never take it into tension. That's solved our first problem. The surfaces of the metal and ceramic pieces must be flat and parallel lest the clamping pressure be uneven and produce a stress concentration. Fatigue is still possible from pre-existing flaws or any generated by the piezoelectric process, so a fine grained and pore-free ceramic must be used.

The metal slabs are dimensioned so the whole acts like an organ pipe, with a node of vibration at the back surface and an anti-node at the front. This ensures that all the sound goes out of one end of the sounder. The graphs in Figure 4.35(b) show the quarter wavelengths and half wavelengths which achieve this. By setting this arrangement of nodes and antinodes we match the mechanical impedance to that of the drive circuit. Of course, the device is now a mechanical resonator and must be designed for an appropriate frequency. Sea noise falls with increasing frequency up to ≈ 50kHz, but attenuation of the transmitted wave rises with frequency. It seems likely that around 10 kHz would be favoured.

The front metal slab of magnesium and the rubber seal help the acoustic-impedance matching of the transducer to the sea water.

Finally, let's choose a PZT. High coupling efficiency and coercivity are the dominant factors. Potassium-doped tetragonal-phase material has the best coercivity; but the composition must also be close to the morphotropic boundary to get a high coupling coefficient. Hot pressing of an ultra-fine powder will give the small-grained, zero-porosity ceramic required; and to suppress the possibility of fatigue, we need to avoid poor stoichiometry at grain boundaries. One way to do this would be to have a furnace atmosphere of lead oxide.

I wonder how many military secrets I've guessed correctly in these paragraphs. One bet is sure: the design must be obsolete or I would not have found it published.

Hydrophone designs are not so easily found, except for rather primitive examples. We shall confine ourselves to the material selection. The fundamental requirement is to get the best voltage signal δV from a tiny impressed stress $\delta\sigma$. As with the burglar detector, the approach is to regard the element in Figure 4.36 as a

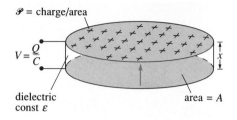

Figure 4.36 Receiving transducer considered as capacitor

capacitor. Since the capacitance is $C = \varepsilon A/x$ and the charge on it due to the polarization of the piezoelectric material is $\mathscr{P}A$,

$$V = \frac{Q}{C} = \frac{\mathscr{P}x}{\varepsilon}$$

The voltage sensitivity then becomes

$$\frac{\mathrm{d}V}{\mathrm{d}\sigma} = \frac{x}{\varepsilon}\frac{\mathrm{d}\mathscr{P}}{\mathrm{d}\sigma} + \frac{\mathscr{P}}{\varepsilon}\frac{\mathrm{d}x}{\mathrm{d}\sigma}$$

Now you can see the voltage sensitivity depends on both the piezoelectric coefficient $\mathrm{d}\mathscr{P}/\mathrm{d}\sigma$ and on the compliance $\mathrm{d}x/\mathrm{d}\sigma$ of the material. Since $\mathscr{P}$ and x both change the same way under stress, the two terms enhance one another. The question is, how to make best use of this. Evidently we should choose a compliant piezoelectric material with a good value of $\mathrm{d}\mathscr{P}/\mathrm{d}\sigma$. That gives us a 'squashy' capacitor for the second term and a useful polarization change. The polymeric ferroelectric PVDF cited in Table 4.1 is one candidate and has been successfully used. In the PZT series, we are led to the morphotropic boundary again for the best coefficients but this time to the rhombohedral side to take advantage of the higher compliance of this crystal.

But a better trick is a composite of stiff piezoelectric rods embedded in a compliant matrix such as an epoxy resin (Figure 4.37). Analysing the voltage sensitivity of

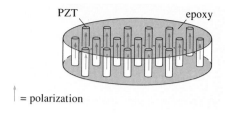

Figure 4.37 PZT rods in matrix composite

this arrangement is much more complex than for the simple case of Figure 4.36. The outcome is that the compliant matrix doesn't just provide a geometrical change of capacitance. In addition, if the difference in compliance of the matrix and PZT is large, and if the area fraction of the rods is about 10% of the total, the voltage sensitivity is increased by an order of magnitude. This is just one example from the quite new field of composite electroceramics.

Objectives for Chapter 4

4.1   Give simple explanations of conduction by O^{2-} diffusion and by electron and hole conduction in ionic materials (oxides). (SAQ 4.3)

4.2 Describe how conductivity in ionic materials may be controlled and give examples of practical methods. (SAQ 4.4)

4.3   Describe the modes of action of voltaic and resistive oxygen sensors and account for the materials selection in each. (SAQ 4.2)

4.4   Describe common features of ferroelectric and ferromagnetic phenomena. (SAQ 4.5)

4.5   Sketch the phase diagram of PZT and know the electrical properties of the various phases. Hence interpret the materials selection for a variety of typical applications. (SAQs 4.7 and 4.8)

4.6   Describe capacitive transduction using a pyroelectric radiant heat detector as an example. (SAQ 4.6)

4.7   Discuss the roles of additives to PZT and explain how they can control a range of technically important characteristics (such as conductivity, Curie temperature, coercivity, ageing, piezoelectric coefficients and transparency). (SAQs 4.9 and 4.10)

4.8   Identify the composition zone in the PLZT composition map where there is strong para-electric behaviour and explain the principles of application of this material as an electro-optic modulator. (SAQ 4.11)

4.9   Distinguish between 'motor' and 'generator' piezoelectric effects, giving examples of practical applications of each. Explain the origins and limitations of piezoelectricity in poled ferroelectric ceramics. (SAQ 4.12)

Answer to exercises

EXERCISE 4.1 The crystal structure of the solution must be the same as that of the host. The amount added must not be so much as to induce the formation of a new phase. For one substance to be extensively soluble in the crystal state of another, the materials should be chemically similar, and their atoms (or molecules or ions) should be of similar size to those of the host. This ensures that not too much crystal distortion is caused.

For an oxygen sensor, as we have already seen, the dissolving cations should carry a different charge from the host-crystal ions. This is our way of making the oxide conduct. Let us not try to make them too different though. In the example of MO_2, the host cations will be M^{4+}. The M' oxides suggested will be $3+$. The other key factor is the size of the substituting ions since this will control which sites in the host lattice are occupied by the solute ions.

EXERCISE 4.2 The fluorite structure of ZrO_2 is the loosest of the three shown.

In the light of the doping argument, you should have chosen as an additive an oxide of a lower-valent metal. This would promote oxygen vacancies. M'_2O_3 or $M'O$ would do the job in zirconia. If you chose a different crystal structure, you should still have chosen a dopant of lower cation valency.

EXERCISE 4.3 The 0,0 state would correspond to unmagnetized ferromagnetic material. Soft materials rest close to this condition.

The 0,R state would correspond to permanently magnetized material.

EXERCISE 4.4 Triangle $A\hat{X}B$ is redrawn in Figure 4.38(a). From this sketch we see that

$$\sin \tfrac{1}{2}A\hat{X}B = \frac{a/2}{a\sqrt{3}/2} = \frac{1}{\sqrt{3}}$$

$$\tfrac{1}{2}A\hat{X}B = 35.25°$$

$$A\hat{X}B = 70.5°$$

Similarly the sketch of triangle $A\hat{X}C$ in Figure 4.38(b) shows that

$$\sin \tfrac{1}{2}A\hat{X}C = \frac{a\sqrt{2}/2}{a\sqrt{3}/2} = \frac{\sqrt{2}}{\sqrt{3}}$$

$$\tfrac{1}{2}A\hat{X}C = 55°$$

$$A\hat{X}C = 110°$$

The angles between adjacent domains in the rhombohedral form are thus 70.5° and 110°. It is not so simple to see the fitting patterns of the crystal structure across these walls as it is for the right angle domains of the tetragonal form.

EXERCISE 4.5

$$CR = \frac{\varepsilon_0\varepsilon_r A}{h} \times \frac{\rho_e h}{A} = \varepsilon_0\varepsilon_r\rho_e$$

Both the thickness h and the area A cancel, so they don't affect the value of CR. To get CR high, both ε_r and ρ_e should be large. The resistivity doesn't feature in the merit index, but the relative permittivity appears in its denominator. A high ε_r will spoil the rate of voltage rise dV/dt.

EXERCISE 4.6 La^{3+} is much larger than the $4+$ ions which inhabit the octahedral sites, so it would cause great distortion if it went there. It goes easily into the larger cuboctahedral sites. Lanthanum oxide La_2O_3 brings in 1.5 O^{2-} ions with every cation, whereas PbO brings just one O^{2-} ion per lead atom. So by substituting into the Pb^{2+} sites, this addition will fill up any vacant sites on the oxygen lattice. If La_2O_3 substituted for ZrO_2 or TiO_2, it would *increase* the deficiency of O^{2-} ions since, as their formulae show, Zr^{4+} and Ti^{4+} bring two oxygen ions per cation.

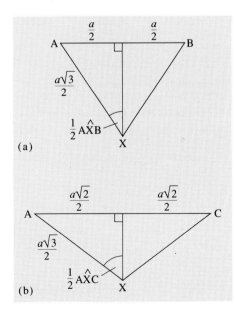

Figure 4.38 For answer to Exercise 4.4

EXERCISE 4.7 Quenching the steel induces shear transformation of the FCC crystals to a body centred tetragonal structure known as martensite. Slower cooling allows time for a diffusion driven transformation to a BCC phase.

EXERCISE 4.8

(a) Ceramics are weak in tension and fracture by sudden crack growth from the 'worst flaw'. A combination of flaw size and stress, determines which is 'worst'.

(b) Fatigue originates from flaws too.

(c) Energy not dissipated mechanically becomes heat. The ceramic could be depoled if the heating is too much.

(d) Impedance matching controls power transfer. The mechanical impedance of the vibrator must present a matched electrical equivalent to the drive circuit.

(e) Impedance matching at the water–transducer interface maximizes power transfer into the sea water.

(f) The drive field must not exceed the coercivity of the PZT.

The above answers imply the following categories of solution (I shall take them in the above order):

(a) Weakness in tension. Solution: material + geometry. Make the ceramic flawless and design to keep it in compression. Avoid stress concentrations.

(b) Fatigue. Solution: as for (a).

(c) Power dissipation. Solution: electronic. Use a pulsed output with a small 'mark to space' ratio (that is, most of the time the device is not transmitting). This gives time for heat to disperse. Use of pulse code modulation suggests pulsing.

(d) Power transfer into ceramic. Solution: geometry. Mechanical impedance is to do with how the vibrator is loaded. We shall look at ideas for optimizing this.

(e) Power transfer into sea. Solution: material. Acoustic impedance is the mathematical product of density and sound velocity. At the boundary between ceramic and sea, the acoustic impedance of the two materials should match as closely as possible.

(f) Drive field and coercivity. We have already discussed additives to PZT to make coercivity high.

Answers to self-assessment questions

SAQ 4.1 The vacancy population is entirely dominated by the dopant concentration. That's what the dopant is there for. The temperature sensitivity therefore reflects the ability of oxygen ions to 'rattle' free of the lattice sites they are bound in. The lines are nearly parallel because they express the activation of the same process: oxygen ions diffusing in zirconia. Remember, it is the gradient of the lines that holds information about the activation energy. The question asks you to revise the calculation of E_a. The gradient of the line is

$$\frac{\Delta(\log_{10}\sigma)}{\Delta(T^{-1})} = \frac{-E_a}{2.3\,k}$$

Putting $k = 86\ \mu eV\ K^{-1}$ and reading from the graph gives

$$E_a \approx \frac{-2.4 \times 2.3 \times 86 \times 10^{-6}}{6 \times 10^{-4}}\,eV$$

$$= 0.8\ eV$$

SAQ 4.2 The function of the oxygen sensor in the car exhaust is to detect sudden changes of oxygen concentration when the engine setting (accelerator) is adjusted. It must respond quickly so that the mixture can be reset to stoichiometry quickly. A large change of oxygen concentration is expected when the engine setting is altered so accuracy is not vital.

The welding application does not need a sensor with fast response: a few seconds wait will not matter in a flushing procedure which will take several seconds (for example, 1 minute) to establish correct conditions. On the other hand accuracy is important because an absolute threshold of concentration has been specified. The furnace is needed first to heat the sensor to its working condition. During flushing the argon is cold, of course. Temperature control of the furnace is essential to make the instrument accurate: its calibration is temperature sensitive.

SAQ 4.3

$$\sigma_e = n_{(Ti^{2+})} \times 2e^- \times \mu_{(Ti^{2+})} + n_{Ovac} \times 2e^- \times \mu_{Ovac}$$

Every oxygen vacancy is a place where titanium ions have been unable to put two electrons. So the populations $n_{(Ti^{2+})}$ and n_{Ovac} are equal. If one mechanism of conduction dominates it must then be controlled by the relative mobilities. The mobility of the electrons is evidently higher than that of the oxygen vacancies.

SAQ 4.4

(a) False. Electron-hopping will be very difficult in this material because both cations are oxidized to inert-gas electron structures and are therefore extremely reluctant to lose more electrons.

(b) True. Electron-hopping is easier for transition elements for the reason stated. However the composition of the oxide must still be adjusted so that there are cations which are not fully oxidized for the effect to occur.

(c) False. Niobia brings *extra* oxygen to the zirconia lattice and therefore fills oxygen vacancies and suppresses this mode of conduction.

(d) True. With the composition on the oxygen-rich side of TiO_2 the cations cannot provide enough electrons to satisfy all the oxygen ions. An electron deficiency in the oxygen constitutes a 'hole' making conduction p-type.

SAQ 4.5 In iron below T_C the exchange interactions between adjacent iron atoms are strong enough to hold their atomic magnetic moments parallel in spite of thermal agitation (rattle) tending to randomize them. In barium titanate, the crystal structure changes from cubic symmetry to a tetragonal (elongated cube) at its T_C, with positive and negative ions displaced oppositely. This confers an electrical polarity on the emerging crystals. Rattle has a randomizing effect, tending to destroy the polarization. The Curie temperature is that beyond which thermal rattle is too vigorous for the alignment forces to keep hold.

SAQ 4.6 The power flux from a point source falls off as an inverse square. At range r the radiant energy is spread over an area $4\pi r^2$ or about 1250 m^2 at the chosen range. The power flux at the detector is thus about 0.05 $W\ m^{-2}$. If the area of the detector is $A\ m^2$, the detector receives $(A \times 0.05)$ W.

For TGS our derived formula gives

$$\frac{dV}{dt}$$

$$\approx \frac{1}{A} \times (A \times 0.05) \times 360\ mV\ s^{-1}$$

$$\approx 18\ mV\ s^{-1}$$

whereas PZ, with a merit index only one-sixth as good, will give 3 mV s^{-1}.

SAQ 4.7 The pyroelectric coefficient Π $(= d\mathscr{P}/dT)$ is large as the Curie temperature is approached (Figure 4.14, PZT curve). This application therefore requires a composition close to the zirconium end of the alloy series where T_C is low. Because of the antiferroelectric phase we must have at least 10% Ti to pole the material. In fact the transformation between two rhombohedral forms shown on the phase diagram also changes the polarization and so enhances the pyroelectric effect. Enough Ti must be present to move this transition out of the danger zone of the element being accidentally depoled by getting warm.

SAQ 4.8

(a) 350°C (read from Figure 4.23).

(b) 70.5° and 110°. Figure 4.23 shows the material to be rhombohedral; see Exercise 4.4.

(c) Tetragonal, that is cubic stretched parallel to one edge.

(d) (i) No polarization. The field, applied above T_C, will have polarized the specimen while it was switched on, but this condition will have rapidly relaxed when the field was switched off.
(ii) Poled parallel to applied field. The field now applied below T_C will have encouraged existing ferroelectric domains nearly parallel to the field to grow at the expense of others. Cooling with the field on allows no relaxation of the new domain arrangement.

(e) The dimensional change from cubic to tetragonal is so large at this end of the composition range that the ceramic is liable to break during poling.

(f) This composition is antiferroelectric — the polarizations of adjacent crystal cells cancel.

(g) Tetragonal. The cubic to tetragonal phase change will have been accomplished at around 350°C but the tetragonal to rhombohedral transition expected at 50°C will be very sluggish at such a low temperature.

SAQ 4.9 Nb^{5+} will substitute into octahedral sites and as Nb_2O_5 will bring extra oxygen to the lattice to fill the vacancies. Mn^{3+} is also small enough to fit the octahedral sites but as Mn_2O_3 it will produce *more* oxygen vacancies, as Exercise 4.6 indicated. The K^+ ion is big, so will go to cuboctahedral sites, but K_2O has half an oxygen per cation so it too is bringing insufficient oxygen to fill vacancies.

SAQ 4.10

(a) (i), (iii);
(b) (i), (vi);
(c) (ii), (v);
(d) (i), (vi);
(e) (vi).

SAQ 4.11   The composition with $y = 0.9$ and $z = 0.65$ is cubic at room temperature but is very close to the Curie transition temperature to a ferroelectric state. The application depends on switching the material to a polarized state by applying an electric field. A composition of $y = 0.06$, $z = 0.65$ would not be suitable because it is already polarized at room temperature (it lies in the rhombohedral phase field). A composition of $y = 0.12$, $z = 0.65$ would not be suitable because it is antiferroelectric.

SAQ 4.12   Stress on block is force/area, so

$$g = \frac{EA\Delta\mathscr{P}}{F}$$

so

$$\frac{gF}{AE} = \Delta\mathscr{P}$$

The change in polarization is related to the voltage generated by the capacitance,

$$A\Delta\mathscr{P} = C\Delta V$$
$$= \frac{\varepsilon_r\varepsilon_0 A}{t}\Delta V$$

So

$$\frac{gF}{AE} = \frac{\varepsilon_r\varepsilon_0}{t}\Delta V$$

Hence $\Delta V = \dfrac{gFt}{\varepsilon_r\varepsilon_0 EA}$

$$= \frac{30 \times 20 \times 10^{-3}}{700\varepsilon_0 \times 70 \times 10^9 \times 4 \times 10^{-6}}V$$

$= 3.4\,kV$

This is just enough for a 1 mm spark.

Chapter 5 Memory Microtechnology

5.1 Introduction

In Chapter 3 we looked at some magnetic memories. A common feature of these magnetic systems is that the bits of stored information can only be read when they pass a 'read head'. Successive bits of information can be read at quite a high rate, but if it's necessary to jump physically from one area of memory to another the rate may be much slower. This is because magnetic systems have moving parts which bring together the data and the read/write head. You can imagine these systems to be electronic equivalents of filing cabinets (especially if they are tape- or disc-based).

The nature of electronically controlled functions is such that there is a great deal of picking up and putting down of scraps of information, and maybe the withholding of actions until further data arrives, or periodically checking and comparing values. For this a filing cabinet is not as well suited as a large table top. What we need is a memory system which can supply and receive data at a steady rate irrespective of where the data were stored last time. The way to escape the delay associated with mechanical moving parts is to avoid such parts altogether. This is what semiconductor-based systems do, pushing electrons hither and thither. This chapter is concerned with just one class of semiconductor memory, the so-called **CMOS** (pronounced 'sea-moss') **technology**. It has evolved as the cheap, volume-production memory close to the heart of many control systems, ranging from those in word processors to those in robot assembly lines.

CMOS stands for Complementary Metal Oxide Semiconductor, which can be further elaborated as follows. A sandwich of insulator between conductors, as you should know, is one way to form a capacitor. If one slice of conducting 'bread' is replaced by a semiconducting slice, the structure can also be used as an electronically controlled switch — a sort of relay but without mechanical motion. We will be investigating these MOS switches in some detail and seeing how to assemble them to form a memory system. The 'complementary' in CMOS comes from the use in this technology of MOS sandwiches with both n-type and p-type semiconductors.

Semiconductor circuits have a certain mystique. They are often seen as being at the prestige end of electronic materials, largely because through miniaturization an awful lot of circuit function can be crammed into a small space. Furthermore, the scale of the manufacturing operation, and the sensitivity of some of the effects and processes to contamination mean that, more than in any other manufacturing process, cleanliness is next to Godliness; and the materials science is closer than ever to physics.

The goal of this chapter is an understanding of the properties, principles and processes behind semiconductor memory. Following this introduction we shall discuss some general semiconductor properties while looking specifically at the control of the electrical nature of the basic materials used. (The basic materials are silicon dioxide insulator, silicon semiconductor and aluminium conductor.) I will then show how progressively more complicated electronic structures are made and how they function. One of the merits of this technology is its potential for mass production; we shall investigate the production processes used. Finally, I want to probe the limits and limitations of all this 'micro' technology.

Let's sneek a preview of the final product. Figure 5.1 shows an optical image and an SEM of an area of semiconductor memory fabricated on a silicon chip. (The chip is a few millimetres square and less than half a millimetre thick.) Using a little elemental surface analysis in an SEM (or similar tool), we could find more than just traces of oxygen and aluminium, as well as the host silicon material. Very carefully resolved surface area analysis and depth profiling, in which a few atom layers at a time are removed between elemental analyses, would show small quantities of phosphorus, boron, arsenic, antimony, tungsten and copper (measured in parts-per-billion to parts-per-thousand) locally accumulated in regions only a few microns in size. Compared with typical metallurgical surveys, the silicon sample of Figure 5.1 would appear astonishingly pure, in spite of the organized dispositions of these foreign atoms.

Figure 5.1 SEM/Optical image of an integrated circuit (CMOS array). The interconnecting tracks are aluminium. Courtesy of Plessey Semiconductors

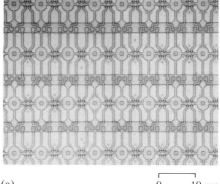

Fortunately we are not in the position of archaeologists trying to work out why the silicon sample is the way it is. In principle we know exactly what it was intended for. It was deliberately engineered to perform a certain function and we still have the plans! In years to come though, if people forget why such devices were made, the micro-archaeologists may find it difficult to distinguish the sort of CMOS technology which we shall investigate from the companion technology of so-called 'bipolar devices', which also uses layers of n-type and p-type semiconductor to a similar end but in a very different way. On the scale of microns they look

(a)

0   10 µm

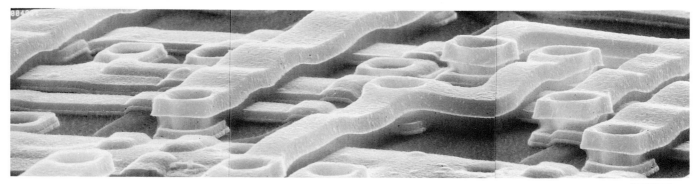

(b)

0   2 µm

much the same. The difference lies in their conception and operation (that is, the principles and the product) rather more than in the properties and processes. So bear in mind that our study of microtechnology is restricted to CMOS only in order to keep the concepts and details manageable. It has a great deal in common with the rest of semiconductor engineering, especially where silicon is involved.

▼Intrinsic and extrinsic semiconductors▲ will reveal our starting point so far as the principles are concerned and the next section begins to look at some properties and processes.

▼Intrinsic and extrinsic semiconductors▲

You should be aware, from your previous studies, of some of the language of semiconductors and their properties. Here I want to revise some basic ideas so that we can build up a consistent picture. The completed statement in SAQ 5.1 should serve to establish our starting point.

SAQ 5.1   (Revision)
Complete the text by supplying missing words and phrases from the list at the end.

Pure crystalline silicon is an   1   semiconductor. Above absolute zero, all four outer electrons of each atom in the bulk are involved in   2   with surrounding atoms. The structure is like that of   3   . Above absolute zero, thermal energy may enable an electron to leave its covalent partnership and hop from atom to atom. It is then termed a   4   as it contributes to conductivity. At the same time, the hole left behind is filled either by an electron from a neighbouring bond or by recombination with another conduction electron. The hole thus moves or is lost. It is effectively a   5   as it is drawn towards regions of negative potential; holes also contribute to conductivity.

In terms of the band model of electron energy states in a solid, the bonding levels (6   band) are separated from the delocalized states (7   band) by a distinct gap in the near continuum of closely spaced energy levels found within the bands. To change its state an electron must receive enough energy to cross the   8   (thereby escaping the bond but not the solid) into one of the available empty states.

In real materials, there are many factors which can upset the ideal picture of   9   . In regions where the highly ordered crystal structure is disrupted the energy gap is not so clearly defined.   10   ,   11   and impurity atoms give rise to locally available energy states within the gap. So what? Well, if a wandering conduction electron encounters a lower energy state which is localized, such as an incomplete bond (effectively a   12   ) or an orbit around an impurity atom, the electron may fall into it. This is bad news for the conductivity, particularly if the energy gap is large compared with thermal energy and if the state associated with the impurity is near the middle of the gap (deep level). See Figure 5.2(a). On the other hand, an electron loosely held by an impurity atom might be easily rattled off by thermal energy to the benefit of conductivity (a   13   impurity). Likewise an impurity which for only a small energy price will take an electron away from a silicon bond creates a hole in so doing (an   14   impurity); see Figure 5.2(b).

15   silicon owes its conductivity to charge carriers provided from shallow donor and/or acceptor levels. If this is done on purpose we refer to the foreign atoms as   16   rather than impurities. The majority carriers in   17   silicon are mobile positive holes. Their charge is balanced by the minority carriers (electrons) and immobile   18   ion charges (acceptor levels). In   19   silicon electrons are the majority carrier and a donor becomes a site of fixed   20   charge on giving up an electron.

Word and phrase list:

valence	conduction
intrinsic	extrinsic
n-type	p-type
conduction	carrier of
electron	positive charge
donor	acceptor
positive	negative
hole	diamond
covalent bonds	defects
energy gap	energy bands
crystal boundaries	dopants

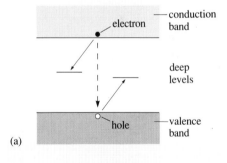

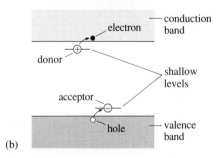

Figure 5.2 (a) Trap and (b) dopant levels in the band gap

SAQ 5.1 refers to silicon, but apart from the specific mention of its valency (four) and its diamond-like crystal structure, the statements would apply to any semiconductors.

We are interested in controlling the properties of semiconductors by measured additions of foreign atoms. Semiconductors so modified are called **extrinsic semiconductors**. I want to describe here some of the ways in which the electrical nature is changed, and I will concentrate on what happens to the charge carriers: electrons and holes.

In intrinsic semiconductors, there are equal numbers of electrons and holes free to participate in electrical conduction. Figure 5.3(a) is a two-dimensional representation of intrinsic silicon, showing schematically the regular silicon crystal lattice and equal numbers of oppositely charged mobile charge carriers. Beware! The carrier concentration is exaggerated. In silicon at room temperature, for example, only 1 in 3×10^{12} atoms is persuaded to release an electron. In n-type extrinsic semiconductor the foreign atoms introduce almost *pro rata* additional electrons. This is because within the solid the n-type dopant atom has five outer-shell electrons, but only four are needed to bond to the surrounding silicon. Now when we add dopant we do this usually so that somewhere between 1 in 10^9 and 1 in 10^2 silicon atoms are replaced by a dopant atom. The extra electron is easily detached from its parent (leaving the parent positively charged) and, even at the higher concentrations, it rarely finds its way back home. So, you can see that the intrinsic free-electron population (those electrons thermally kicked out of *bona fide* silicon–silicon bonds) can be easily swamped by the electrons imported by the dopant. The holes are now very much in the minority. Worse still, they run the constant risk of encountering one of the much more numerous electrons, whereupon they are filled in. For the holes in n-type material, as the doping increases, life expectancy is drastically reduced, so the number of holes dwindles. Figure 5.3(b) attempts to illustrate these points,

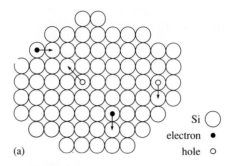

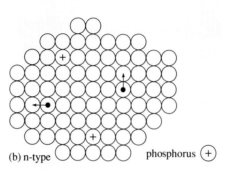

Figure 5.3 Two-dimensional particle picture of (a) intrinsic (b) n-type extrinsic silicon

but again beware. The dopant atoms are not usually as numerous as the figure indicates.

Let's just reiterate with a little algebra, considering first intrinsic and then n-type material.

Intrinsic material has equal concentrations of electrons and holes. Let n and p be respectively the number of electrons and holes per cubic metre.

$$n = p = n_i$$

where n_i is simply the density of either type of carrier in intrinsic material. The product of electron and hole densities (which I will use in a moment) is simply

$$np = n_i^2$$

If donors are added to a concentration N_D, the extra electrons swamp the intrinsic population ($N_D \gg n_i$). A useful

approximation is that the electron density is now entirely due to the dopant so

$$n = N_D$$

Since doubling the electron number doubles the risk of a hole's being annihilated (halving the hole population at a stroke), it turns out that in thermal equilibrium (no currents) we can still say

$$np = n_i^2$$

With electron density (n) determined by the donors, the hole density in n-type material is

$$p = \frac{n_i^2}{N_D}$$

EXAMPLE Calculate the electron and hole densities in silicon doped with 10^{19} donors m^{-3} ($n_i = 1.5 \times 10^{16}$ m^{-3} at room temperature).

ANSWER Using the above conclusions

$$n = 10^{19} \, \text{m}^{-3}$$

$$p = \frac{(1.5 \times 10^{16})^2}{10^{19}} \, \text{m}^{-3}$$

$$= 2.3 \times 10^{13} \, \text{m}^{-3}$$

Electrons are 400 000 times more numerous! This must be n-type silicon.

Have you been keeping an eye on the net charge? Each donor which has donated an electron is a positive charge fixed in the lattice and there are also mobile positive holes: that is $N_D + p$ positive charges per cubic metre. Above I said there would be about N_D electrons per cubic metre and as you have just seen since p is so small I was just about right. A more rigorous analysis would show that there is in fact no net charge.

Try for yourself the equivalent argument for the majority charge carriers and minority charge carriers in p-type material (respectively holes and electrons). You should deduce that p is roughly N_A, that is the concentration of acceptor atoms, which we assume now to be negatively charged.

5.2 Silicon dioxide, silicon and aluminium

Solid-state electronics has made much use of insulators, semiconductors and conductors in a variety of configurations. I want to look at the bulk properties and production routes of some typical materials before we go on to wonder about how we can control their assembly into devices. I will take silicon, silicon dioxide and aluminium as examples of each class. Table 5.1 shows some bulk physical properties for the three materials (the silicon is not intentionally doped). Exercises 5.1 and 5.2 will help you to appreciate the data.

Table 5.1　Properties of silicon dioxide, silicon and aluminium at 300 K ($kT/e = 0.025$ eV)

Material	SiO_2	Si (intrinsic)	Al
structure	amorphous	diamond	FCC
density/(kg m^{-3})	2.2×10^3	2.3×10^3	2.7×10^3
dielectric constant, ε_r (relative permittivity)	4	12	–
resistivity $\rho/(\Omega\ \text{m})$	10^{15}	2.3×10^3	2.7×10^{-8}
melting temp./°C	1600	1415	660
energy gap E_g/eV	9	1.12	–
breakdown strength $\mathscr{E}_B$/(V m^{-1})	10^9	3×10^7	–
atomic density/(atoms m^{-3})			

EXERCISE 5.1

(a) The atomic masses of oxygen, aluminium and silicon are respectively 16, 27 and 28. Use data from Table 5.1 to complete the atomic density (atoms m^{-3}) row in the table. One kg mole (the formula mass expressed in kilograms) contains 6×10^{26} units.

(b) How much less conductive than aluminium is silicon?

(c) How much more conductive than silicon dioxide is silicon?

EXERCISE 5.2　A one-millimetre cube of silicon with a ten-nanometre thick native oxide has one micron of aluminium deposited on a pair of opposite faces. Estimate (a) the capacitance, (b) the resistance and (c) the breakdown voltage between the metal faces, ignoring the contribution of the surface oxide to the dielectric for the capacitance calculation. By the end of this section you may realize that this exercise is over simplified. ($\varepsilon_0 = 9 \times 10^{-12}$ F m^{-1}.)

There are two other things you should know about silicon and aluminium. First they form a eutectic alloy at 11.7% Si which melts at 577 °C so don't go thinking that process temperatures can approach 660 °C (see Table 5.1) once we've made something involving these two elements in contact. We'll come back to metal/semiconductor connections later. Secondly, aluminium like gallium, boron and indium is a well known dopant capable of rendering silicon a p-type semiconductor. We will look at the resistivity of doped silicon shortly but the next thing I want to investigate is the insulating silicon dioxide.

SAQ 5.2 (Objective 5.1)
Briefly describe the electrical roles played by silicon dioxide, aluminium and silicon in integrated circuits.

5.2.1 Growing and depositing silicon dioxide

You are going to find that silicon dioxide is a versatile companion to silicon and every bit as important to microtechnology as silicon itself. As an insulator it is very good but its real merit is that it can be grown into a silicon surface, bonding extremely well to the underlying silicon. Furthermore, it can then be selectively removed to expose regions of the silicon substrate (more of this in Section 5.3). Let's look at the chemistry relevant to the oxide formation.

When oxidation is an undesirable consequence of our oxygen-rich environment we call it corrosion. With rusting in particular in mind, you can guess what conditions promote oxidation: warmth and dampness. In the case of silicon, it is found that at elevated temperature (thermal oxidation) both 'dry' (oxygen) and 'wet' (steam) routes are available,

$$Si + O_2 \xrightarrow{\approx 1000\,°C} SiO_2$$

$$Si + 2H_2O \xrightarrow{\approx 1000\,°C} SiO_2 + 2H_2$$

Both reactions take place at the silicon/silicon dioxide interface. ▼SiO_2: Growth rate and quality▲ looks further into the processes involved.

EXERCISE 5.4 Thermally growing silicon dioxide on silicon consumes surface material. Using data in Table 5.1, calculate the depth of silicon used up in growing a uniform 1 μm layer of silicon dioxide.

Thermally grown oxides are fine at an early stage of the manufacture of microcircuits but, towards the end, it is necessary to find cooler processes to avoid disturbing the results of previous steps. Also it is sometimes necessary to insulate one layer of metal from another, so we should look for ways of making layers of silicon dioxide on non-silicon substrates without the need for high temperatures. **Chemical vapour deposition** (CVD) techniques are useful here, using spontaneous reactions such as that between silane, SiH_4 (silicon's answer to methane, CH_4), and oxygen:

$$SiH_4 + 2O_2 \rightarrow SiO_2 + 2H_2O$$

▼SiO₂: Growth rate and quality▲

When a silicon surface is consumed by a growing silicon dioxide layer, it turns out that the oxidation takes place at the Si–SiO₂ interface. That means that the rate of oxide growth may be limited both by the reaction itself and by the rate of arrival of oxidant through the previously grown material. Chemical reactions and diffusion are generally both accelerated by increasing temperature, so it should be no surprise to find substantial increases in growth rates over the range 800 to 1200 °C.

It would be convenient if either the reaction rate or the diffusion rate would limit oxide growth rate, because then we could easily calculate the required oxidation time for a given set of conditions. Under reaction-rate control, the oxide thickness will increase in proportion (linearly) with time. Under diffusion-rate control, a square-root dependence on time is more likely (from $x \approx \sqrt{Dt}$).

Figure 5.4 shows results for oxide grown using the wet process. Does it show a linear or square root variation with time?

As you can see by looking at the slopes on the graph of log x against log t, after an hour (thickness ≈ 0.1–0.5 μm) it is predominantly diffusion limited (this is true even at lower temperatures). Neither

one limit nor the other fits the region we are interested in (up to 1 μm in a few hours) and in practice experimental data must be used to specify oxidation times.

EXERCISE 5.3   Which of the following might also be expected to affect growth rate?

(a) silicon crystal orientation,
(b) the presence of dopant atoms in the silicon,
(c) the nature of the oxidant species (O₂ or H₂O).

The quality of oxides can be an important factor in determining the performance of semiconductor devices. For example, certain 'field effect' devices require a stable silicon dioxide insulating layer, but early devices failed because of the migration of sodium ions through the oxide. Sodium ions are kept out chiefly by cleanliness (a common source was human perspiration during handling). The addition of chlorine (one per cent or so of HCl or Cl₂ or CHCl₃) to the oxygen gas during dry oxidation also helps to immobilize sodium ions.

Slower grown oxides are denser, because there is more time to get the right atoms to the right places. Hence thin, high-quality oxides are grown in oxygen (plus 1%

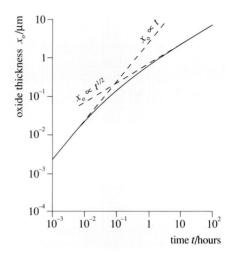

Figure 5.4 Oxide growth during the wet oxidation of silicon at 1000 °C

chlorine) at around atmospheric pressure and at the lower temperatures. This is the sort of material we will need at the heart of our memory devices. Thicker 'ordinary' insulators can be grown much faster using the wet reaction and high temperatures. When lower temperatures are essential, though it is useful to use higher pressures (several atmospheres). This material is a useful 'underlay' for metal tracks and general purpose protection for the silicon wafer.

in which the silicon dioxide is deposited on warmed surfaces. In practice, modest heating to 400 °C gives good uniformity. At atmospheric pressure, about half a micron of silicon dioxide can be deposited in one hour. You can see from Table 5.1 that this would provide insulation up to about 500 volts. But silicon dioxide is more than just an insulator in microtechnology, so watch out for its cropping up again. Meanwhile, after SAQ 5.3, we'll move on to silicon.

SAQ 5.3   (Objective 5.2)
Briefly highlight and account for two ways in which a grown silicon dioxide layer differs from one deposited from a chemical vapour (including process details).

5.2.2. Controlling the resistivity of silicon

Until we enquire into the physics of specific metal/semiconductor/insulator structures, the virtue of doped semiconductors appears to be as a material with variable resistivity. This property is extremely sensitive to composition over an extensive range of values. For example, 0.01% aluminium in silicon decreases the resistivity by seven orders of magnitude (but 1% of silicon in aluminium increases aluminium's resistivity by a mere 3%). We will look at the effects both of composition and of temperature, and then at how the composition can be manipulated in practice.

Composition

Figure 5.5 shows the resistivity of doped silicon. For making a resistor, it doesn't much matter whether p-type or n-type (acceptor or donor) dopant is used, just so long as solubility limits are not exceeded. Why do you think that is?

For a dopant atom to accept or donate an electron to the benefit of conductivity, it must substitute for a silicon atom in the structure. Beyond the limit of solid solubility, further simple substitution does not occur in thermal equilibrium. Figure 5.6 shows solubility data for a number of possible dopants, classified as p-type or n-type.

> EXERCISE 5.5   Suggest dopants for very low resistivity n-type and p-type silicon.

A closer look at Figure 5.5 suggests that the resistivity of p-type silicon is slightly different from that of n-type. To see why, it helps to work with conductivity σ, the inverse of resistivity ρ. As you know from earlier chapters,

$$\frac{1}{\rho} = \sigma = nq\mu \ (\Omega^{-1}\,m^{-1})$$

where n is the number of charge carriers per cubic metre, q is their charge and μ their mobility. What happens if there are both electrons and holes, but in such small concentrations that they don't interfere much with each other?

The situation is rather like commuters being able to take the bus or the train. If both are available more commuters (or charge) can be transported. So with n electrons and p holes per cubic metre

$$\sigma = ne\mu_e + pe\mu_h$$

(Don't forget that although electrons and holes carry opposite charge, they move in opposite directions and so they both make a positive contribution to conductivity.)

Now we can see what controls the conductivity of silicon: n, μ_e, p, μ_h. So, in n-type material conductivity is mainly due to the majority electrons

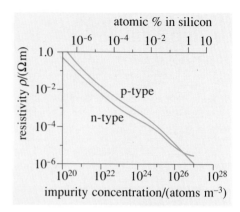

Figure 5.5 Resistivity of p-type and n-type doped silicon (300 K)

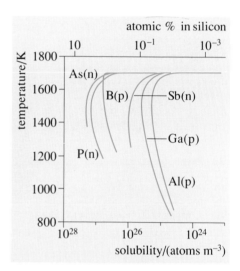

Figure 5.6 Solid solubilities in silicon

and their mobility. Can we now explain the general shape of Figure 5.5 for n-type conductivity?

The argument runs along these lines:

1 For n-type materials we dope with donors.

2 At 300 K the electrons easily escape the atoms and roam the conduction band. They are available for conduction, limited by their mobility μ_e (which is generally a little larger than μ_h).

3 At low doping ($<0.001\%$) the distortions of the atomic structure arising from randomly sited foreign dopant ions are insufficient to affect the electron mobility. So every tenfold increase in dopant causes a tenfold increase in the density of charge carriers and so a tenfold decrease in resistivity (see Figure 5.5).

4 At higher doping ($>0.001\%$) electron mobility is adversely affected by the dopant ions (because they locally distort the structure). The resultant resistivity falls less rapidly than before, the increase in charge carriers being slightly offset by a decrease in their mobility. Notice that in silicon below about 0.001% doping, the mobility must be dominated by thermal effects (phonon interactions).

> SAQ 5.4   (Objective 5.3)
> Account for the changes in resistivity of silicon doped with increasing amounts of acceptor such as boron.

Temperature

Because the electron and hole populations are strongly dependent on temperature the conductivity of an intrinsic semiconductor rises steeply with temperature, in spite of the thermal degradation of mobility. This standard characteristic of semiconductors arises because thermal energy kicks electrons out of bonds (the valence band) creating pairs of electrons and holes located respectively in the conduction and valence bands. For extrinsic conductivity at normal temperature, a given level of doping can swamp this temperature sensitivity — in Figure 5.5 the number and type of charge carriers is in fact almost entirely determined by the doping (and not the temperature). The way extrinsic resistivity varies with temperature is governed therefore by thermal-scattering effects, just as in metals.

But the intrinsic effects won't go away. Eventually, as temperature rises, thermally generated electron-hole pairs become so numerous that the intrinsic behaviour (falling resistivity – rising conductivity) prevails. The intrinsic carrier density in silicon varies with temperature as shown in Figure 5.7. At any temperature the doping density must exceed the intrinsic carrier density if extrinsic behaviour is to dominate.

> EXERCISE 5.6   Figure 5.8 shows the variation of resistivity of a sample of n-type silicon with temperature. Account for the observed temperature dependence.

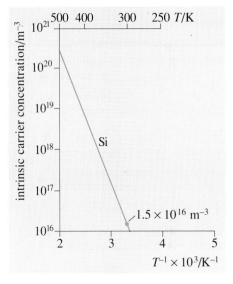

Figure 5.7 Intrinsic carrier concentration against temperature

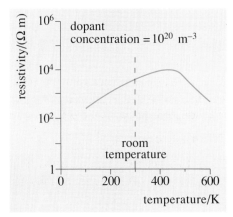

Figure 5.8 Resistivity of an n-type silicon sample against temperature

▼Electronics grade silicon▲

As you should know, the preparation of single-crystal wafers for semiconductor devices is a specialized business. In Table 5.2 there are some of the specifications of a typical 'off the shelf' wafer designed for the CMOS technology to be described later. The intended application calls for a heavily doped substrate with a few microns of lightly doped material at the surface (in which the devices will be made). This structure is achieved by growing a surface layer of silicon on top of a suitably doped, single-crystal substrate. The deposited layer extends the substrate crystal provided atoms have time and sufficient energy to find and settle into vacant lattice sites. Layers grown in this way are said to be **epitaxial** with the substrate. Polycrystalline films result when deposition is either too quick or too cool. One useful chemical vapour deposition (CVD) process is as follows:

$$SiCl_4 \text{ (gas)} + 2H_2 \text{ (gas)}$$
$$\xrightarrow{1200\ C} Si \text{ (solid)} + 4HCl \text{ (gas)}$$

This reaction proceeds at about a micron per minute; but at a temperature a hundred degrees or so hotter or cooler the reverse reaction, silicon etching by hydrogen chloride, predominates. In fact, prior to deposition the substrate is cut back in this way to give a fresh surface onto which the the epitaxial layer is grown.

The new material is only as structurally perfect as its host, from which dislocations and impurities can be inherited. Initially,

Table 5.2 Typical specifications for silicon wafer

Quantity	Specification	Comment
epi. thickness	2–50 μm $\pm$ 5% 5% radial variation	user specified
transition region epi./ substrate	about 2 μm	change from modest to heavy doping
epi. resistivity (user specified)	0.01–0.75 Ω m $\pm$ 15% 0.004–0.75 0.004–0.75	p-type (boron) n-type (phosphorus) n-type (arsenic)
epi. surface quality	< 20 sparkles/wafer scratch free	'Sparkles' are surface defects visible by reflection
substrate thickness	675 μm $\pm$ 4%	
substrate resistivity	5×10^{-5} to 2×10^{-4} Ω m $\pm$ 10% 8×10^{-5} to 3×10^{-4}	p$^+$ (boron) n$^+$ (antimony)

dopant atoms will diffuse out of the substrate into the growing layer and the gas. These will swamp the effect of any hydrides (such as PH_3, B_2H_6) which will have been added to the gas to establish the dopant concentration in the new layer. Only after the first micron or so has sealed the substrate is the transition to light doping complete. Undesirable impurities such as carbon and oxygen, which are inevitable contaminants from the bulk production process, are effectively reduced in the epitaxial layer for similar reasons.

The back surface of the wafer is sealed prior to epitaxial growth with a deposited layer of silicon dioxide. This layer is thick enough to prevent dopants from diffusing out of the bulk of the wafer into the gas. You might guess its thickness from the formula $x \approx \sqrt{Dt}$, where D is the diffusion coefficient for the dopant in the oxide and t is the deposition time. In fact it is not as simple as this because materials like boron and phosphorus are glass formers and so can introduce a liquid phase into the silicon dioxide. Something like half a micron of oxide is reckoned to be enough. The customer can ask for the back seal to be left on so that subsequent processing is also protected, but longer hotter processes may need a thicker oxide layer.

You may be wondering how you buy silicon. Does it come as wafers of intrinsic silicon or can it be had ready doped? ▼Electronics grade silicon▲ is what you should ask for.

5.2.3 Doping — atomic diffusion and ion implantation

Two techniques are available for introducing dopant atoms into a silicon crystal in a controlled way. The first is atomic diffusion, which is thermally activated and driven by density gradients. The second is ion implantation, in which electrically accelerated ions penetrate the crystal structure by brute force. Both are important in semiconductor technology.

Diffusion

In polycrystalline material, grain boundaries tend to channel and to retain foreign atoms. This is one reason why single crystals are used in semiconductor devices, where uniform doping of the bulk is desired. As you should know, diffusion processes are strongly temperature-dependent, a fact which is exploited in many fields of materials processing, not just microelectronics.

At elevated temperatures (>900 °C) boron, for example, can be diffused into a silicon crystal, rendering it a p-type semiconductor at room temperature. Under normal operating conditions (<200 °C) there is no worry that the boron will diffuse out again. To initiate the diffusion, a large concentration of the required dopant must be established at the wafer surface and the temperature raised until 'appreciable' diffusion proceeds. The deepest diffusions need the highest temperature and/or longest diffusion process time. The other variable, the surface concentration, cannot exceed the limit of solid solubility for the particular dopant atoms in silicon (Figure 5.6).

The first task in the diffusion process is to establish a surface concentration of the chosen dopant atoms. The following methods are used, depending on the dopant species:

(a) A suitable gaseous mixture containing the dopant is arranged to flow for a short period of time over the heated wafer surface. This technique provides heavy, but generally not saturated, surface doping.

(b) A solid material saturated with the dopant can be deposited over the surface followed by a short heat cycle which causes some dopant to diffuse into the wafer. The material must not itself interfere with the electronic properties of the wafer: silicon dioxide (SiO_2) is the natural choice as it readily bonds to silicon and is stable at the high temperatures needed for the diffusion process.

A key feature of solid-state diffusion is that it is inherently associated with a non-uniform concentration profile, so the heaviest doping will occur at the surface. This characteristic is not always undesirable; for instance a very high concentration is essential for the surface regions to which metal contacts are to be connected as it avoids an electronic barrier forming between the metal and the semiconductor (we discuss this in Section 5.2.4). On the other hand, it may be that a part of, say, an n-doped region is to receive a second diffusion to convert it locally back to p-type material. In this case the first diffusion must leave a relatively low surface concentration or else we won't be able to swamp it with a second dopant. The diffusion profile of Figure 5.9 is typical of the case when the surface concentration is fixed throughout the process — so-called **constant-source diffusion**. The dopant profile at successive diffusion times is shown.

A different profile can be obtained in a two-stage diffusion process (Figure 5.10). First, a very shallow constant source diffusion is

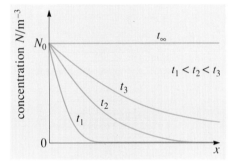

Figure 5.9 Constant source diffusion profiles (x is diffusion depth)

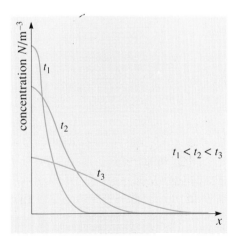

Figure 5.10 Limited source diffusion profiles (x is diffusion depth)

performed, followed by a 'drive-in' diffusion. Only those atoms introduced by the first (pre-deposition) diffusion are available for the second diffusion. The drive-in process is called a **limited-source diffusion** when performed intentionally. As you may imagine, there is a risk that any subsequent thermal cycles of the wafer during processing may also cause a redistribution of the diffused dopant. ▼Designing diffusions▲ looks at how we can predict and follow the dopant distribution.

▼Designing diffusions▲

The mathematics of diffusion should not be unfamiliar to you (but you may be relieved to know we won't be needing it much). The dimensions of semiconductor circuits are critical to device operation, and the equation

$$x \approx \sqrt{Dt}$$

(for characteristic diffusion distance x after time t in a system with diffusion coefficient D) is quite inadequate for the sort of precision required. For instance consider the following problem.

EXAMPLE Phosphorus diffuses into pure crystalline silicon at 1000°C with a diffusion coefficient of 10^{-18} m^2 s^{-1}. If we diffuse phosphorus from a saturated surface into p-type silicon uniformly doped with 10^{23} acceptors per cubic metre, after one hour, where will the silicon change from n-type to p-type?

FOUR SUGGESTED ANSWERS

(a) If $x = \sqrt{Dt}$, then the pn junction is at

$$x = \sqrt{10^{-18} \times 3.6 \times 10^3} \text{ m}$$
$$= 0.06 \text{ } \mu\text{m}.$$

But what we have calculated here is only a characteristic distance and not a precise measure of where the material changes from n-type to p-type. This is inadequate.

(b) Solve Fick's first and second laws for the phosphorus concentration N:

$$\frac{\partial N}{\partial t} + D\frac{\partial^2 N}{\partial x^2} = 0$$

(subject to the solubility-limited boundary condition). I'll do it for you. The junction is where $N = 10^{23}$ (donor concentration exactly equals acceptor concentration) and this turns out to be at a depth of 0.48 µm.

after one hour. But this equation only applies if the diffusion coefficient for phosphorus atoms in silicon (D) is constant. In practice it isn't since it is sensitive to impurities such as the dopant atoms themselves in significant number (the surface is at the solubility limit, remember). This is still inadequate.

(c) One could perform the diffusion experimentally to find the junction, bearing in mind that (a) and (b) suggest that it may be a few tenths of a micron. (I'll say how it might be done shortly.)

(d) CAD. You can buy Computer Aided Design packages which will at least address the right problem. Provided they are supplied with reliable data for the variable diffusion coefficient (so someone's got to do (c) anyway), such packages provide a close guide to process specification.

So, having created a junction how do we locate it empirically?

First, we grind and polish an angled section crossing the junction at a known small angle so that several sideways microns are needed to penetrate one micron down into the bulk (see Figure 5.11). Next we ask our chemists for n-type/p-type selective etchants, for example 0.1% nitric acid in hydrofluoric acid. The hydrofluoric acid etches silicon; the nitric goes for the dopants staining p regions dark relative to n regions. After that it's straightforward optical measurement and geometry. To get the reproducible results needed for a mass production we must reproduce not only the diffusion temperature and dopant species but also the wafer condition (previous treatments, crystal orientation and so on). Local resistivity measurements can also indicate dopant variations across the wafer.

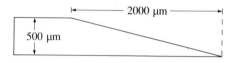

Figure 5.11 An angle lapped wafer

Although algebraic formulae are not capable of guiding process specification, they do give valuable information on feasibility. The following example illustrates this.

EXAMPLE Is it reasonable to consider thermal diffusion for establishing a dopant in silicon 10 µm below the surface? Data: Boron and phosphorus have similar diffusion coefficients, at least at low concentration, which are about 10^{-16} m^2 s^{-1} at 1160°C and 10^{-14} at 1400°C.

SOLUTION Let's give $x = \sqrt{Dt}$ a try to see how long it might take. Higher concentrations in the bulk will need higher surface concentrations to drive them.

At 1160°C

$$t = \frac{x^2}{D} = \frac{10^{-10}}{10^{-16}} \text{ s} = 10^6 \text{ s}$$
$$(\approx 11.5 \text{ days})$$

At 1400°C

$$t = \frac{x^2}{D} = \frac{10^{-10}}{10^{-14}} \text{ s} = 10^4 \text{ s}$$
$$(\approx 2.8 \text{ hours}).$$

The cooler treatment looks prohibitively long: suppose a mistake were made in subsequent processing after this much investment in process time. Provided the higher temperature can be tolerated (silicon melts at 1415°C) it's worth considering a more careful calculation.

EXERCISE 5.7   Figure 5.10 shows diffusion profiles for a so-called limited source diffusion. Identify two characteristic features which distinguish it from the constant source case.

Diffusions are carried out in high purity quartz (SiO_2) tubes surrounded by a resistance-heated furnace winding (Figure 5.12). A batch of wafers (20–30) held in a quartz jig is loaded into the furnace tube and the system is then heated up to the operating temperature with inert gas flowing through the tube. Dopants can be carried by the inert gas flow or else doped silicon dioxide can be first deposited on the wafers to provide a 'solid' source for the diffusion. For the solid source, a doped glass (SiO_2) is deposited from silane and oxygen with a suitable addition such as phosphine, PH_3 (n-type), or of boron trichloride, BCl_3 (p-type), so that dopant (P or B) is incorporated in the silicon dioxide layer as it accumulates. The coated wafers are then ready for a predeposition diffusion stage, transferring the dopant to the wafer. After this diffusion, the glass is removed by etching (see later) and a **drive-in** stage follows in order to push the dopant to the depth specified by the designers.

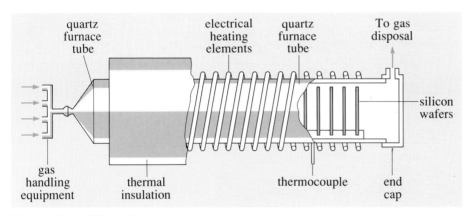

Figure 5.12 A diffusion furnace

Diffusion offers a 'broad brush/paint spray' approach. However, to create microcircuits it is often necessary to perform diffusion into selected areas of the wafer surface. (Stencils or masks can be used to establish patterns of diffusion, and we'll cover these additional techniques in Section 5.3)

Ion implantation

The thermal diffusion process has a major drawback. Complicated structures require more than one diffusion cycle, and the high temperature needed for each diffusion causes re-distributions of other dopants and inevitably risks increasing the levels of background contaminants. This is not to say that diffusion can't be used, but it can restrict device performance, particularly for high-density memory chips.

A more refined technique is to project the dopant atoms at high speed into the wafer surface. This doesn't of itself require high temperatures and even permits 'fine-brush' doping, in which the area of interest is directly addressed by the beam rather than being defined by gaps in a mask. (Such detailed writing is expensive, though, because it is much slower than masking.)

I don't mind the thought of diffusing through a crowd at a football match. I'm less keen on the idea of being blasted through it. But what about thinking of a football as the 'dopant' in a structure of human bodies? Suppose a player kicks a ball waist high into the crowd. How far will it get? Well, it may penetrate a few rows but it will have to dislodge the first two or three people who are in the way if it is to get very far. The same thoughts will apply to dopant atoms projected into a silicon crystal. From this idea we can estimate the energy needed to push a dopant say 1 μm below the surface as follows:

(a) From Exercise 5.1 the atomic density of silicon is 5×10^{28} m^{-3}.

(b) If a dopant atom of diameter 0.2 nm followed a straight 1 μm path it would encounter approximately

$$\pi(1 \times 10^{-10})^2 \times 10^{-6} \times 5 \times 10^{28} \text{ atoms,}$$

or roughly 1600 silicon atoms.

(c) If each atom encountered has to be dislodged, energy will be required to break its bonds. Bond energies are typically of the order of 1 electron volt (1 eV $= 1.6 \times 10^{-19}$ J) per atom, so this means 1.6 keV for the 1 μm path. Let's call it 2 keV to work in round figures.

At least 2 keV of energy (3.2×10^{-16} J) may be needed for each dopant atom. This is kinetic energy, so you can work out the velocity of any particular species.

EXERCISE 5.8   How could a boron atom be given 2 keV of energy?

The figure of 2 keV in fact considerably underestimates the energy required to penetrate one micron of solid because the path is not a straight line and the physics of the collision is not so simple; but it's a start and it provides a sort of minimum. In fact, 100 keV will typically take a dopant only around 0.3 μm below the surface. In general the penetration depths are limited to less than a micron (whereas the thermal diffusion process can be used to provide doping over tens of microns deep). The resulting dopant distribution is non-uniform (▼Range and scatter▲). Several stages of implant at different energies can give a more uniform profile, but at proportionally reduced throughput.

Not surprisingly, this 'breaking and entering' by a dopant atom leaves a trail of damage in its wake. Furthermore a dopant atom may itself come to rest at an interstitial site where it is useless (remember it must be incorporated into the lattice to be useful). So, annealing is necessary to

▼ Range and scatter ▲

When a beam of energetic ions impinges on a solid surface a large number of particle interactions occur. In any one of these the ion may simply scatter off the nuclei of the solid (as in an elastic, billiard-ball collision). Alternatively it may excite or ionize atoms in the solid (inelastic collisions) or even lead to the expulsion of electrons, ions and atoms from the solid (secondary collisions). Each type of collision has an associated probability of occurrence. These random collisions not surprisingly lead to final spatial distributions of ions which is almost Gaussian in shape: a mean depth of penetration, or **range** R, with a standard deviation or **scatter** ΔR. The scatter is usually between $\frac{1}{3}$ and $\frac{1}{2}$ of the range and both scatter and range increase in rough proportion with ion energy. That is the ideal picture. In fact an important omission from the above is the possibility of ions being channelled through a near perfect single crystal, avoiding collisions because along certain directions there are no 'line of sight' atoms. Although the range of channelled ions is much greater

than that of the ordinary scattered ions, exploiting them is difficult because they need accurate shooting. Instead, angles of incidence are deliberately chosen to avoid channelling altogether. Ion implanting into amorphous or polycrystalline material proceeds in much the same way except that deep penetration by channelling is unlikely.

Figure 5.13 shows profiles for boron implantation into silicon; the dosage is the same for each energy. Dose is an

important quantity (in the figure it is the area under the concentration profile) and it is easily related to the total energy delivered by the ion beam. For example, a dose of 10^{16} ions m^{-2} each at 100 keV (that is $10^5 \times 1.6 \times 10^{-19}$ J) introduces 1.6×10^2 J m^{-2} of energy. Since much of this energy will end up as heat, it's important that it doesn't arrive too quickly: 160 J m^{-2} in 100 seconds is 1.6 W m^{-2} (which is quite tolerable).

The Gaussian concentration profile $N(x)$ of an ion implanted dopant is

$$N(x) = \frac{\phi}{\sqrt{2\pi}\Delta R} \exp\left[-\frac{1}{2}\left(\frac{x-R}{\Delta R}\right)^2 \right],$$

where R and ΔR are the range and scatter and ϕ is the dose (if you don't recognize it, try plotting it out for $\phi = 10^{16}$ ions m^{-2}, $R = 0.1$ μm and $\Delta R = 0.03$ μm). If some different profile is required then we could consider adding a second implantation at a different dose and energy (range and scatter). Alternatively, a subsequent diffusion cycle could be used to redistribute implanted ions.

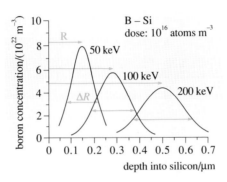

Figure 5.13 Ion implantation profiles

recover the single-crystal structure and to 'activate' the dopant. Some diffusion during annealing is inevitable, but can be restricted by keeping the anneal temperature less than about 900 °C.

There are several reasons why this method for doping semiconductors has become a standard process tool. For example:

(a) The technique offers fine control of dopant dose, density and concentration profile.

(b) As a comparatively low-temperature process it does not excessively disturb previous processing stages.

(c) The range (depth) of implants is shallow and so the technique has become useful for small, low-current devices such as those found in integrated circuits for high-density memory chips.

(d) A wide variety of dopants are readily available using this technique.

A typical implanter system is shown in Figure 5.14. The whole apparatus is inside a vacuum vessel. The dopant source is usually a gas (such as PH$_3$, BCl$_3$). The molecules are broken up and the atoms ionized by electron bombardment in the ion source. Positive ions are extracted by a large electric field and then separated in the magnetic analyser which bends ion trajectories according to their ratio of charge to mass. The required species is selected, accelerated and focused onto the target area. In much the same way that a CRT beam is steered, the ion beam is

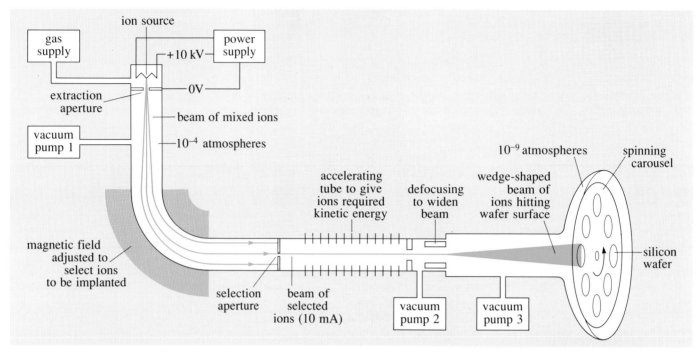

Figure 5.14 Ion implantation equipment

scanned across the target area and across the wafer to implant all the devices on it.

Both ion implantation and atomic diffusion can be used to give a controlled concentration and distribution of charge carriers in silicon. We must look next at how to connect those charge carriers (through currents and voltages) to the outside world.

SAQ 5.5 (Objective 5.4)
A post-implant anneal cycle may require a silicon sample to be heated and kept at 900 °C for several tens of minutes. What happens to the host and dopant atoms during this stage which is different from what they would do during the alternative doping procedure of atomic diffusion?

5.2.4 Connections

The peculiar properties of special dispositions of p- and n-type silicon are only useful if electrical connections can be made both amongst them and with the outside world. In this section we are going to discuss the way in which a thin layer of aluminium is used to provide most of the metallic conductors on most silicon circuits. Aluminium has emerged as the standard material for 'metallization' because it has proved sufficiently satisfactory, reliable and versatile. We will examine the following aspects: deposition, adhesion, resistivity, compatibility, patternability and price.

Deposition

We've discussed thin-film deposition before.

> **EXERCISE 5.9**   In Chapter 3, I mentioned the various ways in which thin films can be deposited. What are they? For an aluminium-deposition process based on each one, give a few words of explanation.

There is no real difficulty in depositing aluminium, although in practice aluminium alloys are often specified for certain metallizations. We shall see why in a moment, but you may care to select one of your answers to Exercise 5.9 as the most suitable for alloy deposition.

Adhesion

Metal–silicon connections are required in certain places, but not everywhere, and for much of its contact area, a metallization layer will be sitting on silicon dioxide. It is fortunate that aluminium adheres well to silicon dioxide through the mediation of the oxygen atoms in the silicon dioxide surface. This also means that once in the surface, the aluminium atoms are tightly held by oxygen atoms and diffusion into the bulk is slow.

Resistivity

The resistivity of aluminium is about 2.7×10^{-2} $\mu\Omega$ m, fourth behind silver, copper and gold (1.6, 1.7 and 2.4×10^{-2} $\mu\Omega$ m respectively). This is going to have to be low enough since we can't get much less, but what layer thicknesses are involved?

> **EXERCISE 5.10**   A one-micron layer of aluminium has a sheet resistance of 2.7×10^{-2} Ω square^{-1}. Suppose a power supply track is to have no more than 4 Ω of resistance along its 3 mm length. Is a track one micron thick, and five microns wide, suitable?

In fact layers of order one micron thick are used (too thick and internal stress can cause layers to peel off). This amount can be sputter deposited in a minute or so.

Compatibility

What happens if aluminium is put on to silicon? Figure 5.15 shows the Al–Si phase diagram. It is pleasantly uncomplicated. You can a see eutectic point at 850 K (577 °C), 11.7% silicon. Given a chance to establish thermal equilibrium, the aluminium will dissolve some silicon and vice versa, and a liquid phase may develop above 577 °C. Does this matter? That depends on the thermal treatments which follow metallization (and on the scale of the underlying electronic device). When aluminium dissolves silicon from the wafer, it swaps aluminium atoms

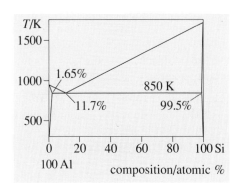

Figure 5.15 Al–Si phase diagram

for silicon atoms. Subsequent diffusion of a nearly pure aluminium band into the silicon can follow, possibly destroying the intended characteristics of this region. Furthermore, aluminium dissolved in silicon contributes holes and is capable of converting n-type semiconductor into p-type. You can probably guess the solution to these particular problems: deposit an aluminium–silicon alloy with enough silicon to satisfy any thermal-equilibrium demands. The increase in resistivity is only slight. In addition, contacts to n-type regions ought to have very strong donor dopant concentrations in the region of the aluminium contact (so much so that it is described as an n^+-type layer). A glance back at Figure 5.6 will confirm that the usual n-type dopants are more soluble than aluminium so we can always outnumber the holes the aluminium may introduce. Another way to prevent aluminium contacts from interfering with the silicon is to use a thin barrier layer which locks up the aluminium (and does not itself have a worse tendency to dissolve silicon). Transition metals are suitable partners for aluminium here in an intermediate compound (for example, Al_3Ta).

That is not the whole story though because we ought to worry a little about how electrons are going to move from the spacious energy-band structure of aluminium to the more confined gapped-structure of silicon. For our memory system based on semiconductor technology, we must make electrical connections to the semiconductor which do not impede the flow of current. Although we can tolerate the junction being a bit more resistive than the contact metal, anything more eccentric, such as non-ohmic behaviour, won't do at all. See ▼Ohmic contacts▲

▼Ohmic contacts▲

You should be familiar with the notion of the contact potential developed on the joining of two dissimilar metals. We will see later that there are different but related effects associated with the juxtaposition of dissimilar semiconductors. The intermediate case of a metal/semiconductor contact can resemble either extreme. The whole story must involve the nature of the semiconductor/metal interface (that is, the details of the band structure near the interface), as well as the Fermi levels of the materials involved.

If the junction is more like that between dissimilar semiconductors then it favours current flow in one or other direction across the junction. This is diode behaviour (we will discuss diodes in Section 5.3.2) and the contact is said to be a rectifying contact. The effect was exploited in the 'cat's-whisker' crystal radio receivers of old.

The case we want to pursue here is where the electrons can pass freely in both directions between metal and semiconductor. We need to make the semiconductor as much like a metal as we can. Any ideas? Well, a distinguishing feature of a metal is that its Fermi-level is surrounded by many closely spaced electron states, about half of which are empty.

EXERCISE 5.11
(a) Where is the Fermi level in intrinsic silicon?

(b) What happens to the Fermi level if n-type dopant is introduced?

To make the semiconductor 'look like a metal' we need as many charge carriers as possible, so we should specify very heavy doping near the junction (n^+ or p^+). This

in fact pushes the Fermi level out of the energy gap, surrounding it with vacant energy states. Under these circumstances we can be assured of an ohmic (non-rectifying) connection between the semiconductor and a metal in intimate contact with it.

For ohmic contacts between p-type silicon and aluminium (itself a p-type donor) we need hardly worry as the tendency towards p^+ doping (by aluminium dissolving in the silicon) is inevitable and anyway this particular junction is inherently non-rectifying.

The other case, aluminium and n-type silicon, needs more attention since dissolved aluminium can go some way to offset our efforts towards heavy (n^+) doping. Fortunately phosphorus is more soluble than aluminium and suitable ohmic contacts can also be established between aluminium and n-type silicon.

Patternability

You can see from Figure 5.1 that the aluminium layer is not just a continuous film. It must be patterned to establish interconnections between one area of silicon and another and to form connections with the outside world. One of the reasons why aluminium was first favoured as a metallization layer was the fact that it could be etched using similar chemicals to those used in other stages of fabrication (chemicals such as aqueous acids). Nowadays, aluminium is so well accepted and understood that ways of producing the necessary pattern of connections on the silicon/silicon dioxide have evolved to cope with the smaller scale required with barely a thought for an alternative material. We will come to how the patterns are defined in Section 5.3.

Price

A simple single integrated circuit, say 3 mm × 3 mm, may have aluminium over half its surface area to a depth of one micron (about 12 µg of aluminium). The original depth of silicon committed to the device is about 0.5 mm so an integrated circuit might use about one thousand times more silicon than aluminium. It's not the raw cost and availability of metal that is of concern: it is the cost of the metallization process (depositions and patterning) which has to be right!

> SAQ 5.6   (Objective 5.5)
> The printed circuits discussed in Chapter 2 used copper tracks between components. Integrated circuits use aluminium tracks. Why might copper be regarded as unsuitable for integrated circuits? (Consider deposition, adhesion, resistivity, compatibility and patternability.)

5.2.5 Summary

Silicon dioxide, silicon and aluminium are the major constituents of silicon-based semiconductor circuits. Together they can be processed satisfactorily to provide exotic structures of p-type and n-type regions of semiconductor, insulator and conductor.

Silicon dioxide can be grown on silicon or deposited from a gaseous silicon compound. Different regions of single-crystal silicon can be doped with donor atoms (for example P, As) or acceptor atoms (for example B, Al) by diffusion or ion implantation. Metallic interconnections can be formed by sputter deposition of aluminium silicon alloys.

We have not yet said how the pattern of p-type and n-type silicon or of the metal tracks are defined. This will come in the next section, which examines some of the electronic properties of the patterns we can produce.

5.3 Electronic structures

In Section 5.2 we looked at our basic materials and some processes used to manipulate them. Now I want to be a little more specific about how the materials behave electrically in response to voltages and currents within a particular device. In this analysis we must always be mindful of the context of our material. It is possible, for instance, for one device, or part of a device, to interact with another in ways that are not simply a combination of their separate behaviours. A pair of pn junctions in close proximity for example could, unexpectedly, act like a transistor (don't worry if you can't see why at the moment). Also, just about any electronic component has some kind of temperature sensitivity and it may unwittingly be more a thermometer than it is, say, a diode. Similarly capacitors have some inductance and resistance and inductors have capacitance and resistance. Our aim is to engineer a design which optimizes one property above others. We will look specifically at how electronic structures can be made to behave as resistors, diodes, capacitors and transistors.

5.3.1 Building an electronic structure

A simple electronic structure to have a go at is an isolated resistor. I am going to design a 1 kΩ resistance as a track 10 $\times$ 100 μm on a p-type silicon wafer using atomic diffusion. Here are some constraints. The resistor must fit within a 3 $\times$ 3 mm surface area of the silicon. As received from the manufacturers, the silicon is 525 μm thick and its resistivity is 10^{-2} Ω m. Diffusive doping will be used to modify the properties of the silicon wafer. (In Exercise 5.14 you can have a go at the ion-implanted version.)

Now the first idea I have is this. A track of extra-doped silicon will be created in the p-type silicon wafer. The track will be more conducting than the base material so that current will preferentially follow the track. I'm worried already. Diffusion may let me get tens of microns into the wafer, which is hundreds of microns thick and thousands wide. In fact the sheet resistance of the wafer is about 20 Ω square^{-1} so any tracks of 1 kΩ will be shorted out by the rest of the material. I may as well solder both ends of a 1 kΩ resistor onto a copper plate!

All is not lost though because I might be able to do something to restrict the current flow to the doped region. I'll make a track doped with phosphorus to convert it locally to n-type silicon so that electrons do the conducting. Also, suppose that I ensure that the p-type substrate (the wafer) is always connected to a more negative potential than that of either of the resistor contacts. That should discourage the electrons (negative charge) from getting too close and influencing the p-type region (and likewise holes from the p-type region will not be encouraged to enter the n-type region). As we shall see in the next section, the track and substrates are effectively isolated by what we call a reverse-biased (discourages current) pn junction. In this way it is all right to work with an island of n-type resistive silicon.

Back to the design. If a 10 μm depth were uniformly doped to give 10^{-3} Ω m of n-type resistivity, then the original p-type behaviour would be swamped. The sheet resistance would be $10^{-3}/10^{-5} = 10^2$ Ω square^{-1}.

In practice we will have to tolerate a non-uniform doping but this figure may be achieved as an average. The required 1 kΩ will be obtained with any track ten times longer than its width; that is, it can be viewed as a series of ten equal squares. I could use the whole surface in a folded 0.9 mm × 9 mm or else as little as a 10 μm × 100 μm track, although I should probably go no narrower than the 10 μm depth of the track. We can't print patterns to this lower size; other techniques such as ▼Etching▲ and ▼Photolithography and masking▲ are needed. To complete the design I ought to specify n$^+$ doping at the ends of the track to allow ohmic contacts to be formed to aluminium contacts. Figure 5.16 illustrates my solution, which leaves plenty of space for other components.

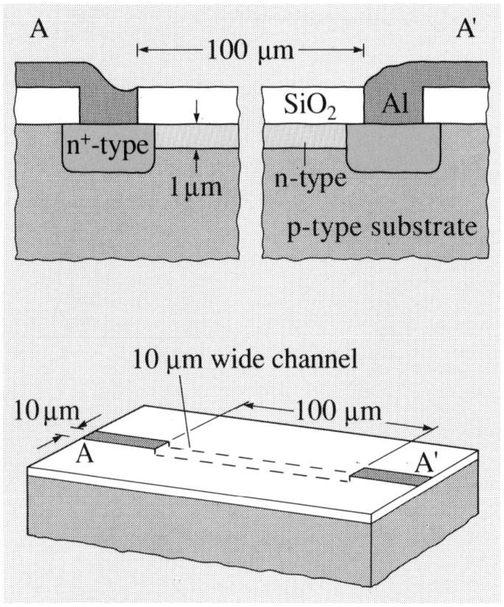

Figure 5.16 An integrated 1 kΩ resistor

▼Etching▲

Figure 5.17 'Philosopher meditating' (1653) by Ferdinand Bol

Etching, the selective removal of material to form patterns, has been around for some time. Figure 5.17 shows an early example. Technological applications of etching include the preparation of metallographic specimens for microscopy and our current interest, which is the fabrication of semiconductor devices. The electronics-grade silicon manufacturer will have already used silicon etchants in preparing the polished wafers.

Here we are interested in removing areas of metallic or insulating layers which have been deposited over the surface of a silicon wafer. The layers are of the order of a few microns thick and the features to be removed or left are of comparable size. Just how we define these features comes next ('Photolithography and masking').· Here I want to deal with how to remove solid a few atomic layers at a time. There

are two extreme approaches: chemistry and physics. The chemical route uses reactions which lure atoms from the solid into forming a molecule which is readily carried away as a gas or in solution in a suitable liquid. The physical route dislodges atoms by the brute force of energetic ion bombardment.

In the early days of silicon technology only wet chemicals were used. Wet etching is more often than not an isotropic process. The development of high-density circuits (more devices per unit area) has required narrower features to be defined. As well as intricate patterns in the semiconductor, metal tracks and contacts have to use less surface area. Eventually a stage is reached when it is necessary to etch down into a surface but not sideways. This is anisotropic erosion, and a different scheme has evolved for it.

Dry etching takes place in a low-pressure gas through which an electric current is passed. In effect we have a glow discharge, not unlike that inside a fluorescent tube. The etching happens by the same process as occurs in sputtering: the gas is ionized to create a plasma of ions (+ and sometimes −) and electrons (−). Ions accelerated from the gas to a few tens of electron volts and above will sputter most things to varying degrees. Bonds are broken by kinetic energy. Chemical energy, which the wet processes rely upon, is also available if chemically reactive gases are included with an inert carrier. For example, aluminium can be etched using a

chlorine-containing gas such as carbon tetrachloride (CCl_4). The electrical discharge plays a dual role here. First it breaks up the gas into highly reactive fragments (or radicals) such as atomic chlorine (Cl). Secondly it bombards the surface with ionic species which are energetic enough to sputter the native oxide from the initial surface and also to help clear other non-volatile debris. The bombarding ions are electrically accelerated perpendicular to the wafer surface so they can encourage etching with a strongly directed component. This technique is often called **reactive ion etching** (RIE) and is suitable for etching fine features of around a micron in size anisotropically.

Dry etching which relies on just chemical processes, without significant ion bombardment, is often distinguished from RIE by dubbing it **plasma etching**. Plasma etching, like wet chemistry, is usually isotropic. Can you see why? (No strong perpendicular bombardment means no preferred direction of erosion: no anisotropy.)

EXERCISE 5.12
(a) What sorts of features will require anisotropic etching? (Try sketching the etching.)

(b) The energy of ions involved in RIE may be tens to hundreds of electron volts. Compare this with ion-implantation energies and comment.

▼ Photolithography and masking ▲

Computer-aided design techniques are used to reduce sophisticated electronic designs (interconnected devices) into a suite of surface processing sequences which will build the structures required. As we shall see, each stage requires some areas of the surface to be exposed to particle fluxes while other areas are protected. Here we must consider how patterns of protection (or exposure) can be achieved. One protective medium which is readily available and widely used is silicon dioxide and by now you should know why. But how do we pattern it?

In Chapter 2 we discussed how optical images could be used to define electrical circuit patterns for printed circuit boards and how screen-printed inks could similarly define thick film conducting patterns. Which of these might be scaled down for use in our microtechnology? As you should already have guessed from the heading, it is the inherently small size of a photon (usually ultra-violet) which extends the lithographic capability to our present requirements.

Figure 5.18 shows a sequence which involves opening a window in the SiO₂ protection layer so that I can make my integrated resistor.

The polymeric photoresist material is applied in an organic solvent as a viscous drop in the centre of the wafer. A fast spinning motion is then used to spread the resist uniformly across the surface (this is **spinning on**). A baking stage follows, to remove solvent. The surface is then illuminated with ultra-violet light through a shadow mask on, or very close to, the surface. Alternatively, the desired pattern may be projected onto the surface with ultra-violet light.

The pattern is developed in the resist by dissolving unwanted material (for example un-crosslinked material in a negative resist). The same pattern is transferred into

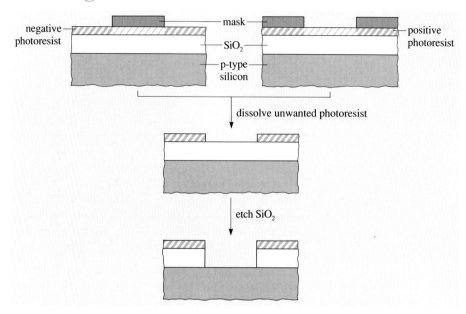

Figure 5.18 Photolithography for opening a window in silicon dioxide

underlying SiO₂ by using the photoresist to protect from the etchant areas which are to remain.

EXERCISE 5.13   Refer to Figure 5.18.

(a) What is the rôle of the photoresist?

(b) What will limit the smallest dimensions achievable (that is, resolution)?

In a subsequent stage I must make ohmic contacts to each end of a doped channel.

(c) What is the photolithographic problem for this step (registration)?

Lithography can also be done using beams of electrons or X-rays to 'write' onto a 'resist'. Electrons can be steered (as in a CRT), but X-rays are simply shaded by an

absorbing mask (which is analogous to the ultra-violet optical method).

Electron-beam resists use similar principles to photoresists, exploiting the fact that low-molecular-mass polymer dissolves more readily than high-molecular-mass material. The electron beam is used to break up 'heavy' chains (positive resist) or else to promote cross-linking between 'light' ones (negative resist). For the negative resist though, the material which remains behind will take up some solvent into any regions which are incompletely cross-linked and this can lead to swelling, to the detriment of resolution.

X-ray absorption by a material ultimately kicks electrons out of atoms so 'electron beam resists' can also be activated by electrons internally liberated during exposure to X-rays. Thus X-ray resists are no different.

EXERCISE 5.14   Specify a 1 kΩ resistor for the same chip, but using ion implantation.

SAQ 5.7   (Objective 5.6)
A chrome pattern on glass is required as the master mask for a photolithographic process. Explain briefly how the mask could be defined, starting with a glass plate sputter-coated with chrome, and using an electron beam controlled by CAD software.

The example of a microtechnology 1 kΩ resistor revealed the need to worry about a little more than just a conducting track. It was suggested that the track could be isolated from its environment by surrounding it with a reverse biased pn junction. The next two subsections look further into how this is achieved and what we can do with it. We shall also investigate the electrical structure of such a junction.

5.3.2 pn isolation

In making a resistor, I introduced n-type dopant to p-type silicon, thereby swamping the acceptors with donors until I had an island of n-type material. To discourage electrons from leaving the island I suggested that a voltage be applied between the two regions making the p-type the more negative. In this section I want to show you how the applied voltage is distributed. It can't be shared equally by all the semiconductor. In fact it's concentrated around the boundary between p-type and n-type materials. To understand this we will look for currents flowing in response to the applied voltage (conduction) and also for currents driven by diffusion; but remember that the idea is to isolate the n-type material so expect the net currents to be small.

Conduction (or drift) currents

You are familiar with current density j (A m^{-2}) flowing in response to an electric field E (V m^{-1}), effectively a potential gradient, in a medium of conductivity σ (Ω^{-1} m^{-1}):

$$j = \sigma E$$

In n-type material electrons are in the majority. There are so few holes that the conduction current is dominated by the flow of electrons and we can say:

$$j_{n \text{ conduction}} = (n_n e \mu_e + p_n e \mu_h)E \approx n_n e \mu_e E$$

where n_n and p_n are respectively the electron and the hole density (in n-type).

EXAMPLE   In a sample of n-type silicon, the doping is such that there are 10^{22} electrons m^{-3} and 2.3×10^{10} holes m^{-3}. What fraction of the conduction current is carried by the minority species (holes)? Mobilities μ_e and μ_h are respectively 0.35 and 0.044 m^2 V^{-1} s^{-1}.

ANSWER   The net conductivity is the sum of an electron conductivity ($n_n e \mu_e$) and a hole conductivity ($p_n e \mu_h$):

$$\sigma = n_n e \mu_e + p_n e \mu_h$$

The conduction current which flows is in proportion to the conductivity so the fraction which is due to minority carriers (holes) is

$$\frac{p_n e \mu_h}{(n_n e \mu_e + p_n e \mu_h)} = \frac{p_n \mu_h}{(n_n \mu_e + p_n \mu_h)}$$

$$= \frac{2.3 \times 10^{10} \times 0.044}{(10^{22} \times 0.35 + 2.3 \times 10^{10} \times 0.044)}$$

$$\approx 3 \times 10^{-13}$$

You can usually forget about the contribution of minority species to conduction currents in extrinsic material.

Note. The subscripts on carrier densities n and p refer to the nature of the host material. Thus n_n is the electron density in n-type material and n_p is the electron density in p-type material. Similarly, p_n is the hole density in n-type material and p_p is the hole density in p-type material.

Diffusion currents

Concentration gradients drive fluxes. Fluxes of charge are currents. For example, if the hole density in n-type material, even though it is small, changes from p_{n0} to p_{n1} over a distance l_n then there is an associated diffusion current:

$$j_{n \text{ diffusion}} \approx - \frac{e D_h (p_{n0} - p_{n1})}{l_n},$$

where D_h is the hole diffusion coefficient in the material.

EXAMPLE   At one end of a sample of silicon in an n-type region, the hole density is maintained at 2.3×10^{10} holes m^{-3} (don't worry yet about how). A distance 10 μm away there are only 10^9 holes m^{-3} because holes are leaking away into a different region (again don't worry yet about how). Estimate the diffusion current density. The hole diffusion coefficient D_h is 1.1×10^{-3} m^2 s^{-1} and the magnitude of the electronic charge is 1.6×10^{-19} C.

ANSWER   The density change is $(2.3 \times 10^{10} - 1 \times 10^9)$ m^{-3} over a distance of 10^{-5} m so the gradient is

$$\frac{(p_{n0} - p_{n1})}{l_n}$$

$$= \frac{(2.3 \times 10^{10} - 1 \times 10^9)}{10^{-5}} \, \text{m}^{-4}$$

$$= 2.2 \times 10^{15} \, \text{m}^{-4}$$

The associated current density is

$$j_{n\,\text{diffusion}} = 1.6 \times 10^{-19} \times 1.1 \times 10^{-3} \times 2.2 \times 10^{15} \, \text{A m}^{-2}$$

$$= 4 \times 10^{-7} \, \text{A m}^{-2}$$

which is about 0.4 pico-amps (or two and a half million holes per second) per square millimetre. This is a small current but if it's all we've got we had better not ignore it.

On its own, a concentration gradient will drive itself away, tending to level out differences in concentration. The gradient can be sustained only if new holes come along from somewhere to replenish the losses. In semiconductors, in the background, pairs of electrons and holes are being thermally generated all the time, so we have a ready source of supply.

Total current

The net steady current at any point is the sum of the diffusion and the conduction currents for both electrons and holes. By good fortune, it is usually the case that one or more of the four possible current terms is negligible.

> EXERCISE 5.15   In words, what are the four terms in the expression for total steady current density in n-type material?

Hole currents

We've just discussed currents in terms of what drives them (gradients of potential and of concentration). Here I want to consider the same currents, but in terms of which type of charge carries them. This will show that, within the bulk of a semiconductor, we cannot have a large potential gradient (or electric field). This is why I said that changes in potential must be concentrated around the junction. When we understand this we can develop a consistent picture of how pn isolation works.

The idea of biasing the n-type island was to hold electrons in the n-type material. What few holes there are in the n-type material will feel the opposite effect and will be encouraged to leave. If the p-type material is held negative it is energetically very attractive to holes. Any hole wandering around the n-type island and near to the edge may be lured away into the more attractive p-type region. In fact because of this there will be hardly any holes left loitering in the edge region of the island; the hole density here is virtually zero. Far from the junction, the hole density might be more like its normal n-type level (p_{n0}), sustained at this level by thermally generated replacements for those leaving. So in the n-type region there is a density gradient of holes and thus a small diffusion current of holes (j_{hn}) driven out into the p-type material. (See Figure 5.19b and the above example where we estimated a diffusion current density.)

Holes therefore arrive in the p-type material at a rate of so many per second. To prevent a build up, the same number of holes per second must be carried away from the junction (as a current density j_{hp}). Here, though, holes are much more numerous, being the majority carrier in p-type. So a very weak electric field is enough to clear the way for the new arrivals. It's rather like a small stream pouring over a waterfall to feed a broad slow moving river (Figure 5.19c).

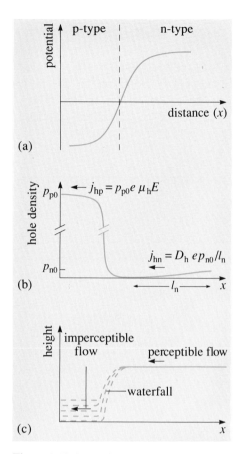

Figure 5.19 Potential (a) and hole density (b) in an extrinsic semiconductor material where the doping abruptly changes from acceptor to donor. (c) An analogy with streams and waterfalls

EXAMPLE   Holes diffuse out of an n-type island at a current density of 4×10^{-7} A m^{-2}. What electric field is necessary to drive the holes as a conduction current through p-type material which has 10^{22} holes m^{-3} (μ_h is again about 0.044 m^2 V^{-1} s^{-1})?

ANSWER   We must use the conduction current density formula

$$j = \sigma E \qquad \text{or} \qquad E = \frac{j}{\sigma}$$

We know the current density. We can calculate the conductivity if we neglect the contribution from electrons (recall our earlier example in which we calculated the minority current to be 3×10^{-13} of the majority current).

$$\sigma \approx p_p e \mu_h = 10^{22} \times 1.6 \times 10^{-19} \times 0.044 \, \Omega^{-1} \, m^{-1}$$
$$= 70.4 \, \Omega^{-1} \, m^{-1}$$

and so

$$E = \frac{4 \times 10^{-7}}{70.4} \, V \, m^{-1}$$
$$\approx 6 \times 10^{-9} \, V \, m^{-1}$$

I said a very weak field would do. There is virtually no electric field needed in the p-type material for the clearing of holes.

Electron currents

> **EXERCISE 5.16**   Account for the steady electron current in the p-type (j_{ep}) and in the n-type (j_{en}) regions.

Voltage and net current

Most of the voltage we apply to isolate our n-type track (it may be a volt or so) isn't needed for the very weak fields in the bulk of the p-type and n-type regions, so it must be concentrated at the junction between p-type and n-type material as illustrated in Figure 5.19(a). There is nowhere else for it to go. In the next section we will look in more detail at just what goes on within this junction region.

The total current crossing the junction is $j_{hn} + j_{ep}$ (Figure 5.19b shows just the hole contribution). As we have just seen, it is independent of voltage, being due to minority carriers leaking (by diffusion) out of the p-type and n-type regions. That is the key to the isolation. Although there is a small leakage current, it is independent of voltage. Figure 5.20 illustrates how the n-type resistor conducts current (along the *x*-axis) without the p-type region knowing about it because the leakage current is immune to voltage changes arising from current flow in the resistive channel. Working like this, the junction (or 'pn junction') is said to be **reverse biased**. So-called pn diodes also exploit the properties of a pn junction, including the opposite case of so-called forward bias, but before we look at these we must consider the nature of the junction proper.

> **EXERCISE 5.17**   Ideally, no current would flow between the isolated n-type region and the negatively biased p-type material surrounding it. What can we do to keep leakage currents small in silicon?

5.3.3 Depleted semiconductor

Before we leave the idea of an isolating pn junction, we ought to look a little more at its electrical structure. I've said that most of the voltage appears across a narrow region where p-type and n-type material meet. Figure 5.21 shows how this can be. (Figure 5.21d is just Figure 5.19a with potential measured from one side rather than the middle.)

In Figure 5.21(a) notice that close to the plane where the dominant doping changes from acceptor (p-type) to donor (n-type) the material is 'depleted' of its majority carriers; the carrier density appears to be almost nothing in this region. This is because near the junction the holes and electrons from their respective sides have paired up and annihilated each other. The 'depletion region' contains dopants which hold fixed positive charge (n-type donor) and negative charge (p-type acceptor), just as in the rest of the material, except that here there is no neutralizing majority

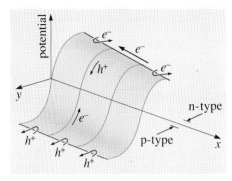

Figure 5.20 The mechanism of pn isolation for the n-type resistor in Figure 5.16

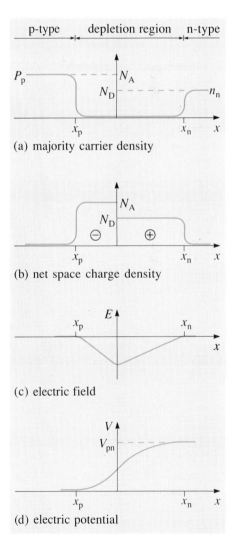

Figure 5.21 Conditions around a pn junction

carrier charge. The depletion region is therefore a pair of charged layers and these are responsible for a local variation of electrical potential. The local potential adjusts to stop further annihilations by holding the mobile species apart with electric fields (Figure 5.21c). Because the material away from the junction has a respectable conductivity, it turns out that the net charge of the depletion layers is zero. Each negative acceptor ion on the p-type side of the depletion region is matched by a positive donor ion on the n-type side.

Figures 5.21(c) and (d) show electric field $E (= -dV/dx)$ and potential V across the junction. Be careful to get the signs of charge and changes of potential right. In going from p-type to n-type, the potential increases across the negatively charged region clearly not because of the surrounding (negative) acceptor ions, but because of the positively charged ions beyond. In ▼pn Diodes▲ and ▼pn Capacitors▲ we look a little deeper.

Concerning the depletion region, you should remember that:
● The amount of charge in the depletion layer depends upon the voltage across it. This means it has a capacitance (it effectively stores charge).
● The width of the depletion region depends upon both the voltage across it and the doping of the p-type and n-type regions. Both the voltage and the doping affect the amount of capacitance (as shown in 'pn Capacitors').
● There is still a contact potential across a depletion region when no net current flows (as shown in 'pn Diodes').

SAQ 5.8 (Objective 5.7)
Starting with a wafer of n-type silicon, say briefly how you would make an array of isolated pn diodes.

SAQ 5.9 (Objective 5.8)
How closely could the devices in SAQ 5.8 be packed if they are isolated by one-volt reverse biased pn junctions and the lightest doping is that of the n-type substrate which has 10^{21} donors m^{-3}? What processing step could we use to reduce the minimum separation between devices? (Assume that $N_{D\ substrate} \ll N_{A\ diode}$ and $N_{D\ diode}$, and we'll suppose that power dissipation is not a problem. Refer to the example in 'pn Capacitors'.)

SAQ 5.10 (Objective 5.9)
Compare and contrast the capacitance of pn junction with that of the conventional capacitors discussed in Chapter 2.

I want to go on to look at another electronic structure where depletion is again important, that of a metal/oxide/semiconductor (MOS) configuration. It is this which will eventually furnish our memory element.

▼ pn Diodes ▲

If we combine our ideas on minority-carrier diffusion currents with details of the depletion region we can begin to see how to make a pn junction behave as a diode. In Section 5.3.2 we considered the flow of minority charges to the edge of the depletion region. What about majority charges? Outside the depletion region these were reckoned to drift in a weak electric field at constant density. Within the depletion region, the potential changes more rapidly and the associated electric field repels majority charge so that their number decreases towards the junction, where the dominant doping changes type. For majority charge, in fact, the resulting density gradient tries to drive a diffusion current across the junction but this is almost exactly matched by the opposing drift current which is driven by the electric field. It can be shown accordingly that the carrier densities follow a Boltzmann law in the depletion region, that is

$$p = p_p \exp\left(-\frac{eV}{kT}\right)$$

describes the hole density at any potential V with respect to the p-type material outside the depletion region (the sign of V will be positive). And

$$n = n_n \exp\left(\frac{e(V - V_{pn})}{kT}\right)$$

describes the electron density at any potential $V - V_{pn}$ (which will be negative) below that of the n-type edge of the depletion region (see Figure 5.21d).

Essentially the band structure of electron energies across the depletion layer is superimposed on the electrical potential energy variation ($-eV$ for electrons, which have high potential energy where V is negative). Here the Boltzmann factor is just the tail of the Fermi–Dirac function, which assigns electrons to available energy states. If you know anything about diodes you'll have noticed that they have an exponential current–voltage characteristic. I want to show how this characteristic arises, but first we must look again at the reverse biased pn junction.

In discussing the isolating pn junction, I said that when the junction was reverse biased, the hole density just outside the depletion region in the n-type material (p_{nl}) was 'virtually zero'. We can now quantify

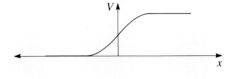

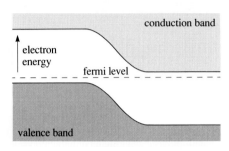

Figure 5.22 The potential and band structure of a pn junction

that. If there were one volt across the junction, at room temperature ($kT/e = 0.025$ volts) the hole density would change from p_p in the p-type to:

$$p_{nl} = p_p \exp\left(\frac{1}{0.025}\right)$$
$$= 4 \times 10^{-18} p_p$$

which is not bad for 'virtually zero', compared with p_p, or even with the (minority) hole density in the n-type material well away from the junction (p_n) as we shall see.

We can also wonder about the conditions under which the net current of holes, which we said was small, is exactly zero. That's how it will be if we just have a pn junction without any attempt to bias it. We are going to find that nevertheless there is still a potential across the junction. It must be true that there is no minority diffusion beyond the depletion region, so outside the depletion region there can be no gradients of minority carrier density. The minority density (of holes in the n-type region) can then be found from the general equilibrium condition that

$$n_{n0} p_{n0} = n_{p0} p_{p0} = n_i^2.$$

In the n-type, the electron density is about N_D and so the equilibrium hole density is

$$p_{n0} = \frac{n_i^2}{N_D}$$

In silicon $n_i = 1.45 \times 10^{16} \text{ m}^{-3}$ at 300 K. If the silicon is doped with donors (and hence electrons) at a concentration of 10^{23} m^{-3}, p_{n0} works out as 2×10^9 holes m^{-3}. To find the potential difference which is consistent with this density of holes just arriving from the p type, we can again use the Boltzmann exponential to get the value of V which must satisfy

$$\frac{n_i^2}{N_D} = p_{n0} = p_p \exp\left(-\frac{eV}{kT}\right)$$

In the p-type, the hole density (p_p) is given by N_A (the acceptor doping) so substituting and taking logarithms:

$$V = \frac{kT}{e} \ln\left(\frac{N_A N_D}{n_i^2}\right)$$

If the acceptor density were also 10^{23} m^{-3} then in silicon at room temperature this would be about 0.8 volts. We call this the **built-in voltage** or **contact potential**. It arises because we have a junction between two different materials, p- and n-type silicon, with different Fermi levels. You will have come across a contact potential between two different metals. It's a similar thing, as Figure 5.22 shows. Basically it's a question of making the Fermi level continuous across the junction.

Incidentally, I can't just measure the built-in voltage with a voltmeter because contact potentials between my copper leads and the two different semiconductor types will just cancel the built in voltage!

We've looked at equilibrium and reverse bias. What if I now use a battery to 'forward bias' the diode? Figure 5.23 shows how the hole density might vary in both p- and n-type regions for the conditions of reverse bias (when there are leakage currents), equilibrium (zero current) and forward bias.

For reverse bias we said that the minority density near the depletion region edge was about zero, whereas in the zero-current condition we have just written it both in terms of the intrinsic carrier density and the n-type doping, and in terms of the p-type doping and the built in voltage:

$$p_{n0} = \frac{n_i^2}{N_D} = N_A \exp\left(\frac{-e\Psi}{kT}\right)$$

where I've put Ψ ('psi') for the built-in voltage. Now I propose to forward bias the

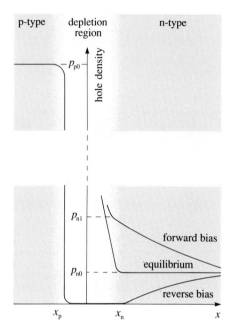

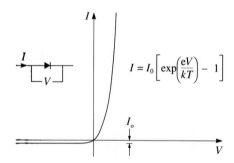

Figure 5.23 Hole density across a pn junction

Figure 5.24 The diode I–V characteristic

$$I = I_0\left[\exp\left(\frac{eV}{kT}\right) - 1\right]$$

junction so that the applied voltage (V_D) decreases the potential difference between p and n regions (from Ψ to $\Psi - V_D$). I will still appeal to the Boltzmann exponential to estimate how many holes from the p-type region have enough energy to get into the n-type, where they can reinforce the native minority charge. At the depletion edge

$$p_{n1} = N_A\exp\left(\frac{-e(\Psi - V_D)}{kT}\right)$$

But our last result ($p_{n0} = N_A\exp(-e\Psi/kT)$) can be used to express the minority hole density at the depletion edge in relation to the native equilibrium hole density (by simple substitution)

$$p_{n1} = p_{n0}\exp\left(\frac{eV_D}{kT}\right)$$

To see how this can account for the current–voltage characteristic of a junction diode,

$$I = I_0\left[\exp\left(\frac{eV_D}{kT}\right) - 1\right]$$

work through the following example. (See also Figure 5.24.)

EXAMPLE

(a) Estimate the diffusion current density of minority charge in the n-type side of a pn junction (assume that the region depletes a distance l_p of the p type and about l_n of the n-type and use the ideas presented in pn isolation).

(b) In the n-type material why do the majority charges also have to carry current?

(c) Estimate by analogy with (a) the majority current density (bearing (b) in mind).

(d) The voltage across the diode is V_D, and the total current at any point must equal the sum of contributions (a) and (b). So what is the current-voltage characteristic for the junction?

ANSWER

(a) From Section 5.3.2, in the n-type,

$$j_{hn} = j_{n\text{ diffusion}}$$

$$= eD_h\frac{p_{n1}-p_{n0}}{l_n}(\text{A m}^{-2})$$

Now we write p_{n1} in term of p_{n0} and V_D:

$$j_{hn} = \frac{eD_h p_{n0}}{l_n}\left[\exp\left(\frac{eV_D}{kT}\right) - 1\right]$$

(b) In the n-type, majority charge (electrons) will flow away from the junction (j_{en}) if the other side (p-type) is injecting electrons (its minority species) into it (j_{ep}). The majority charge drifts at near constant density in a weak electric field, removing the new arrivals as fast as they arrive, so $j_{en} = j_{ep}$.

(c) Since the n-type majority current is just the p-type minority current from elsewhere in its travels, in the n-type

$$j_{en}\,(= j_{ep})$$

$$= \frac{eD_e n_{p0}}{l_n}\left[\exp\left(\frac{eV_D}{kT}\right) - 1\right]$$

(d) The currents in (a) and (b) are both flowing in the n-type. These two current densities are of opposite charges flowing in opposite directions so they add. If there is a uniform cross-sectional area A:

$$I = A(j_{en} + j_{hn})$$

$$= Ae\left[\frac{D_e n_{p0}}{l_p} + \frac{D_h p_{n0}}{l_n}\right]\left[\exp\left(\frac{eV_D}{kT}\right) - 1\right]$$

We can lump all the constants together and write

$$I = I_0\left[\exp\left(\frac{eV_D}{kT}\right) - 1\right]$$

This is the diode equation. (You may well come across more rigorous derivations than this one.)

▼pn Capacitors▲

A parallel-plate capacitor without a dielectric has a constant electric field between the plates ($E = -V/d$). When a dielectric is present, even a polarizable one, because the charges introduced into the gap occur as dipoles, the net charge in any small volume is zero and the field remains constant. A depletion region is a little different. Instead of being on the plates or distributed throughout as dipoles, a depletion region's charge is arranged with all the positive charge spread over one part and all the negative charge occupying the rest. This charge is due to the dopant ions and so we can take it to be dispersed at an approximately constant number of ions per unit volume in each part. Now, an electrostatic field is how charges influence each other at a distance, and the amount of electric field in a capacitor is in proportion to the amount of charge on the plates (see Chapter 2). Let's think a little more about the 'electric field lines' which we draw to represent the interaction of charges. The lines start on positive charge (the source) and end on negative charge (the sink). Figure 5.25 shows two identical capacitors with different charges on the plates of each. The one with more charge has equivalently more electric field lines (a stronger field). This must be so because the greater charge requires the greater potential difference. These formulae should be familiar to you:

$$Q = CV$$

and

$$C = \frac{\varepsilon A}{d}$$

Then

$$E = \frac{V}{d} = \frac{Q}{\varepsilon A}$$

The field is directly associated with the charge on the plates. Next then I want to extend this to describe the distributed charge in the depletion region.

Figure 5.26 shows the electric field emanating from layers of positive charge Δx thick, of area A and uniform charge density ρ. Each layer introduces a charge $\Delta Q \ (= \rho A \Delta x)$ which is matched by a negative charge in layers to the right where the field lines end. Treating each layer as we did the capacitor plate, we can say:

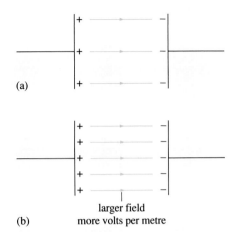

(a)

(b)

larger field
more volts per metre

Figure 5.25 Two identical capacitors storing different charges

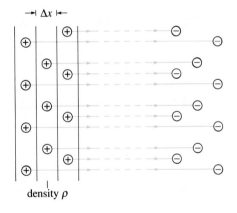

→| Δx |←

density ρ

Figure 5.26 The electric field arising from layers of charge

$$\Delta E = \frac{\Delta Q}{\varepsilon A} = \frac{\rho \Delta x}{\varepsilon}$$

or with proper calculus,

$$\frac{dE}{dx} = \frac{\rho}{\varepsilon}$$

This is one of the fundamental equations of electromagnetism. If ρ and ε are constant the field E increases linearly with x.

To find the way in which potential and charge are related in the depletion region we can integrate to get the field E, and then integrate the field to get the potential. Figure 5.21 indicated a depletion region

having N_A ionized acceptors per cubic metre on the p-type side. In this region therefore:

$$\frac{dE}{dx} = -\frac{N_A e}{\varepsilon}$$

Outside the depletion region the field is zero (all field lines arriving at negative charges come from positive charges), so I can integrate between $x = x_p$ at the p-type edge where $E = 0$, to any point x between x_p and 0:

$$E = \frac{N_A e}{\varepsilon}(x_p - x).$$

The potential gradient $dV/dx \ (= -E)$ can then be integrated from $x = x_p$ to $x = 0$. This gives the change in voltage ΔV_p across the p-type side of the depletion region:

$$\Delta V_p = \frac{N_A e}{\varepsilon} \frac{x_p^2}{2}$$

This voltage develops an energy barrier to encourage the holes to stay in the p-type region.

The depleted n-type material similarly develops a potential owing to its ionized donors, so that the total voltage across the depletion region is

$$V = \frac{N_D e}{\varepsilon} \frac{x_n^2}{2} + \frac{N_A e}{\varepsilon} \frac{x_p^2}{2}$$

$$= \frac{e}{2\varepsilon}(N_D x_n^2 + N_A x_p^2)$$

What we want to know is how much extra charge is involved if the potential increases a bit, so we must find a way to put the charge into the equation. The total charge on each side is the same, and so we can say

$$\frac{Q}{A} = N_D e x_n = N_A e x_p$$

If you now use these to substitute for x_n and x_p we get a relation between potential and charge for the region:

$$V = \frac{Q^2}{2\varepsilon e A^2}\left(\frac{1}{N_D} + \frac{1}{N_A}\right)$$

or

$$\frac{Q}{A} = \sqrt{2\varepsilon e\left(\frac{N_D N_A}{N_D + N_A}\right)V}$$

To get the capacitance we need dQ/dV, which tells us the extra charge dQ associated with an increment of voltage dV. Differentiating the above with respect to V, we find

$$\frac{1}{A}\frac{dQ}{dV} = \frac{C}{A}$$

$$= \sqrt{\frac{\varepsilon e}{2}\left(\frac{N_D N_A}{N_D + N_A}\right)\frac{1}{V}}$$

This expresses the capacitance per unit area for the depletion region of a pn junction. There is about 0.1 millifarad per square metre for silicon with $N_D = 10^{23}$ m^{-3}, $N_A = 10^{21}$ m^{-3} and $V = 1$ volt.

EXAMPLE   The equations above for Q/A can be used to show that the extent of the depletion region on the n-type side of a pn junction is related to the doping on both sides. Do the necessary algebra and decide how to restrict the size of the depletion region on the n-type side. Then estimate the total width of the depletion region at a junction where doping changes from $N_D = 10^{23}$ m^{-3} to $N_A = 10^{21}$ m^{-3}. There

is one volt across the junction and in silicon $\varepsilon_r = 12$. Permittivity of free space ε_0 is 9×10^{-12} F m^{-1}.

ANSWER

From the foregoing

$$\frac{Q}{A} = N_D e x_n$$

and

$$\frac{Q}{A} = \sqrt{2\varepsilon e\left(\frac{N_D N_A}{N_D + N_A}\right)V}$$

so

$$x_n = \sqrt{\frac{2\varepsilon}{eN_D}\left(\frac{N_A}{N_D + N_A}\right)V}$$

To keep x_n small we clearly need to make N_D large (heavy doping). That makes sense because it means we arrange for lots of charge to be around, so we don't need to deplete wide regions of material to find enough charge to build up a change of potential.

Inserting values for N_D, N_A, V and ε_r,

$$x_n =$$

$$\sqrt{\frac{2 \times 12 \times 9 \times 10^{-12}}{1.6 \times 10^{-19} \times 10^{23}}\left(\frac{10^{21}}{10^{23} + 10^{21}}\right)}$$

$$= 11.6 \text{ nm}.$$

The expression for x_p interchanges the doping terms:

$$x_p = \sqrt{\frac{2\varepsilon}{eN_A}\left(\frac{N_D}{N_D + N_A}\right)V}$$

$$= 1.16 \text{ μm}.$$

The total depletion width ($x_n + x_p$) is 1.17 μm.

You can see that the depletion of the more lightly doped material (p-type in this case) dominates the total depletion width. In the equation for x_p above, the factor $N_D/(N_A + N_D)$ is approximately one, and so

$$x_p \approx \sqrt{\frac{2\varepsilon V}{eN_A}}$$

In general, whenever one material is much more lightly doped than the other, the width x_d of the depletion region will be

$$x_d \approx \sqrt{\frac{2\varepsilon V}{eN}}$$

where N is the concentration of donors or acceptors (as appropriate) in the more lightly doped region.

5.3.4 MOS structures

Layers of metal (M), insulator (such as silicon dioxide, O) and semiconductor (S) are as inevitable a part of semiconductor devices as pn junctions (see Figure 5.1). At the very least, an insulator between relatively good conductors will have some capacitance whether we like it or not (▼MOS capacitors▲). Careful adjustment of the dimensions and properties of the layers allows a new electronic effect to be exploited. This new effect, called **inversion**, has electrons locally in the majority at the edge of p-type materials or else holes in the majority at the edge of n-type. It can occur, along with depletion, in the semiconductor adjacent to an insulator in a metal/insulator/semiconductor structure. We will concentrate on the MOS (metal/oxide/silicon) version looking at the silicon-silicon dioxide interface which has a crucial rôle to play, and then at 'inversion' and the conditions necessary to produce it. First you ought to be able to say how we build up MOS layers.

EXERCISE 5.18 Figure 5.27 shows part of an MOS structure. Identify the main process steps necessary to make it starting from a lightly doped p-type substrate.

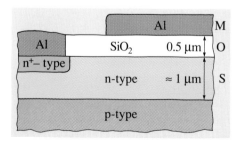

Figure 5.27 Part of a MOS structure

▼MOS capacitors▲

As you know from Chapter 2, there are capacitors (which you intend) and there is capacitance (which you have to live with). The mention of an insulator between conductors (semi- or otherwise) should arouse your curiosity about what we can have and what we must put up with.

EXAMPLE

(a) What is the capacitance per square millimetre between a metal track and a heavily doped (n^+) silicon layer separated by one micron of silicon dioxide?

(b) Specify the dimensions of a 100 pF MOS capacitor capable of withstanding 100 volts.

ANSWER

(a) Using $C = \varepsilon A/d$ with
$\varepsilon = 4 \times 9 \times 10^{-12}$ F m^{-1} (Table 5.1),
$A = 10^{-6}$ m^2 and $d = 10^{-6}$ m,

$$C = 3.6 \times 10^{-11} \text{ F mm}^{-2}$$
$$= 36 \text{ pF mm}^{-2}$$

(b) SiO_2 can withstand 10^9 V m^{-1} (Table 5.1) so we can have 100 V across 0.1 µm ($= 10^{-7}$ m). The dielectric thickness can therefore be decreased by ten from that in part (a). So we can get 360 pF mm^{-2}, thus 100 pF needs about $\frac{1}{3}$ mm^2.

But a semiconductor is not a metal. At the metal/oxide interface a positive charge is built up when a positive voltage is applied here relative to the semiconductor. At the semiconductor/oxide interface negative charge must accumulate to match the positive charge. That's easy for n-type semiconductor. But what would happen if the applied voltage were reversed: negative charge at the metal/oxide interface and positive charge at the semiconductor/oxide interface? The n-type material develops positive charge by withdrawing electrons to expose the fixed positive charge on the ionized donors: that is it develops a depletion region. In very heavily doped material this may be much thinner than the oxide layer as we saw in the previous

section. But when the doping is light, the capacitance of the depletion region must be included. As with pn capacitance, this introduces a voltage dependence. The real situation is yet more complicated because the oxide/semiconductor interface is not as simple as we have just supposed. We'll come back to this soon.

An MOS structure is not the easiest way to make a capacitor if you are in the capacitor business. It becomes a very attractive proposition though if the capacitor is required next to another MOS device on a silicon wafer because, requiring the same processing steps (masking, p^+ or n^+ doping, oxide growth, metallization), its manufacture can be absorbed into the processing sequence. This also means that if you want a compact integrated circuit using MOS structures, the design should minimize the number and value of capacitors as they use a relatively large surface area.

The Si/SiO$_2$ interface

Earlier we looked at thermally grown silicon dioxide and deduced that the 'better' stuff is grown slowly. This is because we want enough oxygen to satisfy the bonding capabilities of silicon atoms after they have forsaken the silicon crystal lattice. But we can't change abruptly from the last layer of crystalline silicon atoms into the first amorphous layer of silicon dioxide. There would be poor adhesion across the interface if we did, but anyway it's not in the nature of atoms at high temperatures to stay in neat lines. So the picture you should have of the interface is of a change from the crystalline order of silicon over several atomic layers (a few nanometres) incorporating progressively more oxygen until the stoichiometry of SiO_2 is achieved. This 'no-man's material' is sometimes said to be SiO_x ($x < 2$). It acts as an adhesive layer and two very important consequences follow from its presence:

● The wonderful regularity of the crystal comes to an end quite sharply, though not abruptly.
● There is certain to be some silicon incompletely bonded in the vicinity of the interface.

The ending of the crystal means that electrons no longer perceive an energy scale consisting of conduction band/gap/valence band. Energy levels associated with silicon atoms close to the Si/SiO_x region are not the entirely delocalized states characteristic of the bulk. As a result, the energy gap near the interface is peppered with localized 'surface states' or energy traps (some shallow, some deep) into which electrons may fall for a time.

In the SiO_x, some silicon atoms have at least one extra bonding electron, which can all too easily be rattled free by thermal vibration to end up in one of the nearby silicon energy levels (some of which may even connect with the conduction band). But this leaves behind static positive charges ('fixed oxide charge') in the SiO_x layer, so the interface holds positive charge as well as extra mid-gap energy levels. If 1% of silicon atoms in the SiO_x are unfortunate enough to lose an electron then in this layer there may be positive charge at a density of around 10^{26} Si^+ ions per cubic metre. Compare that with typical doping levels (Figure 5.5). Silicon adjacent to the interface is going to be different from bulk silicon.

Inversion

Let's think about p-type silicon in an MOS structure but with no externally applied voltage. Holes are the majority charge carrier in p-type silicon, but near the interface with SiO_2 we can expect fewer holes than usual owing to repulsion by the oxide's fixed (positive) charge. Can you see what's going to happen? Holes are going to retreat from the surface, leaving behind their associated fixed acceptor ions (negative) as a depletion layer. Unlike the pn depletion layer, the material on one side is not a conductor and as a result settles down with an electric field in it. Figure 5.28(a) shows a sketch of the potential in this MOS structure.

Any electron (a minority carrier in p-type semiconductor) which wanders into the depletion region will be drawn to the interface where an accumulation of electrons can occur, in part compensating for the oxide charge themselves. In fact if the oxide charge is sufficient, p-type silicon may develop a region at the interface where electrons actually outnumber holes! This is termed **inversion**: an electron-rich (n-type) channel at the edge of a nominally p-type material. Bear in mind that when we look at MOS devices, we will often refer to the polarity of the *channel* rather than that of the *semiconductor* containing the channel.

EXERCISE 5.19 Rework the arguments for n-type substrate material to show that spontaneous inversion to form a p-type channel at the interface will not arise.

If inversion is to be an exploitable property, it had better be easily controlled. The amount of fixed oxide-charge is processing dependent and can to some extent be influenced. Rapidly grown oxide layers will contain more incompletely bonded silicon, for example, and annealing treatments can have some bearing on the fixed charge; but it is usually in the range 10^{14} to 10^{16} charges per square metre. 'Charges per square metre' are also stored on capacitor plates, so let's think about what would happen if a voltage were applied between the metal and the semiconductor, treating the whole assembly like a capacitor. For p-type

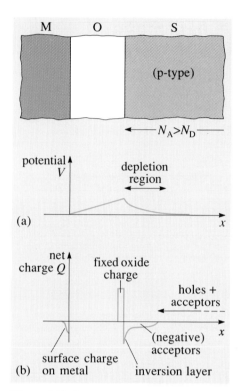

Figure 5.28 Potential and charge in a MOS structure I

semiconductor, negative voltages on the metal will draw holes to the interface. Conversely, positive voltages will drive them away, in just the same way that the positive oxide charge did. We ought to be able, therefore, to exercise electrical control over inversion. Furthermore, we ought to be able to invert n-type material near an oxide interface, in spite of the fixed positive oxide charge.

The electrical threshold to inversion

The next stage effectively includes the oxide charge in the MOS capacitance story (we looked at this in 'MOS capacitors'). Capacitance is to do with charge storage, and we must consider at least five separate sources of charge in an MOS structure (see Figure 5.29).

1 Metal/oxide interface charge Q_m. Electrons are pushed to or withdrawn from the metal surface to give a surface charge density Q_m/A (C m^{-2}).

2 Fixed oxide charge Q_{fo} near the oxide/semiconductor interface ('fo' for fixed oxide). These charges form a thin sheet of charge with a surface charge density Q_{fo}/A (C m^{-2}).

3 Minority bulk semiconductor charge carriers forming a sheet of inversion charge Q_{mi} right at the semiconductor/oxide interface.

4 Semiconductor depletion layer charge Q_D extending into the semiconductor near the semiconductor/oxide interface, arising from fixed ionized dopant atoms.

5 Majority bulk semiconductor charge carrier. This charge has density p or n, depending on whether it's p-type or n-type semiconductor. It keeps the bulk semiconductor neutral.

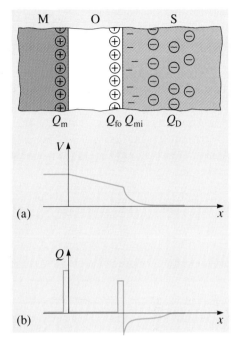

Figure 5.29 Potential and charge in a MOS structure II

EXERCISE 5.20 Which of the charge sources will be altered by applying a voltage across the MOS structure?

You should recognize that if the charging of the MOS capacitor puts positive charges at the metal/oxide interface, they will act in much the same way that the oxide charge did. Think first about a p-type semiconductor layer. Positive charge at the metal/oxide interface will, together with the positive oxide charge, deplete underlying p-type semiconductor to expose negative (acceptor) charges. The exposed acceptor charges must be sufficient in number, together with any electron (minority) accumulation, to match the total positive charge. As the minority charge increases in density, it becomes increasingly important. Less and less additional depletion is needed to match increases in positive charge at the metal/oxide interface as the capacitor voltage is increased. At 'inversion', the electron density at the interface equals the hole density in the bulk and virtually all further negative charge is supplied by this inversion layer (see Figure 5.29b).

What if we reverse the voltage on the MOS capacitor? Negative charge at the metal/oxide interface can itself first match the oxide charge and thereafter call on as many holes as it needs to build up at the semiconductor/oxide interface.

If the MOS structure is made with n-type rather than p-type semiconductor, the positive oxide charge remains the same but otherwise the argument can be reversed. The **threshold voltage** of an MOS structure is defined as the potential difference necessary to cause inversion at the semiconductor-oxide interface (minority interface density = majority bulk density).

It is not too difficult to treat the MOS capacitor properly with depletion layers and oxide charge taken into account. For our purposes, it is sufficient to quote the result for the threshold voltage V_T, at which inversion just occurs. For MOS (n-type substrate, p-type channel):

$$V_T = -\left[\frac{kT}{e}\ln\frac{N_D^2}{n_i^2} + \frac{(Q_{fo}/A) + \sqrt{2\varepsilon_s kT N_D \ln N_D^2/n_i^2}}{\varepsilon_{ox}/d}\right]$$

The relative dielectric constants of the semiconductor and the oxide insulator are respectively ε_s and ε_{ox} and d is the thickness of the oxide. The first term on the right is the voltage across the depletion region just at inversion (remember depletion regions still have voltages across them when no current flows). The rest is associated with the oxide capacitance which has oxide and depletion charges on one side balancing the metal/oxide surface charge.

SAQ 5.11 (Objective 5.10)
An MOS structure is inevitably formed by any metal track insulated by silicon dioxide from underlying silicon. Is there any possibility of inversion occurring in this structure?

SAQ 5.12 (Objective 5.11)
What can you say about the doping of the semiconductor and the dimensions of an MOS structure that has a high resilience to inversion? (Refer to the equation for V_T given above.)

Further charges

Earlier I mentioned the presence of surface traps in the energy gap which arise at the edge of the crystal structure. These traps can fill or empty quite quickly and are sometimes called **fast surface states**. Their influence here is slight because the fixed oxide charge is much more significant at the oxide/semiconductor interface. They do have some influence on the rate at which inversion occurs.

You will recall from Chapter 1 that ionic conduction in insulators gives them a much lower resistivity than electron-based conduction alone would allow. What if there are some ions free to move in the silicon dioxide? Pure SiO_2 is largely covalent but don't forget how readily alkali-metal ions soften silicate glasses. Indeed sodium used to be a nuisance because human perspiration is a rich source of sodium ions. Cleanliness is paramount: even insulators must not be handled! Mobile ions in the oxide can move in steady and low frequency fields, dramatically altering the performance. They are to be avoided.

Finally, a fuller treatment would also worry about contact potentials across the MOS structure, which are responsible for an additional term in the expression for the threshold voltage. It need not concern us further.

5.3.5 MOS switches

Electrical inversion is just what we need for making an electronic switch. Figure 5.30(a) illustrates it. Two regions of n-type semiconductor in a p-type substrate may be isolated by the inherent reverse biased pn junctions. But, because of the MOS configuration between the n-type regions, a connecting n-type inversion channel can be established or destroyed under the influence of a voltage applied to the metal layer. Electrons can wander around within each n-type region and if the inversion layer is formed they can also pass between the once isolated ends. This is a switch with no mechanical moving parts.

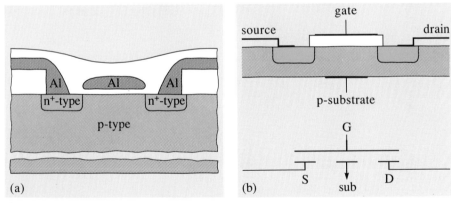

Figure 5.30 An MOS switch structure

To make use of the electronic switch we must make ohmic contact with the extreme n-type regions which is why these are specified to be n^+-type doping. There are now four connections to be made: one to each n^+-type region and the two across the MOS structure. We have in essence an **MOS field effect transistor (MOSFET)**, although the transistor nature is a little more involved. The conventional names for the connections are illustrated in Figure 5.30(b). Charge (electrons here) can flow from source to drain only when the gate-substrate bias is above the inversion threshold.

Sketched in Figure 5.31 are conditions below and above threshold. These situations are discussed further in ▼ MOS transistors ▲.

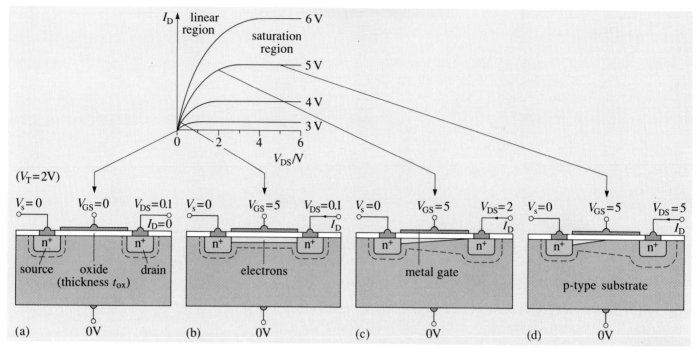

Figure 5.31 A MOSFET structure below and above threshold

SAQ 5.13   (Objective 5.12)
Figure 5.33 shows the characteristic structure for an integrated silicon p-channel MOSFET. Each region uses a particular material, or type of material. During manufacturing, the way these regions are processed is used to determine the specifications of the product. Complete Table 5.3, giving for each region the relevant material and processing factors for the given product features. The first line has been done for you.

Table 5.3

Region	Material	Processing factors	Product features
general insulator	SiO_2	thickness	breakdown voltage
substrate			V_T
source and drain			ohmic contact
p-channel			β (in current–voltage characteristic for MOSFET)
gate dielectric			V_T, β
contacts: gate, source, drain and substrate			electrical connection

▼MOS transistors ▲

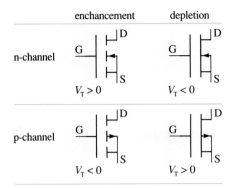

Figure 5.32 Types of MOSFET

Figure 5.32 illustrates four different kinds of MOSFET. The first distinction to make is the nature of the conducting channel when it is formed: n-type or p-type. Thus we have nMOS (sometimes NMOS) and pMOS (sometimes PMOS). The second distinction is whether the conducting channel exists or not with zero gate bias. If it does exist, then the switch is normally on but we can use a gate voltage to deplete (**D**) the channel and so switch it off. **D nMOS** therefore uses a negative gate voltage to disperse the channel, repelling the n-type carriers from the surface. By contrast enhancement devices (normally off) need a gate voltage which encourages a channel to form. The circuit symbols use an arrow on the substrate connection to show n-type (in) or p-type (out). A normally off (enhancement mode) device has a broken path between source and drain. From now on we'll talk only about the enhancement mode device.

The performance of a MOSFET is conveniently summarized in the current–voltage relations of the drain source connections. We need to know what these are so that suitable processes can be specified. Furthermore, the behaviour of any component of an electronic circuit is also crucial to the specifications of others in that circuit, so we have to be able to say how each behaves.

The normal operation of a MOSFET is with the source and substrate contacts at the same potential. This is achieved by patterning the metallization so that the source and substrate contacts are common

(see for example the p-channel MOSFET in Figure 5.33). Remember though that contact potentials between the different materials will leave a built in voltage across the source-substrate pn junction.

When the potential between the gate metal and the substrate (or source) V_{GS} (Figure 5.31) prevents inversion ($V_{GS} < V_T$), there is no conducting channel between source and drain. Virtually no current will flow in response to a drain-source bias V_{DS}. Figure 5.31(a) shows the depletion regions for an n-channel device isolating the source and drain. Only a leakage current between drain and substrate can flow, effectively passing charge between the drain and the source, since the substrate and the source are connected (do you remember how to keep this leakage small?).

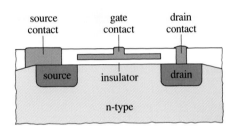

Figure 5.33 Section through a p-channel MOSFET

When the gate–source voltage is sufficient to cause inversion and the formation of a conducting channel, the n$^+$-type source and n$^+$-type drain are linked through an n-type inversion layer. The application of a drain-source voltage causes a current to flow from source to drain. The resistivity of the channel as usual depends upon the carrier density and mobility

$$\sigma = \frac{1}{\rho} = ne\mu_c \quad \text{or} \quad pe\mu_h$$

according to the type of channel. As you can see, the stronger the inversion (the more V_{GS} exceeds V_T) the lower the resistivity. At the drain end, the channel is raised to drain potential so the gate-to-semiconductor bias is effectively the gate–drain voltage. A little careful thought shows that $V_{GS} = V_{GD} + V_{DS}$ (write S, D, G along a voltage scale to convince

yourself) so if the drain end is to exceed threshold ($V_{GD} > V_T$) then

$$V_{DS} < V_{GS} - V_T$$

It can be shown that when this is the case the drain–source current I_{DS} looks something like

$$I_{DS} = \beta[(V_{GS} - V_T)V_{DS} - \tfrac{1}{2}V_{DS}^2]$$

where β is a process-dependent device constant (it is called simply 'beta'; you will see it again in Exercise 5.21).

When the drain end fails to satisfy the ordinary threshold condition, no extra current can be induced to flow. Where the channel is 'pinched off' in this way the resistivity adjusts to keep the current flow independent of the drain–source voltage. In this region the drain-source current virtually saturates at

$$I_{DS} = \tfrac{1}{2}\beta(V_{GS} - V_T)^2.$$

EXERCISE 5.21

(a) Show that for strong inversion ($V_{GS} - V_T \gg V_{DS}$), the MOSFET equation can be made to look like Ohm's law if the channel resistance R_{DS} is $1/[\beta(V_{GS} - V_T)]$.

(b) Complete the missing algebra below to see what quantities determine the constant β for a device with a channel width W and length L.
A capacitance C stores an extra charge ΔQ when the voltage increases by ΔV:

$$\Delta Q =$$

Likewise the extra gate charge which the inversion-layer charge must match is given by:

$$\Delta Q = \varepsilon_0 \varepsilon_{ox} \frac{WL}{t_{ox}} \times$$

If the inversion charge is confined to a thickness t_{inv}, its density is

$$n_{inv}e = \frac{\Delta Q}{LWt_{inv}} =$$

The channel resistance is therefore

$$R_{DS} = \frac{L}{\sigma Wt_{inv}} =$$

The device constant β is then $\beta =$

Now that we have developed the ideas behind a few basic devices I want to show you how these form the building blocks of large memory systems. We can't avoid some more electronics, chiefly because integrated circuit technology thrives on the close interplay between the properties and processing of materials on the one hand and electronic principles on the other. If you are unfamiliar with electronics, take Section 5.4 steadily but superficially right through before re-reading and following the details.

5.3.6 Summary

In integrating an electronic component (resistor, capacitor, diode, transistor) into a silicon wafer, it is important to consider the surrounding material from which it must be isolated. The properties of real electronic structures must be carefully engineered, choosing doping levels and dimensions to provide the basic components.

Regions of semiconductor depleted of majority charge carriers arise wherever the doping switches the material from one type to the other or else where an external potential is able to repel the majority species, such as in MOS structures.

Inversion layers can form in MOS structures providing, for example, an n-type channel at the surface of p-type material. This effect is the key to MOSFET switching elements.

5.4 Memories are made of this

In this section I want to describe, briefly, how simple circuits are used as memory elements. How memory elements in general are organized in such a way that each can be directly addressed in an efficient manner is an important factor in the final design, but it is beyond our scope here. We are simply interested to have an arrangement which is entirely electronic: no mechanical moving parts.

5.4.1 RAM (Random Access Memory)

The main memory in a computer is a short-term storage facility for data to be used in calculations and for text characters. Each memory element has a distinct address so that it can be uniquely accessed. As you know, binary numbers are used to represent data, the 1s and 0s corresponding to voltage levels in microelectronic circuits. Similarly, binary codes are used for characters and punctuation symbols, and typically the code is eight bits or one **byte** long. The addresses used in writing to or reading from memory locations are conveniently associated with bytes. A typical microcomputer of the early 1980s era contained perhaps 128 kbytes (128×10^3 bytes) of 'randomly-accessible-memory' (RAM). By the late 1980s this had risen almost tenfold to 1 Mbyte. Each element of a random access memory can be read from or written to directly, unlike memory locations on magnetic tape or discs where we must wait for the memory element to be aligned with the read/write head.

A desirable requirement of an electronic memory element (which stores one bit) is that its content or state (1 or 0) should remain as set or written after the electronic setting signal to the bit location has been removed. If this is not the case then a slow decline of stored signal can be tolerated if some means of refreshing the memory cells (that is, restoring the set states) is provided within the computer operating system. The former ideal type is known as **static RAM** (SRAM) and can be achieved in so-called bistable circuits which have two clearly defined states (voltage levels). The other type, which needs to be refreshed, is termed **dynamic RAM** (DRAM) and we can consider it equivalent to storing water in leaky fire buckets!

The basis of one particular type of static microelectronic memory cell is a so-called inverter circuit. A low voltage level at the input produces a high level at the output and *vice versa*. Figure 5.34 illustrates the principle. When the input voltage is below the threshold level for the transistor no current flows in the resistor, so the output voltage is 'high' ($+V_s$). If the input is in excess of threshold, current flows and the output voltage falls to 'low' (near zero). However, single inverter circuits do not qualify as static memory cells. Why not? Well, although they have two well defined on/off states, the output state requires the input signal to be maintained for the duration of the required memory state.

If two inverters are connected in series as shown in Figure 5.35, so that the output of the first stage feeds to the input of the second stage, and the output of the whole circuit is fed back to the input of the first (positive feedback), a useful memory cell is formed. This circuit is called a flip-flop and behaves much like an electromechanical switch. Imagine a steady state in which the input is 'low' ($V_{in} \approx 0$). There is no current in the first transistor so the input to the second is 'high' ($V_x = + V_s$). The second is fully on; its output is low ($V_{out} \approx 0$), which fits with its being connected directly to the input. If the input is now given extra charge to try to raise it to the 'high' state, the input gate takes a finite time to charge up and so its voltage, initially zero, does not change immediately to $+V_s$. As the threshold voltage is passed, the first transistor starts to come on so the output of the first inverter stage starts to fall and the second stage output starts to increase as its gate voltage falls. This voltage feeds back to the input and speeds up the input gate charging process. The result is a rapid change of state for both inverters. The original input signal can be removed at any time after the input threshold voltage has been reached: the feedback process ensures the current state of memory is maintained (the circuit is said to *latch* the last input signal).

Well, that's the idea for one memory element. Can we make one? Can we make a million on a chip all at once? Read on.

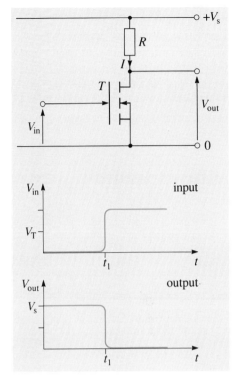

Figure 5.34 A MOSFET inverter

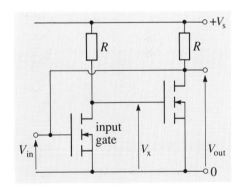

Figure 5.35 A two inverter flip flop

5.4.2 Static RAM

I mentioned earlier that inverter circuits were the basis of an SRAM memory cell. In this section I want to show you how the electronic principles and design must be used to specify the physical design and processing of MOS/resistor inverters. The development of inverters using MOS structures only will then be used to show how the technology must be adapted to meet the low cost mass production requirements of solid-state memory.

Electronic principle

Look back at the nMOS inverter circuit in Figure 5.34. I want this to form part of a memory element on a chip which will house thousands of such elements. The operation of the device was outlined earlier. Table 5.4 is a summary.

Table 5.4   Summary of inverter operation

Input V_{in}	Current	Output V_{out}	Comment
steady and above V_T	current I in R and T	steady low at value of $V_S - IR$	high in/low out
steady and well below V_T	virtually zero current in R and T	steady high at value of V_S	low in/high out
rising past V_T	rising	falling	gate capacitance (MOS) is being charged
falling past V_T	falling	rising	gate capacitance is being discharged

The inverter design relies upon an interaction between currents and voltages associated with a MOSFET and a resistor. Check carefully the following statements which apply to the steady condition (refer to 'MOS transistors' and to Figure 5.34).

(a) For the transistor:

● If the gate–source voltage is above threshold and the drain end is not pinched off, that is

$$V_{in} > V_T \text{ and } V_{out} < V_{in} - V_T$$

then the current flowing between source and drain is given by

$$I = \beta[(V_{in} - V_T)V_{out} - \tfrac{1}{2}V_{out}^2]$$

● If the gate–source voltage is above threshold but the drain end is pinched off, that is

$$V_{in} > V_T \text{ and } V_{out} > V_{in} - V_T$$

then the current is given by

$$I = \tfrac{1}{2}\beta (V_{in} - V_T)^2$$

- But if the gate is below threshold, that is

$$V_{in} < V_T$$

then

$$I = 0 \quad \text{and} \quad V_{out} = V_s$$

(These are just the MOSFET equations with $V_{GS} = V_{in}$ and $V_{DS} = V_{out}$.)

(b) For the resistor:

$$I = \frac{V_s - V_{out}}{R}$$

Electronic design

Next we have to choose values for such things as voltage levels for logic 1 and logic 0 (which may have to be compatible with other systems), transistor parameters (β and V_T) and resistor values. Because this is to be part of an integrated circuit, we may have to accept the devices which the processing can give. But we can express a preference, as we shall see.

- Logic 1. The most positive voltage around is the supply voltage V_s. A popular choice for this is 5 volts, so let's make logic 1 as close to this as we can.
- Logic 0. The lowest voltage in the circuit is zero. This is a good goal for logic 0.

Figure 5.36 gives the curves for I against V_{out} for the transistor and resistor of the basic inverter in Figure 5.34. Check the following statements.

(a) If $R = R_1$, logic 1 is V_D and logic 0 is V_B.

(b) If $R = R_2$, logic 1 is V_C and logic 0 is V_A.
Note: logic 1 output requires $V_{in} \approx 0$ and logic 0 output requires $V_{in} \approx V_s$.

(c) The higher resistance (R_2) dissipates less power than the lower one in both the logic 0 and the logic 1 conditions. (Power supplied $= IV_s$.)

(d) The smaller resistor (R_1) conducts charge at a faster rate than the other.

(e) We should specify $0 < V_T < V_s$ for the threshold voltage, preferably around the middle of the range.

So, the threshold voltage V_T ought to be about mid-range. To get some idea of what value the transistor parameter β should have, we have to investigate the power consumed when the transistor is on. Remember that any current flow I produces a power dissipation $I V_s$ which is shared between the resistor R and the transistor T. We should endeavour to keep the total power down. Just to force a design figure, let's say 0.5 mW

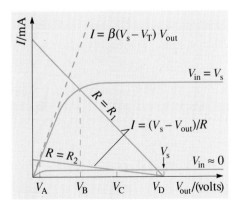

Figure 5.36 I against V_{out} for an nMOS inverter (Figure 5.34)

per inverter. So the current must not exceed 0.1 mA (with a 5 volt supply). This current flows when the inverter input is at logic 1 (high) so the output is at logic 0 (low). Next, we can argue that the logic 0 voltage should be well below V_T to avoid the possibility of a low output, inadvertently increased by noise, triggering a subsequent stage. Let's choose 0.5 volts as a maximum logic 0. These values for maximum current and maximum logic 0 fix β as you can see for yourself:

$$I \approx \beta(V_s - V_T)V_{out} \text{ (for small } V_{out})$$

With $V_T \approx V_s/2$, $V_{out} \approx 0.5$ volts and $I \approx 0.1$ mA,

$$\beta \approx \frac{10^{-4}}{2.5 \times 0.5} = 80 \ \mu\text{A V}^{-2}$$

You should now confirm that to achieve a 0.5 volt output when the transistor conducts 0.1 mA, R must be 45 kΩ. (When 0.1 mA flows through R and T, we are in the logic 0 state and V_{out} is 0.5 volts with 4.5 volts across R.)

EXERCISE 5.22   Follow the consequences of meeting a specification for
(a) a lower power consumption (for the same logic levels),
(b) a lower logic 0 voltage (for the same power),
(c) an existing processing schedule which gives $\beta = 30 \ \mu\text{A V}^{-2}$.

Physical design

We have established specifications for V_T, β and R. By looking at the physical relationships we have established, these can be translated into material and geometric specifications and then into process specifications.

EXERCISE 5.23   The parameters on the left of Table 5.5 can be set by a choice of materials quantities (for example, electron mobility) and geometric quantities (for example, oxide thickness). Fill in the missing quantities relevant to each parameter.

You've probably realized that, since β and V_T are geometrically interdependent, a little CAD may be called for in practice to optimize the MOSFET design. We'll leave it there.

Table 5.5

Parameter	Material quantities	Geometric quantities
V_T		$\left\{ \begin{array}{l} \text{oxide thickness } t_{ox}, \\ \text{gate area} \\ A = W \times L \end{array} \right.$
β	ε_{ox}, μ_e	
R		

The resistor is a little more straightforward. To get high resistivity we don't want to dope too much, but we'll need to swamp the p-type substrate's acceptors. Referring back to the 1 kΩ resistor, you can see that 45 kΩ needs something like an ion implanted track 1 μm $\times$ 5 μm $\times$ 75 μm at 10^{23} donors m^{-3}. That's a lot of surface area (3.75×10^{-10} m^2) for the resistor, so it may be worth looking out for other ways to get the desired effect.

EXERCISE 5.24   If the area of a single inverter is 4×10^{-10} m^2 estimate the number of bits which might be stored on a 3 mm $\times$ 3 mm chip, assuming that isolation between cells and components means that space is filled with only 50% efficiency.

Latch up

An important aspect of the physical design arises when we consider more than just one device. MOS devices are the basis of one style of semiconductor electronics. A different style, called **bipolar technology**, part companion, part rival, also uses layers of n-type and p-type silicon. The electrons and holes are unwitting parties to our technological fancies and, bless them, cannot know what we had in mind.

You can probably imagine what is coming. Suppose a particular area of a MOS circuit ends up, in terms of semiconductor layers and shapes, looking for all the world like one or more bipolar elements. These 'parasitic' (or more kindly 'unintended') devices may well at some time spring to life and foul up the MOS function. The original design may even 'latch up' into a state from which it can not be easily released. Designs which avoid latch up are essential and CAD technology is again useful.

Cheaper, denser, faster

It seems that memory systems are saleable if they are cheaper or denser or faster than their rivals. To see what compromises have to be made, I want to consider three versions of the complete SRAM twin-inverter latch. Figure 5.37 shows them. Notice how the lower pair of components for each latch are the same. Don't worry just now about the way in which they function. The first is based on the nMOS/resistor combination. The second also uses nMOS but instead of a resistance this circuit uses a second, but normally 'on' MOS transistor (that is, a depletion device). The third is referred to as a CMOS (**Complementary MOS**) static memory element since it uses nMOS and pMOS transistors in pairs in the inverter. ▼Memory loads▲ looks into why the choices arise.

What makes a finished semiconductor package cheaper is a reduction in the number of process steps (reduced process time and increased yield) and an increased volume of production. Since the CMOS latch needs substrate material to be of different types (n and p) in different regions there is an additional process step. However it turns out that in use, the CMOS latches consume much less power than the other types and so they are favoured for battery operated devices (where a volume market exists).

On the other hand, the power advantage of CMOS is offset by the geometric area required to establish the different substrate regions and the complementary pairs of transistors. Denser packing means less silicon surface area for each device.

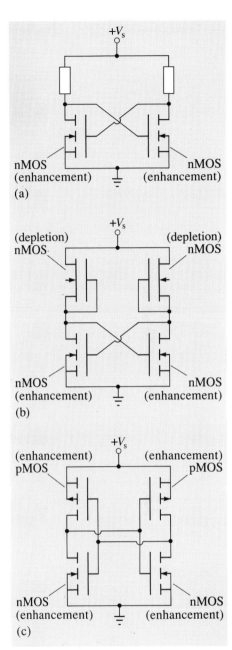

Figure 5.37 Twin-inverter memory latches. (a) *R*/nMOS (b) Enhancement/depletion nMOS (c) CMOS

▼Memory loads▲

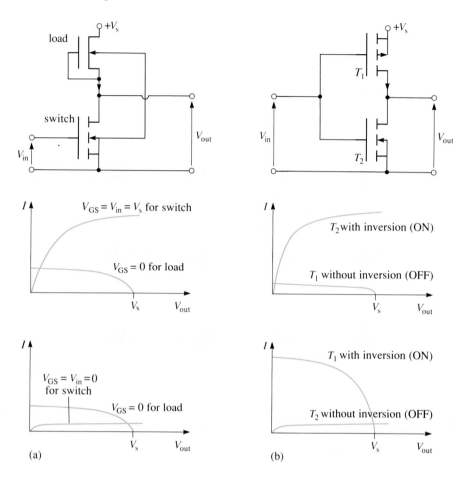

Figure 5.38 Enhancement/depletion nMOS (a) and CMOS (b) inverters

The purpose of the load resistance (the resistor) in the simple inverter circuit is to allow a voltage to be developed across it when it passes current. In fact it need not be an Ohm's-law resistor, provided it gives a well defined voltage from a given current flow. In Figure 5.37 (b) and (c) the load is a second MOS transistor.

For example, a normally 'on' (depletion mode) transistor, kept that way by connecting gate and source together, is the load in Figure 5.37(b). When the switching transistor is off (Figure 5.38a, lower graph), virtually no current needs to flow

in the load, it has virtually no voltage across it and the output voltage is the supply voltage connected through the load. When the switching transistor is on, we can design it so that the load saturates at quite low current (Figure 5.38a, upper graph). In this way it develops sufficient voltage without allowing very much current through and this keeps the power consumption down. To appreciate the advantages of this sort of design we have to look at the processing and design implications. The two MOSFETS must have different threshold voltages, but that's largely a question of doping and geometry,

so we ought to be able to arrange it. Since the material is the same for each, differences in the β parameter which are also needed are easily handled by the channel length-to-width ratio. Again geometry. Notice in Figure 5.38(a) that the two substrate connections have to be joined; the substrate is common to both.

The CMOS circuit Figure 5.38(b) just takes things a bit further. The output voltage can be V_s provided the upper transistor provides a relatively low resistance path while the lower one is off (no inversion). Likewise the output is almost zero volts if the upper transistor is off (no inversion) while the lower one has an inversion channel. In effect the p- and n-channel devices work as switches operated by a common toggle or lever (that is, the gate voltage). When one or other transistor is off there is no significant current drawn from the power supply and so there is negligible stand-by power consumption. Power is consumed during switching when for a short time both transistors may be on.

We'll come to how we make CMOS circuits in due course. In the meantime Exercise 5.25 anticipates some of the problems.

EXERCISE 5.25 The CMOS inverter requires an n-channel and a p-channel device with similar characteristics (β, V_T) together on adjacent areas of silicon, to be operated by a common gate with a smooth changeover (one going off as the other comes on). This will need some careful processing. What control have we over the following?

(a) Substrate material (we need it to be n-type in some parts and p-type in others).

(b) The β parameter (remember one conducts with holes, the other electrons).

(c) The threshold voltage V_T (they must switch together).

(d) Isolation.

The speed of operation of memory circuits is a key parameter in their specification. In the end it is a question of getting charge to accumulate or disperse, moving it along paths with non-zero resistivity. Since electrons usually have a higher mobility than holes, n-channel devices have a slight speed advantage. You probably recognize in this the makings of an *RC* time constant, as though we were charging a capacitance *C* through some resistance *R*. It is unfortunate that neither the *C* nor the *R* are necessarily fixed values (pn capacitance, MOS capacitance and channel resistance). However, clever processing can at least keep down some of the accidental 'stray' capacitance which degrades the performance. See ▼Self-aligned gates▲ .

SAQ 5.14   (Objective 5.13)
The simple resistor/nMOS latch area can be substantially reduced by using a higher resistivity material for the resistor constructed in a deposited layer over the top of the transistor. Suggest a material and relevant processing step.

SAQ 5.15   (Objective 5.14)
Where, in what amount and in what state are silicon atoms to be found in a single element of a CMOS SRAM cell?

5.4.3 Just a wee DRAM

A complete SRAM cell needs four components as a latch to store one bit of information (Figure 5.37) and a further two transistors which are used in addressing it. Rethinking the electronic principles so that fewer components are needed for each bit could have the following consequences: lower production cost per bit of capacity; higher yield of functioning cells; higher packing density. The dynamic RAM (DRAM) follows these ideas using as few as two components per cell: an MOS capacitor to replace the latch and an MOS switch to handle the addressing. The principle is simple: logic levels are represented by the voltage on a storage capacitor which is charged or discharged through the MOSFET switch (Figure 5.39). The stored signal rapidly deteriorates with time but the space saving is more than sufficient for the necessary additional circuitry which refreshes each cell hundreds of times per second. In fact one can buy pseudo- or quasi-SRAM which looks to the outside world like SRAM but internally it is in fact DRAM. Why go to all this trouble?

EXERCISE 5.26
(a) Will a DRAM cell consume more or less power than an SRAM one?
(b) What are the relative merits of SRAM and DRAM?

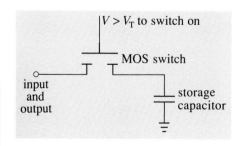

Figure 5.39 A DRAM cell

▼Self-aligned gates▲

A key step in microtechnology is the accurate registration from one lithographic mask to another. Mask alignment is crucial to making smaller faster devices. One of the more critical steps in processing metal–oxide–silicon FETs is the alignment of the gate metallization edges with the source and drain diffusions. If the gate fails to overlap with either source or drain, there will be a break in the channel, as shown in Figure 5.40(a), and the device will not operate. If the gate overlaps too much, an excessive gate-drain or gate-source capacitance will be formed, as shown in Figure 5.40(b). In use (as an inverter, for example) the effective capacitance between gate and drain is increased by the voltage gain of the amplifying circuit. The speed of

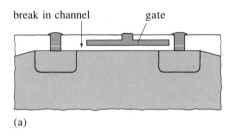

(a)

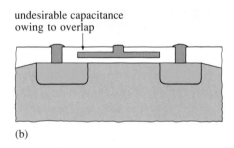

(b)

Figure 5.40 Misaligned gates. (a) Broken channel (b) Excess overlap

the circuit is therefore greatly reduced by the presence of this capacitance. This particular alignment problem has been largely resolved by using the gate itself as a mask to define the position of the source and drain regions.

Most implementations of this technique use polycrystalline silicon (also called polysilicon) instead of metal in the gate electrode, giving a structure like that shown in Figure 5.41. A layer of polycrystalline silicon is laid down on the circuit before the source and drain diffusions take place. The size of the gate electrode is then defined by photolithography and the unwanted polycrystalline silicon is etched away. The remaining polysilicon gate acts as a mask for the subsequent doping of the source and drain. So, as illustrated in Figure 5.41(b), the degree of overlap of the gate with source and drain is precisely controlled, although the actual location of the channel is defined no more accurately than before. This technique is referred to as **self-alignment**.

You should be asking yourself 'why polysilicon?' Well, notice that we want the gate to define n^+ and p^+ regions, so a thermal diffusion or ion implantation will be required after forming the gate. But aluminium melts at 660°C! Even if we stick to ion implantation, the necessary post implant annealing temperatures rules out the use of aluminium gate metallization at this early stage of manufacture. A refractory gate conductor is needed and since silicon deposition apparatus is already on hand, it is a convenient choice.

After being deposited, the polysilicon is heavily doped with phosphorus or another

suitable donor, to make it highly conducting. This can be carried out by the same diffusion that forms the n^+ source and drain regions. The final structure is shown in Figure 5.41(c).

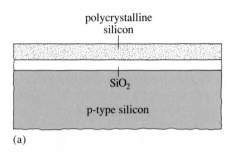

(a)

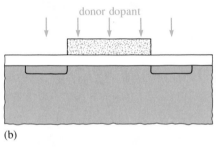

(b)

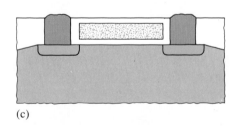

(c)

Figure 5.41 The self aligned gate process

Both SRAM and DRAM memory (and other devices) can be made using the CMOS processing techniques which we will discuss in the next section.

5.4.4 Summary

Random access memory uses MOS structures as switches in various configurations. The basic static RAM cell uses two inverter circuits in cascade to form a latch. The state of any latch (which represents one bit of data) is held until a new setting signal is received provided only that power is supplied to it. In effect, the output of the latch holds its own input as set, except during the transition between one state and the other. By considering the working of one of the inverter arrangements, it is possible to specify values of process-controlled quantities such as threshold voltages and transistor parameters. Geometric and material properties are involved.

The SRAM cell can be designed using different combinations of components, for example: resistors with MOSFETs; enhancement and depletion mode MOSFETs; p-channel and n-channel MOSFETs. The last type is referred to as CMOS. The CMOS device uses least power, has an extra processing step and takes up the most surface area of silicon, but is favoured for battery powered operations. CMOS memory enjoys the benefits of a volume market. Further savings in power consumption and space can be made using dynamic RAM cells but these have the disadvantage that they must be refreshed hundreds of times a second.

5.5 The undoing of CMOS

In this section I want to show you a CMOS processing sequence. The story is based on a process developed by Plessey Semiconductors to produce CMOS circuitry with device features as small as 1.4 μm. The development of a manufacturing sequence is an iterative evolution towards a viable production route, and so there is a good deal of careful locating and arranging before everything 'snaps' into place. In fact I'm going to tell the story backwards so you will see progressively simpler structures and you may see more easily why the various steps are carried out. The story is told in Figures 5.42(a)–(m) and their related text. You can read this material in reverse order if you want to proceed more chronologically.

The process begins with a lightly doped 6 μm epitaxial layer over a heavily doped (n^+) wafer and ends several days later with wells of p-type and n-type silicon chequered across the epitaxial layer into which n-channel and p-channel MOSFETs have been built. Figures 5.42 (a) to (m) follow the sequence back from MOSFETs to wafer, focusing on a region where two wells abut (gate connections are made in a different section of the wafer). The accompanying green text will give process details, but you will be invited from time to time to contribute some ideas.

Figure 5.42(a) is how we want it to finish up. The last treatment was the metallization added to Figure 5.42(b) which provides connections to inputs, outputs and power supplies. The figure shows just one layer of metal. In practice a second layer is added to interconnect the various elements, possibly according to a customer's specification. The metal layer in fact consists of a titanium-tungsten alloy 0.15 µm thick acting as a barrier layer (to stop aluminium migrating into the silicon) and a 0.5 µm layer of aluminium-copper (4% Cu). The copper impedes electro-migration of the aluminium (in which aluminium atoms are literally nudged along by the electron flux when currents flow). Any thoughts about how to do the metallization? Some photolithographic patterning will be needed to establish connections to every device. (See ▼Metallization▲.)

Figure 5.42(b) shows the unmetalled surface. The topmost layer is about a micron of silicon dioxide (dielectric material), with contact holes 1.4 µm × 1.4 µm down to underlying doped silicon (n^+ or p^+ source and drain contacts). It is very difficult to deposit continuous metal tracks over sharp steps, so a desirable feature is that the top of the oxide be much smoother than the silicon surface on which it sits. The dielectric is said to have 'planarized' the silicon surface, and the metal deposition doesn't have to contend with steps over the gates (compare Figures 5.42b and e). The processing here has to cover the whole wafer with the dielectric on which the metal is deposited and through which contact holes must be made to sources, drains and gates. (See ▼Planarization dielectric and contact holes▲.)

Figure 5.42(c) shows n-wells and edges of p-wells capped with photoresist material. (You should know how to do this by now.) Arsenic atoms are introduced by ion implantation into the whole surface deep enough to penetrate the areas covered by the thin oxide, but not deep enough to get through the gates or the resist-covered regions. You should recognize the self-aligned gate technique. (In fact there is scope here for some very smart control over the window through which dopants reach the silicon; we'll return to this in the next section.) The regions of silicon where dopant arrive will form sources and drains for the n-channel devices.

In Figure 5.42(d), source and drain regions are self-aligned on the n-well (p-channel) gates by an analogous sequence which protects the p-wells and implants boron. (See ▼Source and drain implants▲.)

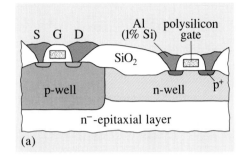

Figure 5.42(a)

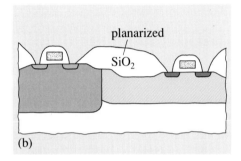

Figure 5.42(b)

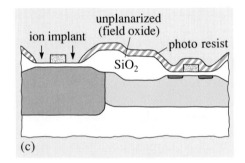

Figure 5.42(c)

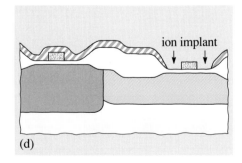

Figure 5.42(d)

▼Metallization▲

Sputtering is ideally suited to provide thin alloy films. Typical deposition rates are $0.1 \ \mu m \ min^{-1}$ so even if wafers are coated singly one station could handle about ten per hour.

The patterning of the metal is a straightforward lithographic job, but did you notice how uneven the surface can be? Projection printing and positive photoresist may be required for high fidelity. Mask alignment is always crucial. Once a photoresist pattern is developed, unwanted metal is etched away using a boron trichloride (BCl_3)/chlorine plasma. A mixture of chemical and physical erosion is necessary here because although aluminium chloride formed on the surface is volatile, copper chloride is not and must be sputtered off. Finally the remaining photoresist is dissolved and washed away. Annealing in a nitrogen-hydrogen atmosphere ensures a good metal-silicon contact and removes damage caused by the sputtering and etching stage.

▼Planarization dielectric and contact holes▲

The area to be insulated is already mostly SiO_2, a thin layer of this having been grown over the entire surface before the gates were defined. The planarizing step is carried out by 'spinning' on an organic silicate in solution in alcohol. This only needs curing at $800°C$ in a nitrogen atmosphere to drive off the organic component to leave a relatively smooth surfaced glass, mainly silica (SiO_2). Over this a further micron of silicon dioxide is deposited by low pressure CVD from silane, oxygen and phosphine. This process requires heating to only to $420°C$ to build up the desired thickness. The phosphine dopes the oxide with phosphorus, which impedes the diffusion of ions (such as sodium) through it.

At this late stage of processing, most atoms are in place so high-temperature cycles are avoided. However the step before this will have ion-implanted dopants to form sources and drains. A 30 minute anneal at $900°C$ is needed to ensure that this dopant is 'activated' (that is, incorporated into the lattice by substitution).

The contact openings to the gates, sources and drains are defined photolithographically and etched out with a low-pressure trifluoro-methane/oxygen plasma at about $0.06 \ \mu m \ min^{-1}$. This scheme produces very anisotropic etching. It cuts deeply, without significant lateral erosion so that the hole is close to $1.4 \ \mu m \ \times \ 1.4 \ \mu m \ \times \ 1.0 \ \mu m$. Silicon is not easily etched by this mixture so an accurate stop is not vital.

The remaining photoresist has been hardened by the etching process so it is first oxidized by exposure to an oxygen plasma and then taken off in a sulphuric acid/hydrogen peroxide dip (which shifts most things!).

▼Source and drain implants▲

Figure 5.43 shows the range R and scatter ΔR of arsenic implanted into silicon or silicon dioxide. At this stage in the processing (Figures 5.42c, d, e) the surface oxide is conveniently that put there to form the gate oxide. The oxide thickness needs to be enough not to break down, but not so much that the inversion threshold is too high. About $0.03 \ \mu m$ is typical. We'll discuss its preparation shortly.

EXERCISE 5.27

(a) Using Figure 5.43, suggest an implantation energy for the arsenic implantation of the n^+ source and drain regions.

(b) What dose (ions m^{-2}) is needed to establish doping of about $10^{27} \ m^{-3}$ in the n^+ regions. (Get an approximate answer by supposing all dopant ends up between $R - \Delta R$ and $R + \Delta R$ at a uniform density.)

The p^+ doping is done using $BF_2{}^+$ ions, which give comparable range and scatter at about 75 keV. After annealing, boron atoms are left in the lattice. During the annealing, the jilted fluorine atoms diffuse out. We have already noted the need to anneal an ion implantation and provision for this was made after the planarization.

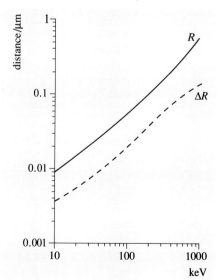

Figure 5.43 Implantation depth of arsenic ions in Si or SiO_2 as function of ion energy

Figure 5.42(e) shows the gates in place in both wells. Self-aligned gates cannot use aluminium to do the aligning because ion-implanted source and drain regions require a 900°C anneal cycle. Polycrystalline silicon is used but it has to be heavily doped to near saturation to get its resistivity down to an acceptable level ($\approx 1 \times 10^{-6}\ \Omega\ m$). This would-be metal acts as the 'M' in the MOS structure. The polysilicon is deposited over a thin oxide layer (which forms the gate oxide) over the whole wafer. Photolithography defines the regions where unwanted material is to be removed, leaving 1.4 μm wide gates. See ▼Gate deposition▲.

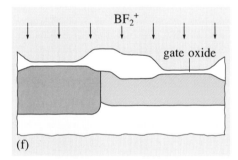

(e)

Figure 5.42(e)

EXERCISE 5.28

(a) Can you suggest a reason why the polysilicon should be much thicker than 0.01 μm?

(b) Can you suggest a reason why the polysilicon should not be much thicker than 1 μm?

Figure 5.42(f) shows the wafer surface after a thin, high-quality oxide has been grown on it. This oxide forms the 'O' of MOS devices. You should now know why its quality is important. The diagram also indicates a blanket dose of ion implantation. This is carried out to establish threshold voltages V_T in the finished device at prescribed levels. (Here V_T is to be less than the $V_s/2$ which we discussed earlier; it will be about ± 0.9 V.)

You will recall that the silicon dioxide/silicon interface region contains fixed positive oxide charges which can to a large extent be locally compensated for by equivalent fixed negative charges just inside the silicon. Burying acceptor atoms in this region does just this. Think of n-type silicon in an MOS structure. Inversion requires mobile positive charges (holes) to be attracted to the silicon-oxide interface. Positive oxide charges discourage this, so 'large' negative gate voltages are needed. Locally compensating for the oxide charge with acceptors will lead to less negative thresholds. Going from say -1.5 to -0.9 V is, mathematically speaking, going 'up' so we say that acceptors raise p-channel thresholds. See ▼Gate oxide and threshold adjustment▲.

Figure 5.42(g) is the wafer at a slightly earlier stage. You will see areas covered with 'thick' oxide and areas covered by 'thin' oxide plus nitride. We'll come to the surface profile presently. Here I want to concentrate on how we can clean off the 'active areas' over which the nitride and oxide has been needed (we're working backwards), ready to grow the high quality gate oxide (which we have just discussed). The problem we face is this: how to remove the nitride (about 0.1 μm thick) and its underlying 0.03 μm of oxide, without removing the other thicker layer of oxide, which is to be used to isolate the surface from other layers (such as metal) and the outside world. (This thicker oxide is termed **field oxide**.) We're getting into the realms of erosive chemistry here; see ▼Removing nitride and oxide layers▲ for the recipes.

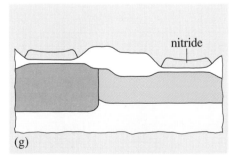

(f)

Figure 5.42 (f)

(g)

Figure 5.42(g)

▼ Gate deposition ▲

Polycrystalline silicon is deposited from silane (SiH_4) at about 0.1 atmospheres onto surfaces at 620°C at a rate of about 3 nm min^{-1}. Under these conditions it takes almost two and a half hours to coat the wafers with 0.44 μm of 0.01 μm grain-sized silicon.

In order to reduce the resistivity and to increase the grain size, the wafers are next heated to 900°C for an hour in a phosphorus oxy-chloride ($POCl_3$) atmosphere. As a result of the initially large crystal surface area of the small-grained layer, the diffusion of phosphorus proceeds apace and dopant gets right through. (In 'Designing diffusions' I estimated only 0.6 μm/hour penetration of phosphorus at 1000°C into single-crystal silicon.) During this heat cycle, silicon atoms also diffuse. The consequent recrystallization into 0.2 μm crystals is driven by the reduction in grain-boundary area.

The etching of the gates needs to be carried out anisotropically, given the feature size (1.4 μm) and layer thickness (0.44 μm). This is done using reactive-ion etching based on chlorine chemistry, reversing some of the possible deposition reactions

$$Si + 2\,HCl \rightarrow SiCl_2 + H_2$$

$$SiCl_2 + HCl \rightarrow SiHCl_3$$

(and others). In reactive-ion etching, the surface is subjected to uniformly directed ion bombardment, so the etching is anisotropic. Afterwards, the photoresist is again stripped by an oxidizing plasma treatment followed by a hot sulphuric acid/hydrogen peroxide wash.

▼ Gate oxide and threshold adjustment ▲

What governs the gate-oxide thickness? If it's too thin it will break down. (The gate voltage will be around a volt, but accidental connection to a 5 volt power supply rail ought to be tolerable. That means a minimum of 5 nm of perfect oxide, so let's say 10 nm for safety. If it's too thick, sufficiently small threshold voltages are not achievable (see the threshold calculations in Section 5.3.4). In fact this CMOS process uses 28 nm of gate oxide, grown in one hour at 950°C from an oxygen atmosphere (with 3% HCl).

At this stage the n-channel devices would have thresholds of about 0.3 volts and the p-channel devices thresholds of about −1.5 volts. If you were developing this process, what would you do to trim these to about +0.9 and −0.9 volts respectively? What you need to do are the following

- Find out how to estimate a rough figure for the required dopant dosage.
- Next estimate implantation energy required to dump it all in the silicon, but close to the surface.
- Next plan some experimental runs which span a range of doses and energies around these estimates.
- Check that the anneal cycles can be tolerated by previously disposed material.
- Do it and measure what you get.
- Iterate towards result.

If you did that, you'd probably find that the following recipe (or a viable alternative) works: 7×10^{15} ions m^{-2} of BF_2^+ at 45 keV (implantation range ≈ 0.04 μm). Allow subsequent thermal cycles above 650°C to anneal it.

In this process, the boron remains as an acceptor, raising both thresholds to more positive or less negative values as appropriate.

SAQ 5.16   (Objective 5.15)
With the electrical inversion of p-type silicon in an MOS structure in mind (that is, n-channel devices):

(a) Will the presence of fixed positive oxide charges raise or lower the threshold for inversion?

(b) What type of dopant in the silicon will promote higher thresholds?

▼ Removing nitride and oxide layers ▲

If you want to 'wash off' a silicon nitride layer after it has done its job, orthophosphoric acid (H_3PO_4) at 170°C is the stuff, especially if you want to leave any silicon dioxide unscathed. Thicker layers need longer treatments.

The removal of 0.03 μm silicon dioxide is also fairly easily done by immersion in wet chemicals for a suitable time. For SiO_2, a blend of hydrofluoric acid (HF) and ammonium fluoride (NH_4F) in water is used. The field oxide is also etched but it is thick enough to sustain this cutting back.

Any wet stage like this is followed by thorough washing in water pure enough to kill you. At a resistivity of 0.18 MΩ m, the water's resistivity is within 0.2% of its theoretical upper limit. It also has no particulates larger than 0.22 μm diameter and it is sterile. In short it has been filtered, demineralized, sterilized and deionized.

Let's look now at how we reach the stage depicted in Figure 5.42(g) from something like that shown in Figure 5.42(h). At this point p-wells and n-wells have been defined and regions of these selected for development as 'active areas' for sources, drains and channels. These areas have been left with a silicon nitride overlayer. The rest of the surface has a relatively thick covering of silicon dioxide. See if you can account for the oxide structure for yourself using Exercise 5.29, then see ▼The bird's beak▲ for some process details.

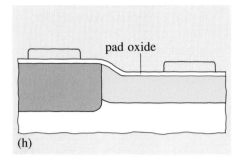

(h)

Figure 5.42(h)

EXERCISE 5.29

(a) A thin oxide layer is grown over the entire wafer before depositing silicon nitride. Which of the following reasons might be behind this strategy.

(i) Amorphous silicon nitride and the deposition process damage the underlying silicon layer.
(ii) Silicon nitride cannot be deposited onto pure silicon.
(iii) The thermal expansivity of the nitride is different from that of silicon.

(b) The thicker oxide regions are grown after the nitride pattern has been established. Look at the structure around the boundary of the active area. Why does this shape arise? (Recall the mechanism of oxide growth.)

Figure 5.42(h) shows the position before field-oxide growth. A uniform oxide layer extends right across the wafer. In addition areas of both wells destined for the actual devices have been capped with silicon nitride.

A particular hazard for n-channel devices arises when some electrons are accelerated in the channel to such high velocities that they shoot through the drain and into other active regions. To inhibit this, n-channel devices are surrounded by a 'guard' of relatively highly doped p-type material in which electrons have a (severely) shortened life expectancy because of the large number of holes into which they may fall. Holes escaping from p-channels are slower and are less troublesome. See ▼p-Guard doping▲.

Figure 5.42(i) is the next stage up stream, in which a thin unpatterned nitride layer has been deposited over a thin oxide. Photolithographic definition of the active areas, followed by etching, will establish the features we needed for Figure 5.42(h). ▼Etching features in silicon nitride▲ tells how. The oxide and nitride layers extend over regions which have already been designated p-well and n-well regions and which go deep down into the epitaxial layer on top of the wafer (Figure 5.42 i). The formation of these layers must either be cool enough and quick enough not to disturb the preformed wells, or else it must be done while diffusion associated with well formation is occurring. ▼Pad oxide and nitride growth▲ discusses one of these options. Before you read that, can you remember why the oxide interlayer is necessary?

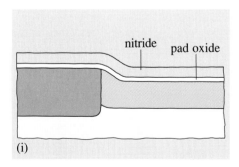

(i)

Figure 5.42(i)

▼The bird's beak▲

The field oxide is grown to about 0.6 μm, consuming silicon as it grows, except under the nitride. This is a fairly early processing stage but the n-well and p-wells will have been diffused into place so this thick oxide is going to need a fairly cool quick growth. Six hours at 900°C in about one atmosphere of water vapour is sufficient. It's important to keep the temperature low for a second reason: ammonia released from the nitride can locally penetrate to the silicon-silicon dioxide interface where a silicon-oxynitride phase disrupts oxidation.

The silicon just under the edge of the nitride is able to oxidize (although less quickly than the areas away from the nitride) owing to the lateral diffusion of oxygen. This reduced growth encroaches a few tenths of a micron under the nitride, diminishing its effective shielding of active areas by this amount. The oxide shape which results is known as a **bird's beak**. (When first encountered in the manufacture of other devices, it looked much more like a gull's head in profile than in Figure 5.42g, but the description remains.)

The surface profile is strongly affected above the birds' beaks where the oxide growth has lifted the edges of the nitride.

▼Etching features in silicon nitride▲

We've already seen how to etch silicon nitride isotropically when we cleaned it off prior to gate-oxide growth. Here though we need some fairly small features (a micron or so) in a thin layer (a tenth of a micron). We can use an isotropic etchant because the width-to-depth ratio is large, but the fineness of the features requires a dry approach, such as gaseous plasma. A carbon tetrafluoride plasma provides free fluorine and fluorinated radicals which decompose the solid nitride into volatiles such as silicon fluoride. A little oxygen is added to oxidize any deposited carbon. It is important to stop the etching before it penetrates the oxide underneath because silicon etches about eight times faster than the nitride. Much effort is expended in developing reliable indicators of the end point for an etching cycle; in dry processes, simply looking at the atomic species in the exhaust gas is a start. In practice wafers are treated singly. The etching takes only a minute or so.

▼Pad oxide and nitride growth▲

The pad oxide provides a useful intermediate layer between the silicon nitride and the silicon where devices are to be made. Its chief function is to prevent the introduction of stress, arising from the silicon-silicon nitride interface, into the active silicon. A secondary advantage, which we have seen, is that it means the nitride etch doesn't end on the more easily etched silicon.

At 950°C, 0.03 μm of oxide is grown in dry oxygen in an hour. This is enough to form the interlayer. The nitride deposition follows and adds a further 0.1 μm of material over the silicon. (Is that enough to mask the p-guard implant?)

The nitride is deposited from a chemical vapour of silane and ammonia. At 720°C the requisite material is laid down in about three hours.

$$3SiH_4 + 4NH_3 \rightarrow Si_3N_4 + 12H_2$$

▼p-Guard doping▲

To make the guard, acceptor atoms must be pushed into non-active regions of the p-wells through the 0.03 μm 'pad' oxide which covers the silicon surface. Ion implantation is the obvious method. The nitride over the p-wells automatically screens the n-channel active areas but the whole of the n-well needs masking with a layer of photoresist. At 25 keV, boron ions can be placed at about 0.1 ± 0.04 μm below the top of the oxide and this is found to give suitable results with a dose of 2.5×10^{11} m^{-2}. Nitride and photoresist masks must be more than 0.1 μm thick to be effective.

▼Twin well drive-in diffusion▲

Ion implantation is used to load n-type and p-type dopant just below the silicon surface. To establish the dopants on lattice sites and drive the dopants down to form deeper wells, a long heat treatment is carried out. After 18 hours at 1050°C the wells are half a micron or so deep. This is the longest and hottest heat cycle in the entire process.

Figure 5.42(j) shows the wafer, before pad oxide growth. The p- and n-wells are clearly almost complete, having just been diffused into the place from ion implanted predepositions, illustrated in Figures 5.42(k) and (m). ▼Twin well drive-in diffusion▲ will tell you how this bit is done but after that you're on your own, back to the 6 μm, n⁻ epitaxial layer on an n⁺ substrate, as supplied by a wafer manufacturer.

SAQ 5.17 (Objective 5.16)
Table 5.6 lists the first page of the manufacturer's schedule for the CMOS technology being described. Suggest process details in the comment column based on what you've already seen. (Note: if you continue to 'undo' the processing, start at the bottom of the table. I've done the twin-well inspection and the twin-well drive-in for you.)

Summary

We have just undone a CMOS fabrication sequence, working back from finished circuits to the wafer of electronics grade silicon. I'll summarize in the other direction. Wafers as supplied are n-type silicon having a lightly doped epitaxial layer on a heavily doped substrate. The twin wells required for CMOS circuitry are established using two separate ion implantations into the silicon surface. Subsequent heat cycles allow the implants to diffuse further to establish the desired well structure, after which shorter, cooler processes are chosen to prevent further movement.

Areas where transistors are to be located (active ares) are capped with nitride while the remaining surface is oxidized to isolate devices and connecting tracks. Then after removal of the nitride overlayer, the high-quality gate oxide is grown using a slow, dry thermal oxidation. The threshold voltage of the transistors is determined next by a controlled ion implant through the gate oxide just into the underlying silicon. A deposition of polysilicon is patterned to form the transistors' gates, which also provide an ion implantation mask for sources and drains (self-aligned gates). In the final stages the wafer surface is smoothed by flowing a glass (silica) over it before masking and etching contact holes for the metallization layer, which is itself then patterned. From start to finish takes several days. These techinques are often referred to as **VLSI** (very large scale integration).

5.6 Sharpening the tools

In this final section I want to explore the scope for finer scale circuitry. We have just discussed a process which involves scale lengths of around one micron across the surface and tenths of a microns down into it. There will come a time if I shrink the size of devices, by whatever means, when there just aren't enough atoms in what remains to enable the intended solid-state wizardry to continue. We'll probe this limit shortly.

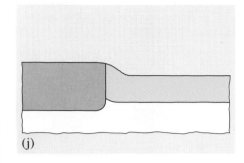

(j)

Figure 5.42 (j)

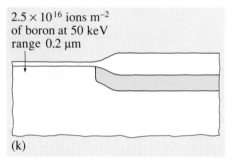

2.5×10^{16} ions m⁻² of boron at 50 keV range 0.2 μm

(k)

Figure 5.42(k)

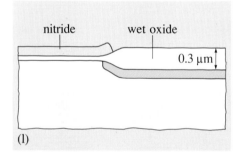

nitride   wet oxide

0.3 μm

(l)

Figure 5.42(l)

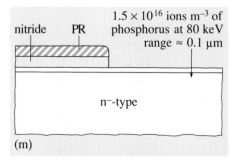

nitride   PR

1.5×10^{16} ions m⁻³ of phosphorus at 80 keV range ≈ 0.1 μm

n⁻-type

(m)

Figure 5.42(m)

Table 5.6

										PAGE:01		BATCH NO.
	MASK:											
	REVISION:											
	PROD CONT:–											

OPERATION			IN	OUT	OPERATOR	DATE	COMMENTS
Provide unique identification for each wafer	LASER MARK	PROC					
		MARK					
		MEGASONIC					
		INSPECT					
		PROC					
Grow pad oxide	INIT OX	PROC					
		CLEAN					
		OXIDISE					
		MEASURE					
		PROC					
Deposit nitride	NITRIDE	PROC					
		NITRIDE					
		MEASURE					
		PROC					
Mask off the p-wells	TWIN WELL PHOTO	PROC					
		COAT					
		ALIGN					
		DEVELOP					
		PROC					
Check. CD = critical dimension	TWIN WELL INSP	PROC					Optical measurement
		INSPECT					
		MEASURE CD					
		REWORK					
		PROC					
Etch nitride from n-well areas	TWIN WELL ETCH	PROC					
		ETCH					
		INSPECT					
		MEASURE OX					
		PROC					
Ion implant n-wells	N WELL IMP	PROC					
		PROC					
Strip off the remaining photoresist	TWIN WELL RESIST STRIP	PROC					
		STRIP (DRY)					
		STRIP (WET)					
		MEASURE CD					
		FINAL INSP					
		PROC					
Thicken up oxide over n-wells (to mask p-well implant)	REOX (AND N WELL DRIVE)	PROC					
		CLEAN					
		NW DRIVE					
		MEASURE					
		PROC					
Remove the nitride over p-wells	ON ETCH TWIN WELL	PROC					
		OX ETCH					
		NIT ETCH					
		CLEAN					
		INSPECT					
		MEASURE OX					
		PROC					
Ion implant p-wells	P WELL IMPLANT	PROC					
		PROC					
Drive in the n and p-wells	TWIN WELL DRIVE	PROC					18 hours at 1050 °C drives wells in half a micron or so
		CLEAN					
		TW DRIVE					
		MEASURE					
		PROC					

In reducing the size of devices, the benefits to be gained are increases in speed and packing density (very saleable attributes) and a more effective use of the materials and processing involved. We will look at one or two refinements which fall into this category.

5.6.1 Sufficient

By now you should be ready to take a hint and estimate a rough limit to the size of MOS devices as we have learnt about them. Exercise 5.30 will set you going.

EXERCISE 5.30

(a) How many silicon atoms are there in a cube of side 0.01 µm (10^{-8} m)? Refer back to Table 5.1

(b) If the silicon is doped to a concentration of 10^{24} atoms m^{-3}, how many charge carriers are introduced into the cube?

(c) What characteristic lengths should we be concerned about?

5.6.2 Assessing the changes

Here we will look at the effects of size changes on MOSFET devices, presuming rather optimistically that we will find a way to make them if the benefits are good enough.

Physical pattern size

To begin with I'm going to suppose that all linear dimensions are decreased by a factor α. (That is x becomes x/α, so if $\alpha = 2$, then x becomes $x/2$.) In this case the area of a device is decreased by α^2 (that is $xy \rightarrow xy/\alpha^2$). The packing density (the number of devices on a chip) will increase by α^2. That result reflects the two-dimensional nature of this technology; the distance we go down into the wafer hardly matters. We must find out what must happen to voltages, currents, resistances, capacitances, and so on, when linear distances change by this factor.

Voltages and currents

If vertical dimensions are to be reduced the gate oxide will be thinner and its breakdown voltage will be proportionately less. To keep the electric field the same in this critical region we must consider reducing the supply voltage by the same factor α. To see what happens to current, we have to look at the MOSFET characteristic equation (from 'MOS transistors'). The saturation current is

$$I_{DS} = \frac{\beta}{2}(V_{GS} - V_T)^2$$

where

$$\beta = \frac{\varepsilon_0 \varepsilon_{ox}}{t_{ox}} \mu \frac{W}{L}$$

Now, just supposing that the carrier mobility μ is unaffected by the diminishing scale, the β parameter will scale as follows

$$\beta = \varepsilon_0 \varepsilon_{ox} \mu \frac{W}{t_{ox}L} \rightarrow \varepsilon_0 \varepsilon_{ox} \mu \frac{W\alpha^2}{\alpha \, t_{ox} \, L} = \alpha\beta.$$

The drain–source current in its turn, if all voltages are decreased by α, will behave as follows

$$I_{DS} \rightarrow \frac{\alpha\beta}{2}\left(\frac{V_{GS}}{\alpha} - \frac{V_T}{\alpha}\right)^2 = \frac{I_{DS}}{\alpha}$$

so the current would also decrease by α. This is only approximate since the mobility doesn't remain constant.

EXERCISE 5.31 Show that if device dimensions are all reduced by a factor α, a 'chip' need not dissipate any extra power in spite of the increased density.

Depletion region widths

In 'pn Capacitors' I showed that the width of the depletion region of pn junction is approximately.

$$x_d \approx \sqrt{\frac{2\varepsilon_0\varepsilon_r V}{eN}}$$

So if the depletion widths are to scale with other physical sizes, the doping levels may have to go up by α: I've got to put $N \rightarrow \alpha N$ and $V \rightarrow V/\alpha$ in the depletion formula, so

$$x_d \rightarrow \sqrt{2\varepsilon_0\varepsilon_r \frac{V/\alpha}{e\alpha N}} = \frac{x_d}{\alpha}$$

Capacitances

The time taken to switch on an MOS device is associated with the loading of charge onto the gate electrode. The gate capacitance measures how many coulombs are needed for each volt of potential established. Charging and threshold voltages, as well as charge flow rates (currents), have all been reduced together so what matters to the speed of operation is the net gate capacitance. What has happened to it?

$$C = \varepsilon_0\varepsilon_r \frac{A}{d} \rightarrow \varepsilon_0\varepsilon_r \frac{A}{\alpha^2} \frac{\alpha}{d} = \frac{C}{\alpha}$$

Capacitance decreases by α, so we might expect faster operation. (But if vertical scaling were abandoned, so that d remains unchanged, then we could reduce the capacitance further.)

Connection resistance and capacitance

If a 'chip' remains the same size but features are reduced by α to pack more per unit area, then some connecting tracks (such as power rails) will not be any shorter. They must still stretch from one side of the chip to the other. The resistance is $(\rho l / A)$. If ρ and l are unchanged, it will increase by α^2 because the cross-sectional area A of the track decreases. The capacitance between the track and the substrate, on the other hand, is unchanged (the area of track presented to the substrate is reduced by α, since the length is unchanged but the width reduced, and the thickness of the inulating layer is reduced by α). Thus the RC time constant may be longer by α^2. Notice therefore that although scaling down may help a device to operate faster, the resulting system of devices may have inherent speed reductions. Clearly a scaling algorithm is called for which does something a litte more careful that just reducing all the lengths, voltages and currents and increasing all the doping. The detail of size reductions is a complicated blend of models for device operation and models and data for material behaviour. A little CAD here is again invaluable.

5.6.3 Secondary effects

The above discussion on scaling supposed that the relevant materials properties can be extrapolated to smaller dimensions without any adverse consequences on device behaviour. Effects other than those we designed for ('second-order effects') can be significant for channel lengths less than about 5 µm, so clearly we need to investigate this question further before ordering the production line for ever denser memories.

The modelling of electron transport in semiconductor materials involves several concepts of length. For example there's the mean free path between collisions (associated with mobility) and the depletion-region width. Figure 5.44 is a collection of such data related to dopant concentration. If scaling involves very large concentrations then our simple transport models no longer apply. For instance, mobility is related to how far electrons move without colliding; but in very short channels, electrons may pass through without a scattering event. This 'ballistic' motion is not necessarily undesirable, but the operating models for such devices will be different from those for the standard MOSFET. Mobility is degraded, even at normal dopant concentrations, as devices are reduced in size. First, the electron mobility in a shallow channel is decreased by the transverse electric field between the gate and the substrate. Secondly, the use of short channels gives rise to large longitudinal fields between the source and the drain. A feature of all semiconducting materials is that at low fields the drift velocity increases linearly with the accelerating field strength (mobility is constant). This cannot continue indefinitely. Drift velocity tends to saturate as it approaches the background random thermal velocity (see Figure 5.45). Both effects mean that transistors must be larger than expected from the elementary first-order design. On the other hand if ballistic motion is

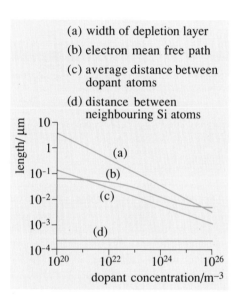

Figure 5.44 Some characteristic lengths in extrinsic silicon at 300 K

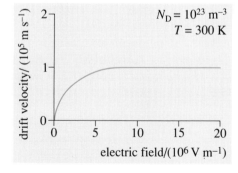

Figure 5.45 Drift velocity and electric field in silicon

involved, then carrier velocities can exceed thermal velocities, offering exciting possibilities for future device innovations.

Strong electric fields may also cause impact ionization. This phenomenon is well known in a reverse-biased pn junction, where the large fields in the depletion region accelerate electrons to high velocities. These charges can easily gain enough kinetic energy to ionize some of the lattice atoms, resulting in a runaway increase in reverse current (that is, breakdown). In a MOSFET channel the result of breakdown is an uncontrolled rapid rise in drain current.

A related phenomenon is the injection of carriers into the gate oxide, as a result of their gaining enough energy to overcome the SiO_2–Si interface barrier potential. This charge can be trapped in the insulator at defect sites (rather like fixed oxide charge), causing long term drift of threshold voltage.

The final feature of scaled systems to be considered is that of the metal tracks for interconnections and contacts. Aluminium has the lowest resistivity of the materials available (a factor of 1000 better than polysilicon). However, there are problems for the power supply lines at increased current density as feature sizes are reduced. We saw earlier that copper must be added to slow the electromigration of aluminium in CMOS circuitry at 1.4 μm scales. For smaller scales the problem becomes worse.

The evolution of conventional memory technology continues as technology is refined to make smaller feature sizes possible. The chief result is that the cell density can be increased but without too much change in the performance of the final product.

5.6.4 Devices and systems

There are two aspects to memory technology which can be tightened up, both with the aid of materials science and engineering. As we have seen, smaller devices may be faster, but systems of such devices may be unable to capitalize on the whole of the potential gain.

It's not solely a question of electronics or system design. The ability to engineer materials to higher specifications is paramount. Try Exercise 5.32 to see if you can see where the scope lies.

As illustrations of how ingenuity enables technological evolution, see ▼Accuracy for speed▲ and ▼Layer upon layer▲ and then try SAQ 5.18.

EXERCISE 5.32

(a) Give one example of a processing step which ought to be minimized if finer scale engineering is to be successful.

(b) How much of the original silicon wafer is used directly for its semiconducting properties in the final CMOS memory product?

- Less than 1%?
- Between 1 and 5%?
- Between 5 and 50%?

▼Accuracy for speed▲

In the processing sequence based on the Plessey Semiconductor CMOS technology I omitted a detail regarding the gate alignment so that I could introduce it here. It is an excellent example of processing ingenuity enabling superior specification.

In a real MOSFET, the gate and the drain are close enough for some of the charges on the gate to induce opposite charges in the drain region. (The gate and source interact similarly.) This behaviour is just what we design into a capacitor and, as we have seen, such a gate-substrate interaction is crucial to the formation of the inversion channel in a MOSFET. Capacitance is not intended, but is inevitable in the gate-drain and gate-source interaction. Generally it is dubbed 'stray capacitance'. I've already indicated (in 'Self-aligned gates') that gate-drain capacitance is especially undesirable as it acts as a feedback path. Gate-drain capacitance can limit the speed at which a MOSFET will switch. Let's see why the MOSFET structure has stray capacitance.

EXERCISE 5.33

(a) What geometric factors tend to decrease capacitance?

(b) What geometric factors ensure the formation of conducting channel between source and drain?

(c) What compromise can you suggest to keep stray capacitance low and still ensure a channel forms?

Exercise 5.33 should show you that we need some careful process control to prevent the source and drain regions extending under the gate. But the self-aligned gate is supposed to do just that. Look back at the processing sequence to see why we might be concerned that ion-implanted dopant finds its way sideways under the gate. (I'll explain why shortly.)

The polysilicon gate acts as a mask for the ion implant which will form the drain and source regions (Figure 5.42d). However, during the anneal cycle following implantation, the ions introduced by implantation will diffuse sideways under the gate, thereby increasing significantly the gate-source and gate-drain capacitances. In the CMOS cycle described, this lateral diffusion extends 0.2 μm under the gate. In effect, the gate is too big. Now at this stage the gate is heavily doped polysilicon, and there is a 0.03 μm layer of oxide over the lightly doped epitaxial regions that will form the sources and drains (Figure 5.46a). Suppose before the ion implantation takes place, we were to put in an oxidation stage. Heavily doped polysilicon can be oxidized much faster than the lightly doped epitaxial material. It turns out that 0.2 μm of gate material can be oxidized while a mere 0.002 μm is added to the oxide elsewhere (Figure 5.46b). Now when the ion implantation takes place, the source and drain regions are masked by the gate and its oxide out to 0.2 μm wider than the gate (Figure 5.46c). In the subsequent annealing, lateral diffusion closes the gap and a complete channel can form, but with a reduced stray capacitance (Figure 5.46d). In this way switching speeds can be reduced from 300 picoseconds to 200 picoseconds, without any new processing equipment. In fact processing designed to establish 1.4 μm gates has been sharpened to form them just 1.2 μm wide.

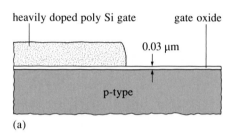

(a)

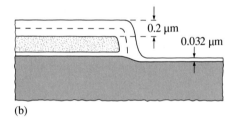

(b)

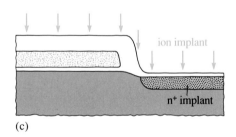

(c)

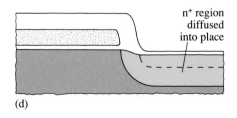

(d)

Figure 5.46 Improving self-aligned technology

▼Layer upon layer▲

One route to increased performance in high density systems is to make more use of the vertical dimension. For example using many layers of active devices with vertical interconnections would allow the maximum interconnection lengths to be reduced to tens of microns rather than several millimetres. (Multiple metallization layers are a well established technology.)

If we are to build circuits in layers of silicon separated by layers of insulator, the first problem we have to address is that of how to lay down the necessary high quality material. Let's begin with an insulating substrate. The p- and n-channel devices must then be built up as islands on top of the insulating substrate, rather than as wells into the substrate (Figure 5.47). The advantages of making devices this way has been appreciated for many years; for example, there are no problems in maintaining electrical isolation between devices.

A suitable single-crystal substrate exists: sapphire, Al_2O_3. It can be grown in large crystals, and cut into wafers and polished. It is not particularly reactive, even at high temperatures. One particular plane of its hexagonal lattice provides a reasonable lattice match to silicon, so an epitaxial silicon layer can be grown on it. An added advantage of this silicon-on-sapphire technology (SOS, or more generally SOI for silicon-on-insulator) is that the resulting chips are **radiation hard**. That is, because the substrate is a good insulator, the creation of electron–hole pairs by

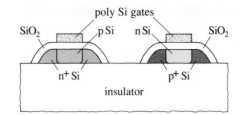

Figure 5.47 Silicon on insulator — a CMOS pair

absorption of radiation is unlikely to corrupt information held in storage cells and logic circuits. Radiation hardness is a very important quality for military and space equipment.

Two-dimensional SOS systems have been in production for some time. Such devices, of course, have the same long-interconnection problem as more conventional devices. Also, the silicon overgrowths contain misfit dislocations and thus lack the high crystallinity of bulk silicon. (Defective structures reduce carrier mobility.) And an additional problem for this technology is its costliness. Artificial sapphire wafers are approximately ten times the price of silicon wafers.

With SOI devices packing density can be increased because:

• The separation between devices is reduced to the minimum feature size. Conventional layouts need field oxides and well diffusion, which increase the space between devices.

• The surface layout area is further reduced by the absence of substrate connections and well-bias connections.

A further advantage of SOI is that during processing, fewer masks are needed than for comparable, high-performance conventional devices, to the benefit of reliability and processing cost.

We won't pursue it further here, but you can probably imagine some of the problems of extending the use of SOI to three-dimensional systems; that is, systems where devices are constructed over other devices. The production of such multi-level active systems is going to require low-temperature process technology.

EXERCISE 5.34 To minimize the surface area needed for a **DRAM** cell, an integrated capacitor has been designed to sit vertically in a silicon surface layer so that both plates extend down into the wafer, like the sides of a trench. Using the CMOS processing as a guide, what can you say about the following?

(a) Capacitor dimensions. How deep can we go? (This affects plate area A.) How close can the 'plates' be? (This affects dielectric thickness d.)

(b) What would be suitable techniques and materials for making the dielectric and the 'plates'?

SAQ 5.18 (Objective 5.17)
'Accuracy for speed' described a refinement in the CMOS processing that enabled devices to be made with an enhanced specification.

(a) What materials *property* underlies this refinement?

(b) How is this property exploited in the *processing* of the device?

5.7 Epilogue

In this chapter we have been discussing just some of the microtechnology which lies behind electronic memory. I haven't even reached a single finished product, packaged and ready to use in the form which may be familiar to you. I have concentrated on the 'micro' scale of memories.

It is a fast moving field, but by far the most rapid change is in the products — their design and disposition at a systems level. Our focus on CMOS SRAM has revealed a materials story with a more stately evolution and a relevance wider than the angle we have taken. Semiconductors, conductors and insulators will continue to be depleted and inverted or patterned and etched or deposited or doped for some purpose long after these memories have been forgotten.

Objectives for Chapter 5

You should now be able to do the following:

5.1 Discuss the basic functions required of insulators, conductors and semiconductors in integrated circuits. (SAQ 5.2)

5.2 Describe the means of deploying silicon dioxide in the fabrication of silicon microcircuits by chemical deposition and thermal growth. (SAQ 5.3)

5.3 Explain the control of resistivity in semiconductors through the introduction of certain foreign atoms. (SAQ 5.4)

5.4 Describe the mechanisms and techniques of doping semiconductors by atomic diffusion and ion implantation. (SAQ 5.5)

5.5 Discuss the use of aluminium as the major conductor in silicon microcircuits. (SAQ 5.6)

5.6 Describe the key steps in the selective removal of material on the scale of microns through the use of photolithography. (SAQ 5.7)

5.7 Explain how pn isolation is achieved and discuss the associated effects. (SAQ 5.8)

5.8 Describe the nature of depletion regions and carry out basic calculations given appropriate formulae. (SAQ 5.9)

5.9 Explain how the depletion of charge carriers in semiconductors contributes to capacitance. (SAQ 5.10)

5.10 Explain how the inversion of a layer of semiconductor in a MOS structure can be brought about. (SAQ 5.11)

5.11 Carry out basic calculations on inversion thresholds given appropriate formulae. (SAQ 5.12)

5.12 Discuss the engineering of materials properties through composition and geometry to make basic MOS switch devices. (SAQ 5.13)

5.13 Appreciate the operation of basic memory cells sufficiently to be able to challenge the constraints on space and speed of operation. (SAQ 5.14)

5.14 Describe the nature of materials used in MOS memory devices. (SAQ 5.15)

5.15 Give scientific explanations of various steps in a MOS processing sequence. (SAQ 5.16)

5.16 Discuss the materials and processing requirements of MOS-based semiconductor RAM microcircuits. (SAQ 5.17)

5.17 Explain some of the property-processing links of MOS circuit fabrication. (SAQ 5.18)

5.18 Expand the following shorthand:

MOS	p^+/n^+-type
nMOS	pn junction
pMOS	polysilicon
CMOS	epi-layer
RAM	p-well
SRAM	n-well
DRAM	VLSI
p/n-type	

Answers to exercises

EXERCISE 5.1

(a) 28 kg of Si have 6×10^{26} atoms, so 1 kg has

$$\frac{6 \times 10^{26}}{28}\text{atoms}$$

From Table 5.1, 1 m³ of Si weighs 2.3×10^3 kg, so 1 m³ has

$$\frac{2.3 \times 10^3 \times 6 \times 10^{26}}{28}\text{atoms}$$

$$= 5 \times 10^{28} \text{ atoms}$$

Hence atomic density is 5×10^{28} atoms m⁻³.

$(28 + 32)$ kg of SiO_2 have 6×10^{26} formula units, so 1 kg has

$$\frac{6 \times 10^{26}}{60}\text{formula units}$$

From Table 5.1, 1 m³ of SiO_2 weighs 2.2×10^3 kg, so 1 m³ has

$$\frac{2.2 \times 10^3 \times 6 \times 10^{26}}{60}\text{formula units}$$

$$= 2.2 \times 10^{28} \text{ formula units}$$

Each formula unit has two oxygen atoms and one silicon atom, so we have 4.4×10^{28} oxygen atoms m⁻³ and 2.2×10^{28} silicon atoms m⁻³.

Similarly, 1 kg of Al has

$$\frac{6 \times 10^{26}}{27}\text{atoms}$$

so 1 m³ of Al has

$$\frac{2.7 \times 10^3 \times 6 \times 10^{26}}{27}\text{atoms}$$

Atomic density of aluminium is 6×10^{28} atoms m⁻³.

(b) Silicon is 11 orders of magnitude (10^{11}) less conductive than aluminium.

(c) Silicon is 12 orders of magnitude more conductive than its oxide.

EXERCISE 5.2

(a) For capacitor plates of area A, separation d and insulator of relative permittivity ε_r,

$$\text{capacitance } C = \frac{\varepsilon_0 \varepsilon_r A}{d}$$

Ignoring the thin oxide dielectric

$$C = \frac{9 \times 10^{-12} \times 12 \times 10^{-6}}{10^{-3}}\text{F}$$

$$\approx 10^{-13} \text{ F}$$

This is not a very useful amount. Electronic circuits need capacitances of nano- and microfarads. Under some circumstances it could, nevertheless, be a nuisance in parts of a circuit where capacitance is undesirable.

(b) If the oxide is intact, we have two oxide resistances in series with one silicon resistance. Using resistance $R = \rho l / A$

$$R = \frac{(2 \times 10^{15} \times 10^{-8})}{10^{-6}}\Omega$$

$$+ \frac{(2.3 \times 10^3 \times 10^{-3})}{10^{-6}}\Omega$$

$$= (2 \times 10^{13}) + (2.3 \times 10^6)\,\Omega$$

As an insulator it's not bad, but it's not a very useful amount of resistance, even if the oxide is absent. Electronics circuits need Ω and $k\Omega$.

(c) The total thickness of the oxide layer is 2×10^{-8} m. So breakdown voltage V_B is

$$V_B = 2 \times 10^{-8} \times 10^9 \text{ V} = 20 \text{ V}$$

For the silicon $V_B = 10^{-3} \times 3 \times 10^7$ V or 30 kV. So beyond 20 volts, the resistance is that of the silicon.

EXERCISE 5.3

(a) Diffusion to the growing interface is through amorphous silica. If silicon crystal orientation is important, it must be through its influence on reactivity. (A weak effect is observed.)

(b) As you probably know from the behaviour of silica-base glass, dopant atoms in silica will weaken its structure, enhancing diffusion a little.

(c) Growth rate depends on diffusion rate and reaction rate, both of which are species dependent; so there could be an effect. (In fact, H_2O diffuses and reacts faster than O_2.)

EXERCISE 5.4 From Table 5.1 (and Exercise 5.1),

1 m³ of Si contains 5×10^{28} Si atoms

and

1 m³ of SiO_2 contains 2.2×10^{28} Si atoms

Therefore a volume of 1 μm $\times$ 1 m² of SiO_2 contains

$$2.2 \times 10^{28} \times 10^{-6}$$

$$= 2.2 \times 10^{22} \text{ Si atoms}$$

There are 5×10^{28} silicon atoms in 1 m³ of pure silicon so 2.2×10^{22} atoms of unoxidized Si occupy

$$\frac{2.2 \times 10^{22}}{5 \times 10^{28}}\text{m}^3$$

$$= 0.44 \times 10^{-6} \text{ m}^3 \text{ of Si.}$$

Each square metre of Si is therefore eroded to a depth of 0.44 μm to produce 1 μm of SiO_2.

EXERCISE 5.5
The lower resistivities call for higher dopant concentration. The highest solubility donors (n-type) are phosphorus and arsenic. The highest solubility acceptor (p-type) is boron. All may achieve around 1% solubility at room temperature.

EXERCISE 5.6
At high temperature, thermally generated carriers and their temperature dependence have swamped the steady carrier concentration established by the dopant. At lower temperatures when the dopant is dominant in determining the charge carrier population, the resistivity rises with temperature just as for metals.

EXERCISE 5.7
You may have spotted some of the following contrasting points during limited-source diffusion.

(a) The concentration at $x = 0$ falls.
(b) The concentration gradient at $x = 0$ is always zero.
(c) At any point the concentration rises before eventually falling.
(d) After a very long time, there is very little diffusant everywhere.

EXERCISE 5.8
If we ionize the atom and accelerate it electrically, each volt gives each electronic charge 1 eV of energy. So a singly ionized boron atom (B^+) in passing between regions differing in potential by 2 kV gains 2 keV of energy.

EXERCISE 5.9 They are CVD (Chemical Vapour Deposition), evaporation and sputtering.

For CVD of aluminium, aluminium atoms must be produced at or just above the substrate through an appropriate chemical reaction.

For evaporation of aluminium, a refractory filament or other means (electron bombardment, induction) is used to vaporize pure aluminium to generate a thermal flux of aluminium atoms towards the substrate.

For sputtering of aluminium, ions of an inert gas (e.g. argon) are accelerated onto an aluminium target causing a non-thermal flux of dislodged target atoms (mainly aluminium!) towards the substrate.

EXERCISE 5.10 A 5 μm track 3000 μm (3 mm) long is made up of 600 × 5 μm squares in series. If each one contributes $2.7 \times 10^{-2}\ \Omega$ the total track has a resistance of

$$(6 \times 10^2 \times 2.7 \times 10^{-2})\Omega$$
$$= 16.2\ \Omega$$

This is four times too big so a track four times wider (or else four times deeper) must be specified.

EXERCISE 5.11

(a) The Fermi level lies near the middle of the gap between valence and conduction band.
(b) If n-type dopant is added, the Fermi level moves towards the conduction band.

EXERCISE 5.12

(a) Isotropic etching will tend to produce spherical pits, eroding sideways as much as downwards. Closely spaced features may not be able to accommodate the lateral spread and so require anisotropic etching.

(b) Ion implantation uses tens of keV, that is hundreds of times more energy than RIE, yet it only penetrates about one micron. RIE may implant ions but to considerably less depth, perhaps a hundredth of a micron.

EXERCISE 5.13

(a) The photoresist defines areas of SiO_2 which are to be removed right back to the Si surface. It protects the SiO_2 which is to stay.

(b) *Resolution* is limited by the optical system, the mask and how its image is formed on the photoresist. The transfer of this resolution to the silicon surface depends on how the SiO_2 is removed.
(c) The mask for the contacts must be *registered* (aligned) correctly with respect to the doped channel.

EXERCISE 5.14 As before, consider an island of n-type material, electrically isolated from the p-type substrate. Using ion implantation restricts the penetration to 1 μm. If we want the same sheet resistance as employed for the diffused resistance ($10^2\ \Omega$ square^{-1}), the resistivity will have to be less (because the shallower channel is 'geometrically' more resistive). $10^{-4}\ \Omega$ m of 1 μm depth materials gives a sheet resistance of $10^2\ \Omega$ square^{-1}. An n-type track doped with 3×10^{24} phosphorus atoms per cubic metre over 1 μm depth, and again ten times longer than its width (5 μm × 50 μm for example), will provide 1 kΩ of resistance. There need to be n$^+$ regions at the ends to allow aluminium contacts to be formed.

EXERCISE 5.15

• Hole conduction current (small in n-type semiconductor).
• Electron conduction current.
• Hole diffusion current.
• Electron diffusion current.

EXERCISE 5.16 Electrons in p-type are in the minority. Any near the boundary will be swept onto the island so the boundary density is virtually zero. Far away, the electron density may be near its characteristic p-type value n_{p0}. A small diffusion current (j_{ep}) of electrons is driven by the density gradient.

Once electrons are in n-type material, a very weak field is enough to move newcomers away from the edges as a conduction current (j_{en}). It must be the case that $j_{en} = j_{ep}$ or charge would accumulate.

EXERCISE 5.17 The leakage currents depend on minority-carrier concentrations n_{p0} and p_{n0}. If these are kept low the leakage will be low. In extrinsic material, as we increase the majority-carrier concentration, we decrease the minority-carrier concentration. (For example, electrons in n-type semiconductor are almost matched in charge by positively ionized donors, so we don't need so many

holes to achieve overall charge neutrality. What holes there are have an increased risk of being dropped into by an electron as the electron density increases.) So we should use the heaviest doping we can to keep the minority charge density down.

Note: Since
$$n_n p_n = n_i^2$$
and in n-type
$$n_n \approx N_D,$$
$$p_n = n_i^2/N_D.$$
In silicon
$$n_i \approx 10^{16}\ \text{m}^{-3}$$
whereas in germanium
$$n_i \approx 10^{19}\ \text{m}^{-3}.$$
You can see why silicon is preferred!

EXERCISE 5.18

1 Grow n-type epitaxial layer by (i) CVD or by (ii) implanting donor ions and then heating so that donor ions diffuse and swamp acceptors, converting p$^-$ region to n-type. The n-type layer must be at least 1.25 μm thick to allow for silicon consumption during oxide growth.

2 Grow 0.5 μm of oxide. A good quality of oxide is going to be needed, especially since the Si–SiO$_2$ interface has been highlighted as important.

3 Etch a contact hole to the n-layer and implant extra donor to make n$^+$ regions for an ohmic contact.

4 Deposit aluminium.

5 Pattern the aluminium using lithography.

EXERCISE 5.19 The oxide charge is positive. This will draw negative charge (mobile electrons) from the n-type material, making it even more n-type. To invert n-type would require negative charge in, or beyond the oxide.

EXERCISE 5.20

• Q_m is how the metal holds its part of the capacitively stored charge. Clearly it is voltage dependent.
• Q_{fo} was fixed during processing. No change.
• Mobile 'minority' charge can accumulate near the semiconductor/oxide interface if encouraged to do so by opposite-polarity charge either in the oxide or further away on the metal–oxide interface; equally it could be driven away by like charges. Since more or less Q_m will influence Q_{mi}, it is voltage dependent.

• If Q_m repels majority charge, the ionized dopant left behind in the crystal structure forms a depletion layer. The degree to which Q_m repels majority charge and drives the depletion is voltage dependent and so too therefore is Q_D.

EXERCISE 5.21

(a) If $(V_{GS} - V_T) \gg V_{DS}$ we can do some approximating and say

$$I_{DS} \approx \beta(V_{GS} - V_T)V_{DS}$$

Comparing this with Ohm's law $I = V/R$, we can identify the term

$$\frac{1}{\beta(V_{GS} - V_T)}$$

with the channel resistance.

(b) A capacitance C stores an extra charge ΔQ when the voltage increases by ΔV:

$$\Delta Q = C\Delta V$$

Likewise the extra gate charge which the inversion-layer charge must match is given by:

$$\Delta Q = \varepsilon_0 \varepsilon_{ox} \frac{WL}{t_{ox}}(V_{GS} - V_T)$$

If the inversion charge is confined to a thickness t_{inv}, its density is

$$n_{inv}e = \frac{\Delta Q}{LWt_{inv}}$$

$$= \varepsilon_0 \varepsilon_{ox} \frac{1}{t_{ox}t_{inv}}(V_{GS} - V_T)$$

since $\sigma = n_{inv}e\mu$. The channel resistance is therefore

$$R_{DS} = \frac{L}{\sigma W t_{inv}}$$

$$= \frac{Lt_{ox}}{\varepsilon_0 \varepsilon_{ox}\mu(V_{GS} - V_T)W}$$

The device constant β is then

$$\beta = \frac{\varepsilon_0 \varepsilon_{ox}}{t_{ox}} \frac{\mu}{L} \frac{W}{L}$$

EXERCISE 5.22

(a) If logic levels remain $+5$ and $+0.5$, the power is reduced by operating at lower current in the logic zero state. This is most strongly influenced by β, which should be lowered. (The threshold voltage ought to be left mid-range.) The load resistor R

must then be raised to ensure the lower current still reduced the output to logic 0).

(b) The logic zero is established by current in R dropping potential. If logic zero is to be lowered, a larger current must flow or else a larger resistance must be used. The former is ruled out if the power is to remain unchanged. (β can therefore be raised.)

(c) Using the existing process leads directly to (a).

EXERCISE 5.23   See Table 5.7.

Table 5.7

Parameter	Material quantities	Geometric quantities
V_T	N_A, n_i, ε_s Q_{fo}, ε_{ox}	oxide thickness t_{ox} gate area $A = W \times L$
β	ε_{ox}, μ_e	oxide thickness t_{ox} channel aspect ratio W/L
R	N_D, μ_e	diffusion depth and channel aspect ratio L/W

EXERCISE 5.24   The area of a single memory element (1 bit) is at least that of two inverters. Really we should allow for extra components (transistors) to address the individual cells, but since the inverter area is dominated by the resistor we can neglect these. So each bit needs

$$8 \times 10^{-10} \text{ m}^2$$

for the components. Including space for isolation etc. (in which we must allow a factor of two for inefficient usage of area) this rises to

$$2 \times 8 \times 10^{-10} \text{ m}^2.$$

A 9×10^{-6} m^2 chip can therefore accommodate about

$$(9 \times 10^{-6}/2 \times 8 \times 10^{-10}) \text{ bits}$$

$$= 6000 \text{ bits.}$$

EXERCISE 5.25

(a) The p-channel device needs an n-type substrate. The n-channel one needs a p-type substrate. We must develop n-type islands (or 'wells') in a p-type wafer to hold the p-channel transistors.

(b) β depends on geometry and carrier mobility. To have the same β, p- and n-

channel devices differ in W/L ratio to compensate for the different mobilities μ_n and μ_p.

(c) The substrate doping can be used to establish the required threshold voltage.

(d) If n wells are formed in a p substrate there is inherent isolation here but it may be necessary to improve this isolation if higher packing density is required.

EXERCISE 5.26

(a) A DRAM cell will consume power every time it is refreshed as well as when a setting operation changes its state. An SRAM cell based on CMOS consumes power on switching from one state to the other but has a negligible standby power and will hold its state so long as the supply voltage is uninterrupted. Unless the application calls for repeated changings of state, a DRAM cell will consume more power.

(b) SRAM is economical to run. The cell is accessible any time after setting (DRAM is repeatedly 'closed for refreshing'). SRAM does not need complicated refresh circuit and cycles. DRAM uses less space per bit than SRAM. It uses simpler cells, which ought to imply a higher production yield.

EXERCISE 5.27

(a) We need the oxide thickness to correspond with $R - \Delta R$, so that dopant (most of which ends up between $R - \Delta R$ and $R + \Delta R$) penetrates the silicon. Constructing the $R - \Delta R$ line on Figure 5.43 (taking care with the scales) I estimate this to require between 70 keV and 90 keV.

(b) A concentration of 10^{27} dopant atoms m^{-3} spread over a depth $2\Delta R$ corresponds with a dose of $2\Delta R \times 10^{27}$. Taking ΔR from Figure 5.43, the dose is

$$2 \times 1.8 \times 10^{-8} \times 10^{27} \text{ ions m}^{-2}$$

$$= 3.6 \times 10^{19} \text{ ions m}^{-2}$$

EXERCISE 5.28

(a) The self-aligned process uses the gate (polycrystalline Si) as the mask for the source and drain implantations. During those implantations, ions will have to pass through the oxide layer into the underlying silicon. The oxide and the polysilicon gate are comparable in their screening ability, so the gate thickness will have to exceed that of the oxide to prevent implantation into the silicon beneath the gate oxide.

(b) The gates are 1.4 μm wide and have to be defined by etching unwanted material from a deposited layer. If the layer thickness approaches the feature size, more care has to be taken with the sides of the features. Therefore it is wise to keep the gate thickness well below 1 μm.

EXERCISE 5.29

(a) Although they all might be reasons, it is unlikely that a thin film of a silicon compound cannot be successfully deposited on to silicon. Thermal mismatch between thin films and substrates can be a problem but in fact (i) is the major cause for concern. We're going to want high quality material in the active regions so don't let's damage it in trying to protect it!

(b) This is the bird's beak feature. Diffusion in silicon nitride is generally much slower than in silicon dioxide and there is virtually no oxidation beneath the nitride-covered region. However the edges of the nitride-covered region are permeated by laterally diffusing oxygen; here oxidation takes place but at a slower rate than away from the nitride.

EXERCISE 5.30

(a) There are 5×10^{28} silicon atoms m^{-3}. In 10^{-24} m^3 therefore there will be 5×10^4 silicon atoms.

(b) At 10^{24} dopant atoms m^{-3}, each dopant has about 10^{-24} m^3 to call its own. So the dopant introduces one charge carrier into the 0.01 μm cube.

(c) We should consider depletion-region widths, distances between collisions for electrons, channel widths and lengths, and so on.

EXERCISE 5.31
As we have seen, linear scaling by α requires voltages to scale by the same factor and, as a result, currents too. The area of a device reduces by α^2 and

at the same time the power it dissipates (volts × amps) also decreases by α^2. If twice the number of devices each use half the power, the total consumption is unchanged.

EXERCISE 5.32

(a) Thermal cycles inevitably encourage the spreading of dopants. If smaller scale features are to be retained, high temperature processes must be avoided, or at any rate be as short as possible, to minimize the smearing out of dopants towards a uniform thermal-equilibrium state.

(b) The CMOS devices we have discussed use less than a micron depth of silicon in their active regions (channel, source and drain). The original wafer was over 500 μm thick. Even if all of the surface were active, much less than 1% of the original electronics-grade silicon is finally exploited for its electrical properties. The rest has proved useful in maintaining the quality of the vital layer and in constructing it, to say nothing of providing silicon for much of the insulating material and mechanical support for the whole.

EXERCISE 5.33

(a) Capacitance decreases with decreasing area and increasing separation.

(b) Continuous conducting channels between source and drain in a MOSFET are ensured if the gate electrode 'overlaps' the source and drain regions.

(c) The gate should be very precisely fitted between source and drain, being no wider than necessary to form the channel, if possible without overlap.

EXERCISE 5.34

(a) Dimensions. The CMOS process was able to cope with surface features which were close to a micron in linear dimension, and was used from time to time to etch several tenths of a micron into various materials. For the trench capacitor it will probably be necessary to cut features deeper than their width. To obtain useful amounts of capacitance we must trade plate separation (which consumes surface area) for plate area (width and depth) which is why we call it a trench.

There is scope for a little modelling here to see where the advantages lie. We want to maximize the number of farads per square metre of surface. Say the surface area is $a \times b$. The 'vertical' dimension is c. A 'horizontal' capacitor would provide $\varepsilon ab/c$ farads, or ε/c farads per square metre of surface. A vertical capacitor gives $(\varepsilon cb/a)$ farads, or per square metre of surface:

$$\frac{(\varepsilon cb/a)}{ab} = \varepsilon c/a^2 (\mathrm{F\ m^{-2}})$$

Thus the 'vertical' option provides $(c/a)^2$ more capacitance for the same surface area. So we have to go deeper than the plate width to benefit.

(b) A 'deep' trench will need a strongly anisotropic etch, such as ion-assisted etching or even just ion sputtering. The sides of the trench have to be 'metallized'. We might try heavy doping of the side walls, but the geometry militates against ion implantation so it will have to be thermal diffusion. The dielectric between the plates could be a deposited silicon dioxide (CVD).

Note. We're only just beginning to see some new possibilities here. It's not unreasonable to contemplate 'putting' a switching capacitor on the side wall of the trench (by diffusion, for example) before filling it with dielectric to make a complete DRAM cell.

Answer to self-assessment questions

SAQ 5.1 1 intrinsic, 2 covalent bonds, 3 diamond, 4 conduction electron, 5 carrier of positive charge, 6 valence, 7 conduction, 8 energy gap, 9 energy bands, 10, 11 defects, crystal boundaries, 12 hole, 13 donor, 14 acceptor, 15 extrinsic, 16 dopants, 17 p-type, 18 negative, 19 n-type, 20 positive.

SAQ 5.2 Silicon dioxide. Insulator: isolation between layers of conductor and semiconductor. Ideally electrically inert.

Aluminium. Conductor: electrical interconnection between active semiconductor regions and connections to the outside world.

Silicon. Semiconductor: areas in which electrically sensitive properties allow the development of active regions of material capable of modulating conducting properties.

SAQ 5.3 You may have picked some of the following:

1 Grown layers (wet or dry) require a high-temperature process. Deposited layers are formed at much lower temperatures — hence CVD for cooler processes.

2 Grown layers consume substrate silicon. CVD imports its own silicon—hence CVD for putting silicon dioxide over non-silicon surfaces and for filling-in over silicon material already committed to a device.

3 The quality of slow-grown oxides is superior to that of deposited material — thermal energy aids atomic diffusion so the higher temperature processes can get more of the right atoms in the right place.

SAQ 5.4

(a) Acceptor atoms, such as boron, in a silicon structure render it p-type material.

(b) At 300 K a few electrons easily escape from silicon bonds (the valence band) onto the acceptors leaving holes free to roam the valence band. They are available for conduction limited by mobility μ_h.

(c) At low doping ($<0.001\%$) the distortions of the atomic structure arising from randomly sited foreign dopant ions is insufficient to affect the hole mobility. So every tenfold increase in dopant causes a tenfold increase in the density of charge carriers and so a tenfold decrease in resistivity.

(d) At higher doping ($>0.001\%$) mobility is adversely affected by the dopant ions (they locally distort the structure) so the resultant resistivity falls less rapidly than before — the increase in charge carriers being slightly offset by a decrease in their mobility.

SAQ 5.5 During a post-implant anneal, dopant and host atoms move enough to allow the dopant to join the regular atomic structure by substituting for the host at lattice sites. The substitution is driven by the same thermal diffusion mechanism that is exploited in doping by atomic diffusion, but being a cooler and shorter process the distance moved by any one atom is much smaller.

SAQ 5.6 Copper can be sputter deposited and although it is much more noble than aluminium, it may be persuaded to adhere to silicon dioxide. It can also be patterned, as PCB technology clearly demonstrates. Its resistivity is lower than aluminium's, so thinner layers could be used. Its short-comings must be connected with:

(a) the scale of patternability
(b) its compatibility with silicon.

In fact there is no problem with (a). The trouble is with (b), because copper atoms put deep levels (see SAQ 5.1) in the silicon band gap.

SAQ 5.7 We want to define areas of chrome to be removed, but we don't yet have a photomask to do it with. Now, an electron beam directly steered under computer control could selectively expose an electron resist coated over the chrome. The developed resist would allow selected access to the chrome layer which might then be removed by an acid etchant. The sequence would be as follows. Spin on resist. Bake to remove solvent. Expose to directly steered electron beam. Dissolve resist from areas where chrome is to be removed. Etch through exposed areas to glass substrate. Remove remaining resist. (Note: Steered electron beams are much slower than optical systems when patterning silicon but are well suited to the making of the master mask.)

SAQ 5.8 The isolation is afforded by reverse biased pn junctions. Each diode must therefore be placed in its own well of p-type material. The parent n-type substrate must be always biased to hold electrons in it and to repel holes from the various p-type wells (that is, the substrate must be connected to the most positive voltage around). The p-type material can form the p-side of the pn junctions. Each p-well must contain an island of n-type. This region of silicon will have been initially n-type, then converted by stronger doping to p-type, and must now be reconverted by even stronger doping to n-type. The initial doping must be sufficiently light to permit the reconversion.

SAQ 5.9 The worked example in 'pn Capacitors' is a useful starting point. It shows that the total depletion-region width is dominated by the lighter-doping. In this case that is the donor doping of the n-type semiconductor, so we need to consider the expression for x_n, with appropriate values. Notice, though, that if $N_{D\,\text{substrate}} \ll N_{A\,\text{diode}}$ then the ratio $N_A/(N_D + N_A)$ is almost equal to one. In fact the numbers here are exactly as in the example and the result is that the depletion region takes up about 1 μm of the substrate around each p-well. The p-well and the diode it forms could account for about as much surface area, so we can guess that at least 2 μm $\times$ 2 μm of substrate is needed for a diode. So we could get between 2.5×10^{11} and 10^{12} diodes m^{-2} (divide by 10^6 for mm^{-2}). Clearly the higher the substrate doping the denser we can pack the isolation wells. You might also have suggested placing several devices in a common well.

SAQ 5.10 A pn junction capacitance exploits charge storage in abutting regions which are depleted of mobile charge carriers. A conventional capacitor stores charge on the surface of metal plates.

For a given voltage, a pn junction stores more charge in widely depleted region of heavy doping than in narrowly depleted regions of light doping. In a conventional capacitor, for a given voltage, charge storage can be increased by increasing the polarizability of the dielectric and by increasing the plate separation.

The amount of pn capacitance for given dimensions is voltage dependent (the depletion width changes with voltage) whereas conventional capacitance is not.

SAQ 5.11   Yes. A metal track on insulator over silicon will form an 'accidental' MOS structure. There will therefore be some level of voltage between the track and the silicon which will lead to inversion.

SAQ 5.12   To discourage inversion, semiconductor near the oxide must be heavily doped so that depletion will expose copious fixed dopant charges to match charges at the metal/oxide interface. Thicker oxide layers will discourage inversion, requiring higher threshold voltages. According to the formula, V_T gets bigger as N_D and d increase.

SAQ 5.13   See Table 5.8.

SAQ 5.14   The need is for a material which we can deposit and over which we can exercise control of resistivity. An important consideration must be that this is proposed as a surface layer over an integrated circuit, so we're talking about cool processes in the final steps of a fabrication sequence. Silicon should suggest itself not only because we're already in the business of controlling its resistivity, but also because we've already considered silicon deposition for laying down the polysilicon gate (using a relatively cool CVD process). It doesn't need to be single crystal for a simple resistor application. The layer will have to be deposited on an insulator (silicon dioxide) and will need to be patterned.

SAQ 5.15   The cell (and its addressing transistors) is comprised of MOS devices. The latch proper has two n-channel and two p-channel transistors. Each device is built into the substrate in an appropriate well of doped single-crystal silicon (p-type for n-channel devices). Doping is strongest in the connection regions (n$^+$-type for n-channel) which are established within the well. Electrical inversion of the surface

Table 5.8

Region	Material	Processing factors	Product features
general insulator	SiO$_2$	thickness	breakdown voltage
substrate	n-type silicon	doping under gate oxide	V_T
source and drain	heavily doped silicon (p$^+$)	doping	ohmic contact
p-channel	inverted n-type	geometry (photo-lithography) W/L	β (in current–voltage characteristic for MOSFET)
gate dielectric	SiO$_2$	thickness and quality	V_T, β
contacts: gate, source, drain and substrate	AlSi alloy or poly-crystalline Si	geometry	electrical connection

occurs where a gate voltage exceeds a critical value. The gate is isolated from the substrate and channel by a layer of amorphous silicon dioxide. The gate itself is polycrystalline silicon doped with phosphorus. Interconnecting aluminium contains a few per cent of dissolved silicon.

SAQ 5.16   (a) An n-channel device requires a population of electrons to be established near the silicon/silicon dioxide interface under the influence of charges on the gate electrode.

The presence of positive charges fixed in the oxide near the silicon dioxide/silicon interface will encourage the build up of the inversion layer. Positive gate charge and the positive fixed oxide charge act in the same direction. This means that the threshold voltage decreases (becomes less positive) with increasing amounts of positive oxide-charge.

(b) To raise the threshold voltage, we must install some negative charge to compensate

for that of the oxide and so offset its effects. This can be achieved by implanting acceptor dopant in the silicon near the interface. The resulting negatively charged acceptors add to those which are already there (the material is p-type), increasing the margin somewhat in favour of its surface staying depleted for longer.

SAQ 5.17   See Table 5.9 (overleaf).

SAQ 5.18

(a) The materials property is the oxidation rate of silicon, which is dependent on its purity and form (that is, whether it is polycrystalline or single crystal).

(b) The property in (a) is translated into a processing step because it means that the heavily doped polycrystalline gate silicon can be consumed laterally by oxidation, without adding significantly to the oxide thickness through which implanted ions must penetrate.

Table 5.9

OPERATION			IN	OUT	OPERATOR	DATE	COMMENTS
	MASK:				PAGE:01		BATCH NO.
	REVISION:						
	PROD CONT:–						
Provide unique identification for each wafer	LASER MARK	PROC					
		MARK					
		MEGASONIC					
		INSPECT					
		PROC					
Grow pad oxide	INIT OX	PROC					0.03 μm 950 °C Dry oxygen
		CLEAN					
		OXIDISE					
		MEASURE					
		PROC					
Deposit nitride	NITRIDE	PROC					$SiH_4 + NH_3$ at 720 °C
		NITRIDE					
		MEASURE					
		PROC					
Mask off the p-wells	TWIN WELL PHOTO	PROC					Use positive photoresist
		COAT					
		ALIGN					
		DEVELOP					
		PROC					
Check. CD = critical dimension	TWIN WELL INSP	PROC					Optical measurement
		INSPECT					
		MEASURE CD					
		REWORK					
		PROC					
Etch nitride from n-well areas	TWIN WELL ETCH	PROC					CF_4/O_2 plasma etching
		ETCH					
		INSPECT					
		MEASURE OX					
		PROC					
Ion implant n-wells	N WELL IMP	PROC					Penetrate pad oxide with sufficient dose
		PROC					
Strip off the remaining photoresist	TWIN WELL RESIST STRIP	PROC					Oxygen plasma & sulphuric/peroxide wash
		STRIP (DRY)					
		STRIP (WET)					
		MEASURE CD					
		FINAL INSP					
		PROC					
Thicken up oxide over n-wells (to mask p-well implant)	REOX (AND N WELL DRIVE)	PROC					Cool, fast growth 900 °C/H_2O 6 hours for 0.6 μm so 2–3 should do for 0.3 μm
		CLEAN					
		NW DRIVE					
		MEASURE					
		PROC					
Remove the nitride over p-wells	ON ETCH TWIN WELL	PROC					Orthophosphoric acid at 170 °C
		OX ETCH					
		NIT ETCH					
		CLEAN					
		INSPECT					
		MEASURE OX					
		PROC					
Ion implant p-wells	P WELL IMPLANT	PROC					Penetrate pad oxide with sufficient predeposition material
		PROC					
Drive in the n and p-wells	TWIN WELL DRIVE	PROC					18 hours at 1050 °C drives wells in half a micron or so
		CLEAN					
		TW DRIVE					
		MEASURE					
		PROC					

Chapter 6 Displays

6.1 Introduction

Visual information — words, numbers, graphics and pictures — comes to us from innumerable sources. Newspapers, paintings, films (still and moving), hoardings and indicator boards, car instrument panels and digital watches are just a few examples. These display devices vary enormously. For instance, a page of print is very different from the cathode ray tube (CRT) in a television set. One is thin and the information it displays fairly permanent, whereas the other is heavy and bulky, but can carry quickly changing information. Both have been enormously influential. In the UK, Caxton's invention of the printing press in the fifteenth century initiated profound social and cultural changes. The arrival of the CRT (1903) and its subsequent evolution has similarly transformed our lives. It is found in television, computers, radar, oscilloscopes and many other applications (Figure 6.1), and could be said to have assisted the twentieth-century 'information explosion'.

The revolution in information technology begun by the CRT continues. Manufacturers believe there are large markets for novel 'information' devices such as pocket-sized (and even wrist-sized) televisions and computers, windscreen displays in cars and aeroplanes, wall-sized television screens and video hoardings, and so on. The incentive to produce thin, flat, electronically rewritable displays is thus enormous.

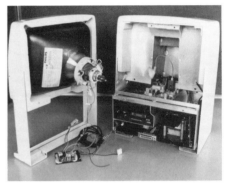

Already, a great deal of time, money and effort has been invested in what in the 1960s was thought to be short-term research and development, but the task has proved to be difficult and complex. In those early days, many companies eventually abandoned their plans and settled for recouping part of their investment in relatively simple displays for pocket calculators and digital watches. ▼An alpha-numeric display▲ describes the simplicity of the format.

The difficulty with the display of general information (words and pictures) stems from the 'human' side of the design specification. Our own vision system has evolved to gather information from three-dimensional space, using only a narrow band of the electromagnetic spectrum together with intelligent image-processing and recognition schemes. For portability and compactness, electronic devices must use a two-dimensional format, and their displays must match the eye's spectral response. Our animal instincts give us heightened sensitivity to flickering peripheral events, but have not required us to be able to follow events faster than a tenth of a second or so. These and other factors complicate the task of communicating information from the electronic world to the brain. We will look at the viewer's specification in more detail in Section 6.3.

Figure 6.1 Cathode ray tubes

▼An alpha-numeric display▲

A simple and flexible way of displaying numbers electronically is by the so-called seven-segment array; a version is illustrated in Figure 6.2(a). By switching on the appropriate segments the numbers 0 to 9 can be generated as shown in Figure 6.2(b). For each digit there are seven elements which need to be separately addressed (switched on or off) by suitable electrical connections.

Displays that show both letters and numbers are called **alpha-numeric**. Letters are a bit more complicated and an alpha-numeric display needs a seven-by-five array of dots. Figure 6.2(c) shows an example. But, even an array of 35 elements is simple compared with the thousands required for a large complex display. We look at this problem in Section 6.3.

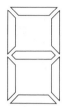

(a) Seven-segment array

(b) Digits generated from (a)

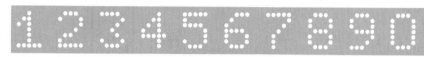

(c) Digits generated by a 7 by 5 matrix

Figure 6.2 (a) Seven-segment array. (b) Digits displayed by seven-segment display. (c) A 7 × 5 matrix

The main aim of this chapter is to investigate some of the materials problems associated with meeting the display function, with flat screens particularly in mind. In order to keep the task manageable we will spend most of our time on just two television technologies: the CRT, representing the standard multipurpose display against which the performance of others are judged; and the LCD (liquid crystal display), representing the novel flat-device end of the market. Before concentrating on these, though, we will take a broad look at what the market has to offer us by way of display technology.

6.2 Devices, mechanisms and materials

An electronic display of any kind converts electronic signals into visual images in 'real-time', that is as the signals actually arrive at the display. Thus the response of the display must be so fast as to appear instantaneous. As we saw in Chapter 1, it is useful to divide display technologies into two classes: light emitting and light modulating. In light-emitting displays, the display material converts the signal directly into light. In light modulation the signal is not used to generate light. Instead, it produces microstructural changes in the display materials, and these changes modulate ambient light or secondary light as it passes through.

▼ Photons and colours▲

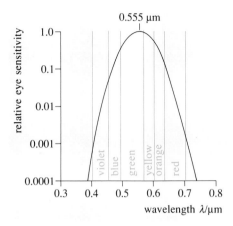

Figure 6.3 Spectral sensitivity of human eye

You should be familiar with the idea, postulated by Planck in 1900, that electromagnetic radiation (radiowaves, microwaves, light, X-rays and so on) can only be absorbed or emitted by atoms in small, discrete units or **quanta**. The quanta are called **photons** and their energy E_p is given by the equation

$$E_p = h\nu$$

where ν is the frequency of the radiation (in hertz) and h is Planck's constant (6.63×10^{-34} J s). To express E_p in electron volts, it is necessary to divide by electronic charge e,

$$E_p(\text{in eV}) = \frac{h\nu}{e}$$

$$E_p = \frac{6.6 \times 10^{-34}}{1.6 \times 10^{-19}} \times \nu \ (\text{eV})$$

$$= 4.1 \times 10^{-15} \times \nu \ (\text{eV})$$

So when an electron 'falls' from an energy state E_1, say, to a lower energy state, E_2, and a photon is radiated, then

$$E_1 - E_2 = h\nu$$

The form of the radiation depends on ν. If ν is roughly in the range 4.3×10^{14} to 7.5×10^{14} Hz it is visible to the human eye.

We are also accustomed to thinking about light in terms of wavelength λ. The energy of a photon is also given by

$$E_p = \frac{hc}{\lambda}$$

where c is the speed of light in air (3.0×10^8 m s^{-1}) and λ is the wavelength in metres.

EXERCISE 6.1   The visible part of the electromagnetic spectrum extends from about 400 nm (violet light) to 700 nm (red light). What are the photon energies (in electron volts) corresponding to violet and red light?

Energy differences are what we have to manipulate in the emitting medium in order to change the photon energies and hence the perceived colour. Figure 6.3 shows the relative spectral sensitivity of the eye. Notice that our eyes do not perceive all colours equally; more about this later.

The main types of device in each category were listed in Tables 1.2 and 1.3, along with the phenomena they rely on and some current applications.

Consider the phenomena on which light-emitting technologies in Table 1.2 rely. At the electron level, what do they have in common?

In each case electrons are excited to a higher energy state and, as they 'decay' to a lower state, light is emitted. The colour of the light depends on the energy difference between the two electron states. See ▼ Photons and colours▲. With the exception of plasma panels, the light-emitting materials in Table 1.2 are solids and employ related phenomena: apart from the filament, they all emit by some form of luminescence. The next subsection explains this.

6.2.1 Luminescence

First some definitions. If they get sufficiently hot, all materials emit light; they become **incandescent**. Light emission stimulated by any other means is called **luminescence**. In general, the radiation can be anywhere from ultra-violet through the visible spectrum to infra-red although, of course, for display materials only the visible part of the range is of interest.

Luminescence in which photons are emitted during or only a short time after excitation (less than 10^{-8} s) is called **fluorescence**; emission after longer times is **phosphorescence**. Despite this distinction, all materials which usefully luminesce are called **phosphors** (from the element phosphorus, which the Greeks called 'light-bearer' because it glows in the dark as it oxidizes).

There are three different kinds of luminescence, depending on how electrons are excited into higher energy levels. The following are of interest for display technologies.

● **Photoluminescence** is produced by bombardment with photons. It is used in 'brilliant white' and 'day glow' paints and in washing powders. It is also the basis of the fluorescent lighting tube. Here the phosphor coating on the inside of the tube is excited to fluorescence by ultraviolet photons from a mercury discharge. My dictionary notes that the name fluorescence derives from fluorite (calcium fluoride), itself so named because it is used as a *flux*. Coloured (that is, impure) fluorite has the property of emitting visible light after exposure to ultraviolet light. That conversion from ultraviolet to visible light is the key to the brilliance of paints and laundry and is also the source of visible light from a fluorescent tube.

● **Cathodoluminescence** is created by bombardment with electrons (or 'cathode rays'). The CRT is the paramount example of its use. Section 6.4 will say more about this. Table 1.2 gives some common examples.

● **Electroluminescence** is generated by currents and electric fields within the solid. Table 1.2 gives some common examples.

So what do we need for visible luminescence? Exercises 6.2 and 6.3 should help you to answer this.

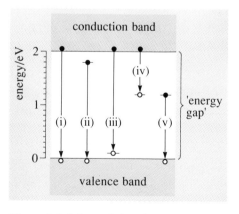

Figure 6.4 Schematic band diagrams for luminescent materials (i) to (v)

EXERCISE 6.2

(a) Figure 6.4 shows various pairs of energy states in the band structure of some imaginary non-metal material. Which pairs could be directly associated with visible-light emission?

(b) Give one other qualification for a useful light emitter.

EXERCISE 6.3 Which classes of material are associated with each of the paired levels in Figure 6.4?

6.2.2 Energy, momentum and efficiency

Besides thinking about whether a transition is of the right energy change to emit visible light, we must also think about whether the transition is possible or 'allowed'. This means thinking about momentum and energy. In the naive view of electrons as particles whizzing about a nucleus, we envisage them having both energy and momentum. In the more sophisticated quantum story, momentum and energy are associated with energy states rather than electrons. In either case, as you should remember from basic physics, momentum and energy must be conserved. We've already suggested that energy is conserved because

electron (energy E_1) $\rightarrow$ electron (energy E_2)
$$+ \text{ photon energy } (E_1 - E_2).$$

If the electron (or the energy state) has different momenta at E_1 and E_2, then we should also be insisting that

electron (momentum p_1) $\rightarrow$ electron (momentum p_2)
$$+ \text{ photon (momentum } p_1 - p_2)$$

Transitions are only allowed if both energy and momentum are conserved. In general we can't conserve both energy and momentum with just two particles (electron and photon) and we could do with a third particle to balance the books. In fact photons don't have very much momentum. However, you will recall that the lattice's thermal energy is sometimes described as being conveyed by phonons. Compared with photons (energy $h\nu \approx 2$ eV), phonons have virtually no energy ($kT/e \approx 0.025$ eV), but they do have lots of momentum. As a result, allowable transitions may conserve energy with a photon while simultaneously conserving momentum with one or more phonons if necessary.

Cases (ii) and (iii) in Figure 6.4 are quite useful here because the levels in the gap are due to dopant. Recombination at these sites involves a third particle (the dopant). Practically all useful phophors are crystalline solids which in the pure state do not luminesce. They rely on the presence of small concentrations of alloying elements, usually called **activators** (or sometimes **dopants**, as with extrinsic semiconductors). For instance, a common phosphor for CRTs is zinc sulphide (ZnS) containing minute amounts of copper. We look at how they work later.

With all this talk about dopants and semiconductors, you may well be thinking about silicon and wondering what its luminescent properties are. The band gap in silicon (1.12 eV) is too small for it to be a light emitter, but other semiconductor materials have proved to be useful. ▼LEDs▲ looks a little deeper.

▼LEDs▲

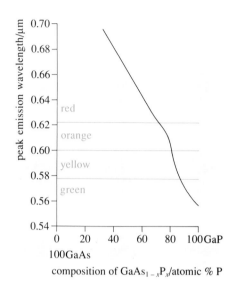

Figure 6.5 Colours of GaAs GaP(N) alloy LEDs

For a display material usefully to emit light energy, there must be some way of supplying it with electrical energy. The light often comes from electrons from a conduction band recombining, via dopant levels, with holes from a valence band. So, the electrical energy input has to fuel the supply of electrons to the conduction band and holes to the valence band.

As we saw in Chapter 5, pn junctions form a diode. When biased one way virtually no current flows. When biased the other way (so-called forward bias) current flows as holes from the p-type are allowed to escape into the n-type and *vice versa*. A hole in n-type materials doesn't last long. Drastically outnumbered by electrons, it is soon filled in by an electron recombining with it. Likewise an electron in p-type recombines with one of the majority holes. In the right material this recombination involves the emission of photons. Here then is one way to generate light: a forward-biased diode.

Light-emitting diodes can be made from semiconductors with band gaps in the range 1.8–3.0 eV, provided the energy and momentum rules allow photons to be involved. One particularly useful material is an alloy of gallium arsenide (GaAs) and gallium phosphide (GaP). Pure GaAs diodes emit in the infrared range (1.42 eV) and are used in television remote controls to communicate between the handset and the main unit. But the addition of gallium phosphide changes the band structure. By adjusting composition, the band gap can be varied between 1.42–2.26 eV.

Although diodes of pure GaP and alloys with more than 60% GaP are very inefficient, the addition of a little nitrogen to trap electrons first before they recombine radiatively greatly increases their efficiency. Figure 6.5 shows how the colour changes with composition.

SAQ 6.1   (Objective 6.1)
Identify entries in Table 1.2 in which the light emission is associated with energy-band structure (giving a brief explanation of the processes involved).

6.2.3 Electromodulation

The other main display mechanism is the one that uses electric fields to modify microstructure and hence to modulate light originating from some other source. Liquid crystals (▼Whence liquid crystals?▲) and ferroelectric ceramics (Chapter 4) are useful materials here.

EXERCISE 6.4

(a) At optical frequencies, what parts of matter interact with electric fields?

(b) What parts of matter interact with lower-frequency electric fields and with d.c. electric fields?

The interaction of light with matter involves electrons. When light is absorbed, electrons are excited into higher energy states (which may lead to photoluminescence). When light is reflected and refracted at crystal boundaries a substance appears opaque. Metals, with copious electrons

▼Whence liquid crystals?▲

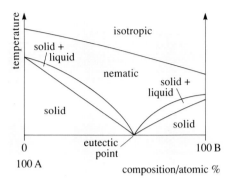

Figure 6.6 Phase diagram of liquid-crystal system

The first observation of liquid crystal behaviour is usually attributed to Reinitzer, an Austrian biologist who in 1888 found two distinct melting points in a derivative of cholesterol. Liquid crystals are more formally called **mesophases** (*mesos* is Greek for middle) because they arise between the solid and liquid phases. There are two families of mesophase, and they are distinguished from each other by how they are described in relation to the phase diagram. **Lyotropic mesophases** generally correspond to a particular concentration, or range of concentrations, of solute in an ordinary liquid solvent. For example, soap and water can form a variety of liquid-crystal mesophases as the water penetrates between layers of soap molecules, weakening the bonding between their polar groups and allowing new structural arrangements to form. A large number of biological systems exhibit this sort of behaviour and it is invoked to describe some of the complex chemical transport through cell walls in living organisms. This type of material is being used in certain cosmetics where their properties are claimed to aid the penetration of active ingredients to underlying cells, but at present none is used in display devices.

The liquid crystals of interest to us are called **thermotropic mesophases**. Here the crystal phase is associated primarily with a temperature rather than a composition. To be useful, thermotropic phases should be stable over a range of temperatures — this is a particularly important requirement for displays. One of the crucial aspects of the development of liquid crystals has been the extension of their range of stability by alloying. Figure 6.6 shows the phase diagram of a typical liquid-crystal system. The **nematic phase** is the liquid-crystal region. In this phase the material is a liquid with relatively long-range order, owing to the shape of the molecules which favours at least parallel alignments. Clearly, the eutectic composition is stable over the widest temperature range. The temperature above which the mesophase becomes a normal liquid is called the **clearing point**.

available for conduction, prevent electric fields from penetrating and so reflect light.

None of these can be much affected by a lower-frequency electrical control signal, so they do not offer the possibility of light modulation. In Chapter 1 I suggested that it is the polarization of light which holds the key to most electromodulation schemes. We have already seen flash goggles at work in Chapter 4 and Section 6.5 includes a description of how thin layers of liquid crystals can be used to twist the polarization. So what have liquid crystals and **PLZT** got which makes them so special? In a word: anisotropy. More fully, it is anisotropy in their dielectric properties — in one case straight, stick-shaped molecules, which can be electrically reoriented and in the other, an ionic crystalline structure that can be electrically stretched. In both cases the anisotropy favours certain orientations of polarized light.

We turn next to the specifications which a flat display device must meet.

6.3 The requirements of a flat panel display

Fairly early on in the personal-computer age it became clear that monitors (VDUs) and television sets have different specifications. For a start, people don't want to watch television from a few centimetres. Also, the most elaborate software rarely needs a display giving the high fidelity colour rendition expected of ten-part serializations of Dickens, gardening programmes, or even real life. So there are two clear markets for versatile display screens: that of the computer monitor and that of the domestic television. For the computer monitor, the incentive for flatness is portability. Were space and weight not at a premium, the CRT would reign supreme — because it works well otherwise. For TVs the position is similar. Bigger screens with higher resolution (1250 lines for European 'high-definition' television) may be a viable commercial and artistic proposition provided the TV set does not then become too heavy and too bulky.

In the product design specification, along with economic considerations and technical functionality, 'viewability' for the user is of prime importance.

> EXERCISE 6.5   Imagine that a new type of television screen has come on the market. In assessing the visual quality of the picture produced give at least three characteristics you would consider.

In essence, the main problem is one of matching the set of optical signals transmitted to the display to the characteristics of the human eye and brain.

6.3.1 Viewing angle, resolution and speed

First some bare necessities concerning viewability. In Chapter 1 I indicated that a television screen ought to be viewable from about 45° off the perpendicular (head on). A simple liquid-crystal digital clock display on my desk (*c.* 1985) is unreadable beyond 35°; this would be barely acceptable for a television screen.

The question of resolution was also introduced in Chapter 1.

> EXERCISE 6.6   From information given in Chapter 1, how many black lines per millimetre on a white background will look like a uniform grey tone, 'at normal viewing distances'.

Resolution is concerned with the spatial detail of an image. A static image on a computer graphics screen may need much finer resolution than a television picture.

Figure 6.7 Blurring to give the impression of speed

Most television images are not static. They are changing, either because the scene is being scanned (for example, the camera is panning) or else because objects and figures in the scene are in motion. Moving images do not need to be as sharp as static ones. Indeed, a common way to imply motion in a stationary image is through fuzziness — deliberate loss of resolution (Figure 6.7). This introduces a number of questions of speed:

• At what rate does a sequence of static images have to change before motion appears smooth?
• How long does any area of the display have to hold its image?
• How fast does any area of the display have to respond to the driving signal?

Let's deal with these in turn. The human vision system has limited temporal resolution such that something like twenty-five pictures per second blend into a continuum. This is because the system's response time is about a tenth (0.1) of a second. Events which are shorter than this are still perceived, but are averaged over the response time. So a sequence of images each one slightly different from the previous one and each one lasting a twenty-fifth (0.04) of a second, form a smoothly moving image. Figure 6.8 shows how a sequence of short step changes in say the brightness (see later) of an image are integrated by the eye into a smooth trend.

The second question, to do with persistence of the image, concerns the response of material used to display the images. Can you foresee the compromise, given the eye's insensitivity to quick changes? If the time between new images is going to be 0.04 second (British television), but the eye integrates over 0.1 second, we do not have to specify a screen capable of holding a fixed image for 0.04 second since the eye will 'hold' it for twice that period anyway. Figure 6.9 shows a possible strategy which exploits phosphorescence — the continued emission of light for a period after stimulation. The screen material is over-stimulated (a short, bright flash) every 0.04 second, then allowed to relax according to the relatively fast decay of phosphors, but the average (or integrated) stimulation of the eye is as before. This would be satisfactory for most purposes were it not for our peripheral sensitivity to flicker — an animal reflex clearly useful for providing early warning of attacks. For the screen image it means that the edges may appear to have a distracting flicker. Projected cinematic film (24 frames per second), for which the flat display is simply a white (reflective) screen, is similarly susceptible to flicker. Each frame of a film is brought in turn in front of the projection gate where it is held for about 0.04 of a second. To avoid flicker, the projector beam flashes on at least twice during this time. The screen therefore shows every frame for a fraction of the 0.04 second period at least twice. This suppresses the flicker sensation because it is now too fast even for the peripheral field. Likewise a CRT television picture is in practice changed every 0.04 second, but lines 1, 3, 5, 7, . . . are shown in the first 0.02 second and 2, 4, 6, 8, . . . are shown during the next 0.02 second. Figure 6.10 shows how the brightness of a region of screen (several lines wide) might change with time (more about this in the next section).

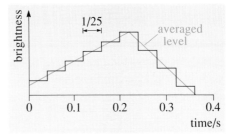

Figure 6.8 A staircase of light levels and its average value

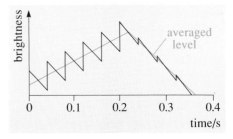

Figure 6.9 A sawtooth staircase

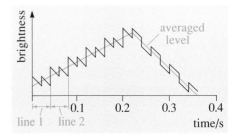

Figure 6.10 A double sawtooth staircase

The final question of speed also concerns the material response. For television displays, luminescent and electro-optic materials must respond fast enough to allow the necessary 25 pictures per second to be presented. The following exercises illustrate two of the options.

> **EXERCISE 6.7** A 625 line television picture has to be changed every 0.04 second. In each line there are 833 elements (picture cells or **pixels**). How long have we got to switch each element in turn, and for how long does each element wait before it is again addressed?

> **EXERCISE 6.8** A 625 line television picture is changed every 0.04 second by 'rewriting' a whole line at a time. How long have we got to switch the material at each element and for how long does each line of elements wait before it is again addressed?

The ratio of on-time to off-time for a device is normally called the **duty factor**. Low duty factors are appropriate when a short burst of activity is sufficient to sustain an effect for a much longer period. The example in Exercise 6.7 involves pixels with a duty factor of

$$\frac{7.7 \times 10^{-8}}{4 \times 10^{-2}} \approx 2 \times 10^{-6}$$

whereas in Exercise 6.8 the duty factor is

$$\frac{64 \times 10^{-6}}{39.9 \times 10^{-3}} \approx 2 \times 10^{-3}$$

For television display materials we can expect to find luminescent processes (only electrons move) used for the lower-duty-factor systems and electro-optic processes (ions move or molecules re-orient) used for the higher-duty-factor systems.

As usual, different classes of material involve different design strategies — with far reaching consequences. This question of response is bound up with how we address the screen when sending information to it (see ▼Forms of address▲).

▼Forms of address▲

In the basic seven-segment digit display we have to be able to get at each individual segment to turn it on and off. That means seven separate connections to each digit. What about an eight digit calculator display — does that have $7 \times 8 = 56$ individual connections?

Well, it could but since a major cause of electronic system failure is faulty connection, and since all those connections require a track and something to drive signals along it, it makes sense to employ a more economical addressing scheme. By connecting the eight digits in a matrix arrangement of columns and rows the driver circuits can be shared between all the digits. Thus, fewer drivers and tracks are necessary. In fact, just $15 (= 8 + 7)$ are needed. Such a scheme is said to be **multiplexed**. Figure 6.11 shows an arrangement for three seven-segment digits. Each segment is represented by a solid square, labelled with a letter A–G and a number of 1–3. So C2 means segment C of digit 2. The seven-segment arrays for each digit in (a) are redrawn in (b) to highlight the three-column seven-row matrix arrangement. You can appreciate that for displays with, say, hundreds of thousands of elements (pixels), matrix-addressing dramatically decreases the complexity of the circuitry. For some types of display it is essential.

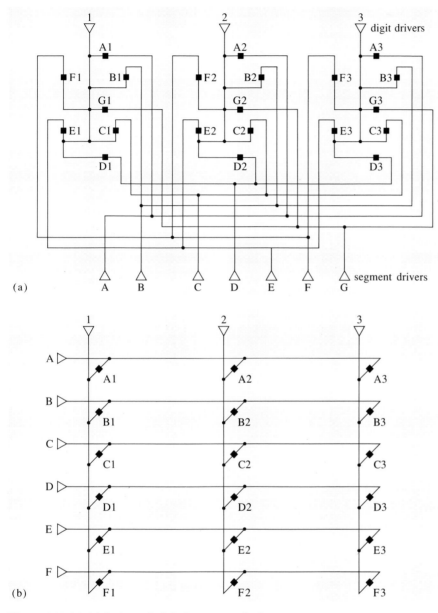

(a)

(b)

Figure 6.11 Multiplexing a 3-digit 7- segment display

should be off. This problem is called **crosstalk** and is clearly a very important factor in determining the effectiveness of a matrix-addressing technique. In particular, it puts a stringent requirement on the electro-optical properties of a display material, especially for a television display with a matrix with something like 800×600 pixels.

EXERCISE 6.9   The element on row F and column 3 is to be switched by imposing a potential difference of 5 volts across it. To do this, row F is connected to $+ 2.5$ volts and column 3 is connected to $- 2.5$ volts. All other rows and columns remain at zero. Three different materials are available for the display elements which vary in brightness as shown in Figure 6.12. Comment on the use of each one.

The conventional CRT television screen gets around the connection problem in a completely different way. An electron beam is scanned across the screen delivering information, in the form of a stream of energetic electrons, along a narrow path as it strikes the screen. For a black and white display one single electron beam delivers the message to all points on the screen in turn. This style of addressing is often described as **raster scanning**.

A distinct advantage of the raster scan is that the beam power can carry information. For example a smooth (grey) scale from pure black to pure white is easily translated into beam power. Simple matrix addressed schemes are more suited to on/off (binary) data so greyness has to be achieved in some other way.

The display matrix looks just like the two dimensional arrays of cells used in solid-state memories. It is accessed in the same way. To operate the display, each row is selected in succession and the appropriate information for the elements in a row is sent along the columns (which are addressed simultaneously) during the time that the row is being addressed. If we want, say, to excite a particular segment of one digit, we send (drive) a signal down that column (say V_c) and across that row (say V_r).

Each segment is adjacent to an intersection of a column with a row, so the segment can be arranged to receive the sum of the voltages from the two drivers ($V_c + V_r$). The other segments connected to the driven column and row will obviously also experience some voltage (at least V_c or V_r) while data is passed to the one selected by this combination. If this voltage level causes light to escape from these other display segments, the viewer may have difficulty in distinguishing between segments that should be on and those that

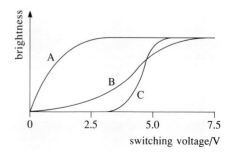

Figure 6.12 Three characteristic brightness/switching responses

277

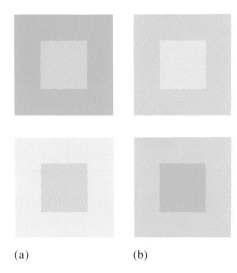

(a)   (b)

Figure 6.13 (a) Central areas are the same 'brightness' but appear different because of the different borders. (b) Outer areas are equal 'brightness'

6.3.2 Brightness, colour and contrast

There are five basic controls on my television set that I can adjust. The function of the off-on/volume is obvious. The station selector seems to have a little influence on programme quality. The remaining three are devoted to picture quality and offer brightness, colour and contrast control. Can't these three be factory-set? The answer is no. All three are subjective: the response of your eyes is almost certainly different from that of mine and all three controls need to accommodate a variety of ambient lighting conditions from strong daylight to cinema-like darkness.

In Chapter 1 I introduced the three qualities, brightness, colour and contrast, and here I want to hint at how they can in fact be quantified. Brightness is associated with the amount of radiated power. If a television screen is against a white wall (reflecting so many watts per square metre of ambient light) its brightness relative to that of the wall is crucial to its viewability. But we have to be careful to account for the eye's own sensitivity because not all colours are perceived to the same degree: green watts are more effective than red or blue watts. ▼The eye▲ looks into more detail of this and the nature of colour response in general.

Brightness is less significant than relative brightness. Figure 6.13(a) shows two squares of colour at 50% print density surrounded by in one case 75% colour and the other 33% colour. Clearly although the inner squares are equally 'bright', their contrasting surroundings give them different appearances. (Figure 6.13b shows the inverse patterns.) Contrast is a measure of relative brightness and can be quantified between zero and one as simply

$$\frac{(\text{maximum} - \text{minimum}) \text{ brightness}}{(\text{maximum} + \text{minimum}) \text{ brightness}}$$

In Figure 6.13 all pictures have 20% contrast. Heightening the contrast brightens the brighter bits while darkening the less bright.

▼The eye▲

The human eye is an amazing transducer. It is able to sense a broad range of colours and accommodate light levels covering a range of seven order of magnitude (that is, the brightest is 10^7 times brighter than the faintest). It does this, however, with some compromises which art, entertainment and nature have turned to advantage. First there is the spectral response of the eye. The eye is most sensitive to green light: there is more nerve activity per watt of incident radiation when it is green. The eye's relative responses to different wavelengths was shown in Figure 6.3.

At very low light levels different parts of the retina (the rods) take over from the main sensors (the cones). Under low light levels, colour sensitivity shifts towards the blue so that reds and yellows are not well perceived.

There are three subjective attributes to colour vision: brightness, hue (that is, distinguishable colour) and saturation (low saturation implies paleness or dilution by white).

The eye apparently senses colour by combining the response of three types of sensor: a generally blue-sensitive one, a generally green-sensitive one and a generally red-sensitive one. So, a yellow light with a wavelength between green and red will partially stimulate both green and red sensors, producing the 'yellow' sensation. But we can also simultaneously stimulate the same green and red sensors with green light and red light mixed together. With the right amount of each the brain gets the yellow sensation again.

Seurat (1859–91) and other *pointilliste* painters exploited similar effects using fine dots of pure colours to create, from a distance, the impression of other hues. This is of course the principle of colour television, which uses red, green and blue dots to produce the sensation of full-colour images.

By dividing the contrast range into distinguishable levels, brightness can be expressed as a binary code. It is convenient to choose the number of levels as 8 (2^3), 16 (2^4), 32 (2^5), 64 (2^6) and so on, being 3, 4, 5 and 6 bits of information. In fact for each pixel at least six bits are needed to encode brightness to give a satisfactory 'grey-scale'.

6.3.3 Power and efficiency

The power consumed by a television set is not a major consideration. Mine (with a CRT) is fused at 3 A and so only requires a few hundred watts. More significant is the efficiency with which electrical power is converted into the light radiated by the image on the screen. A watt per square metre is not an unreasonable optical power output for a screen, so efficiencies may be low (less than one per cent). The power which doesn't escape as light escapes as heat, so a flat-panel system (having a low volume) will·need to be more efficient than a larger-volume equivalent if it is to operate without special cooling requirements. If you want to get a feel for the light and heat involved, remember that a 100 W incandescent bulb produces about 6 watts of light (of which we see only a fraction when we view it directly) and 94 watts of heat, over an area of about 10^{-2} m^2, and these are more than enough to dazzle and burn.

SAQ 6.2   (Objective 6.2)
A large flat screen display (6 m × 4.5 m) is to be used for video action replay at an outdoor sports stadium. Compare the requirements of the display with those of a domestic television by completing Table 6.1.

There is another, more general, aspect to efficiency. As well as operating economically, the display should have a life expectancy long enough to make its manufacture economically worthwhile. Rather than putting a capital cost per hour on it though, let's just say it ought to last ten years at around 2–3 hours per day: that is, 10 000 hours.

Table 6.1   Comparison of large outdoor display with domestic television

	Same as domestic TV? yes/no	How different?
Viewing angle		
Resolution		
Speed		
Brightness		
Colour		
Contrast		
Power		
Efficiency		

6.4 The CRT

6.4.1 Basic operation

The big advantages of the CRT are that it is a well established technology and that it is versatile. Because it has been used for so long and has had no significant competitors in the general market, it is in a pre-eminent position in the industry. As a result, although a relatively complicated device compared with, say, a light bulb (see Figure 6.14), it has proved amenable to mass production methods and therefore has a low product cost.

The versatility of the CRT stems largely from the different mechanisms and materials that can be used to display colour, and from the 'trade-offs' possible between colour quality and resolution for different applications. We shall look at just one type of colour CRT, the so-called **shadow-mask-tube**.

The picture tube in the familiar colour television set is a shadow-mask (or tricolour) tube. It contains three electron guns, each of which produces a separate electron beam. The three simultaneously scan the viewing screen and produce a red, a green and a blue image respectively. The (curved glass) screen is covered with three separate sets of uniformly distributed phosphor dots and each set glows in a different colour when excited by an electron beam. Electrons from the 'red' gun, that is the gun controlled by the red-primary colour signal transmitted from the television camera, impinge only on the red-glowing phosphor dots; they are prevented from striking the green-glowing and blue-glowing dots by the shadow-mask.

The shadow-mask is a metal sheet containing a large number of tiny holes (0.35 mm diameter) each of which is accurately aligned with the phosphor dots on the screen. Similarly, as Figure 6.14(c) illustrates, the

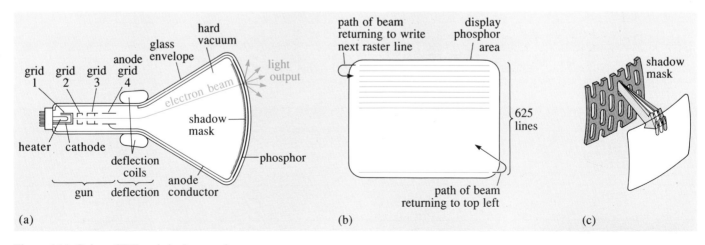

(a)　(b)　(c)

Figure 6.14 Colour CRT and shadow mask

▼ Dotting phosphors ▲

Phosphors are prepared from extremely pure starting materials because, just as with semiconductors, it is carefully controlled dopant additions which give them their key properties. The phosphor coating on CRTs has to be thin enough to allow light to escape, but thick enough to absorb all the energy of the incident electron beam. This requires a coating several microns thick. Particles are around 10 μm diameter and consist of grains a micron or so across.

The usual process route is to precipate the phosphor (or a near precursor) from an appropriate solution. After this dopants are mixed and fired in at 1000°C, during which the final crystal structure is developed. Ideally particles are not allowed to sinter as subsequent grinding to reduce

the particle size could harm the phosphor's properties. However, mechanical damage can be used deliberately to modify the phosphor characteristics.

The coatings are deposited from liquid suspensions. For a monochrome tube the process is straightforward. The suspension settles out onto the inside surface of the screen in about twenty minutes during which the coating is stabilized by a gelling agent. The excess liquid is then poured off. For a colour CRT there are additional problems. Can you guess what these are?

The first problem is that we have to put down three different phosphors (R, G, B) and the second is that we need a sequence of RGB dots, a hundred or so microns for each colour, accurately located so that

each electron beam can address its appropriate phosphor through the shadow mask.

In Chapter 2 we saw photolithography and screen printing as useful patterning techniques with electronic materials. The clue as to which process is more suited here lies in the need for extremely accurate matching of the dots to the shadow mask. Exercise 6.10 (and its answer) complete the story.

EXERCISE 6.10   Using photolithography, how can we establish a dot pattern that will register accurately with the electron beam after it has passed through the shadow mask? Consider one set of dots at a time.

electrons from the other two guns fall only on the green and blue phosphor dots respectively. In this way three separate colour images are formed simultaneously on the screen. The dots are close together and so small that they cannot be resolved by the human eye at normal viewing distances. Hence, we have the impression of continuous colour.

The principle of the tube is simple, but its production and operation are complex. For instance, various adjustments are necessary to compensate for inevitable inaccuracies and distortions — particularly in convergence of the beams and the 'purity' of the primary colour signals. The shadow-mask of a 25 inch tube has over 500 000 holes and, obviously, three times that number of phosphor dots on the screen. As ▼ Dotting phosphors ▲ indicates, making a television screen is tricky.

The electron beam carries a current of about 1 mA and bombards the screen with electrons which have been accelerated through about 20 kV (depending on brightness). The beam power (current × voltage) is therefore about 20 W, which is distributed over the whole screen. Under the spot, the power density is 6.7×10^7 W m^{-2} so it's as well the spot keeps moving.

The whole CRT, its electrodes, filament, steering coils and shadow mask, would be a fit study for another book on electronic materials. Here we shall concentrate on just the phosphors.

6.4.2 Phosphors

Earlier I introduced the nature of luminescence. Now we will look in a little more detail at the chief characteristics of cathodoluminescent materials and their use in television tubes (see ▼Cathodoluminescence▲). The three properties which describe a phosphor are:

- efficiency,
- emission spectrum (colour),
- persistence.

Efficiency

A cathodoluminescent phosphor converts the bombardment energy of a stream of external electrons into cascades of energetic internal electrons. These in turn lose energy to photons (light) and phonons (heat). The energy efficiency of a phosphor is simply the ratio of luminous emitted energy to incident energy.

Zinc and cadmium sulphide phosphors are among the most efficient at around 20% for 20 keV electron bombardment. At lower electron energy, the efficiency declines. The following exercise probes further.

EXERCISE 6.12   Figure 6.15 shows typical phosphor efficiency data with a schematic indication of how far the electron beam penetrates at different energies. Suggest one reason for poor efficiency at low voltage.

At the higher-voltage end of Figure 6.15, efficiency falls if the photons are produced so far within the body of the phosphor that they are reabsorbed before they make good their escape.

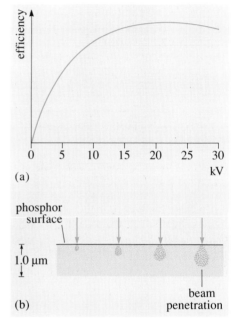

(a)

(b)

Figure 6.15 The effect of voltage on (a) phosphor efficiency and (b) penetration of the electron beam

▼Cathodoluminescence▲

Cathodoluminescence is light emission stimulated by electron bombardment. We are concerned with the luminescence following the impact on a thin layer of material of high-energy electrons, accelerated from a hot filament.

In general the harder the phosphor is hit, the brighter the light it will emit. Why? Because as an incident electron blasts its way into the solid, it knocks several host electrons out of their accustomed positions. These in turn knock others out, so a cascade of electrons is generated, sharing out the incident energy. The result is lots of holes (electrons knocked out of

bonds) in the valance band and an equivalent number of electrons in the conduction band. Order is only restored when the holes are eventually filled, during which time the initial incident energy is dissipated as light or heat.

The more intense the bombardment, that is, the more electrons arrive per second (current), the more light is emitted. However there comes a time when the rate of recombination is limited not by the availability of electrons and holes but by the availability of the paths associated with light emission (provided by the presence of the activator). Beyond a certain current limit the phosphor is saturated.

EXERCISE 6.11   Which of the following expressions best describes the light emission (L) from phosphor excited by an electron beam (current i and voltage V).

(a) $L \propto \dfrac{i}{V}$

(b) $L \propto \dfrac{V}{i}$

(c) $L \propto iV$

(Important comments in answer.)

Emission spectrum

EXERCISE 6.13   How many phosphors are needed for a colour CRT?

In the simpler sorts of cathodoluminescence, electrons from the conduction band recombine with holes in the valence band via levels deep in the energy gap. These deep levels arise from the presence of deliberately added activators, and the emission spectrum of a phosphor depends chiefly upon the types of activator present. So, for example, crystalline ZnS with about 0.1% of zinc sites being replaced by silver atoms will emit blue light, whereas a copper activator (instead of the silver) gives a greener emission (see Figure 6.16). With activator atoms so thinly spread, their action is virtually independent of each other; so the 'colour' of a phosphor could be adjusted by a mixture of activators.

The emission spectrum also depends slightly on the crystal structure. This is important for compounds like ZnS because it exists in two forms, wurtzite and blende. Why do you suppose crystal structure matters?

The band structure arises from the interaction of the outer energy states of neighbouring atoms. The precise detail of the interaction (for example, the size of energy gaps) will depend upon the precise position of each atom. Furthermore, the foreign activator atoms will perturb one structure slightly differently from another. Together these amount to about a 5% wavelength shift between the two forms.

Different colours can also be produced by alloying different phosphors. Zinc cadmium sulphides can be prepared which will give emissions anywhere from blue to red (see Figure 6.17). The band structure is again being modified, this time to much greater effect by changing the nature of the host lattice.

Lastly there is one further emission mechanism which can be important, in particular when manganese is the activator. Manganese has five 3d-shell electrons which would normally have all parallel spins: this is the

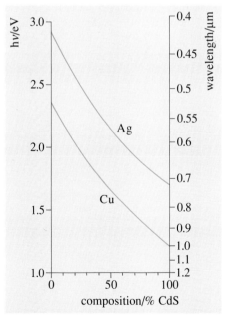

Figure 6.17 Emission colours of zinc cadmium sulphides alloys with silver and with copper activators

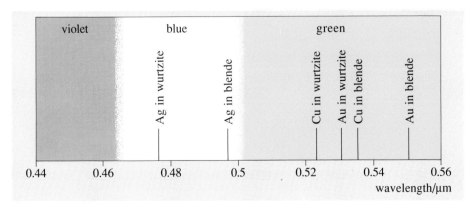

Figure 6.16 Luminescence spectra for activated zinc sulphide phosphors

lowest energy condition. Rearrangements to, say, four parallel spins and one anti-parallel cost energy, and, when manganese atoms are in a suitable crystal lattice, the energy difference lies right in the visible spectrum. So manganese atoms can activate a phosphor simply by allowing themselves to be excited (by electron bombardment for example).

Persistence

The persistence is quantified by the time for the light emission to decay to 10% intensity. A clue as to what controls persistence comes from noting that in most phosphors light arises from a part of a recombination process in which a conduction band electron is reunited with a valence band hole. While electrons are in the conduction band (for example following a burst of electron bombardment) a phosphor has a greatly enhanced conductivity, provided the electrons have a respectable mobility. It is this same mobility which takes the electrons in search of recombination paths. Although we are not interested in currents, mobility μ and number n of charge carriers are clearly implicated. If you want lots of light you need lots of electrons (and holes). If you want it to last a long time you need low-mobility material (zinc sulphide has some of the lowest electron and hole mobilities of the phosphors, and it is a 'natural' phosphor).

Long persistence is associated with the obstruction of the recombination path. This is best done by taking some of the available conduction-band electrons out of circulation for a while — that is, trapping them in a localized state. Eventually they will escape, released by a passing phonon, and, delayed but otherwise little affected, continue their search for a recombination path. The trap could be some kind of defect level just inside the energy gap. The deeper the level the longer it is occupied (phonons energetic enough to release a trapped charge are rarer).

EXERCISE 6.14 **Suggest an example of a defect which might act as a trap.**

Table 6.2 characterizes some zinc based phosphors.

Table 6.2 Some cathodoluminescent phosphor characteristics (20 keV bombardment)

Phosphor	Activator	Colour	Efficiency/%	Decay time/seconds
ZnS	0.01% Ag	Blue	21	
ZnS:CdS (50:50)	0.005% Ag	Yellow	20	$\approx 10^{-4}$
ZnS	0.001% Cu	Green	23	
ZnS	1.5% Mn	Yellow		4×10^{-4}
Zn$_2$SiO$_4$	0.4% Mn	Green		1.3×10^{-2}
Zn$_3$(PO$_4$)$_2$	3% Mn	Red	8	—

6.4.3 Flattening the CRT

The CRT has remained the major electronic screen technology for televisions, computer terminals and video monitors. Its champions have found cathodoluminescence fast enough, fine enough and bright enough. Phosphor displays are well suited to working with the eye. Add to this the relative simplicity of steered-electron-beam addressing and you can begin to see why the CRT takes some beating. But the high voltages needed to excite the phosphors efficiently causes problems for flat panel CRTs.

> EXERCISE 6.15 In a conventional television tube, the beam is accelerated (up to 20 kV) focused and scanned (about $\pm$ 45° up and down and side to side). What are the consequences of shortening the front-to-back dimension of the tube?

Nevertheless, small flat CRTs have evolved for the flat-panel market. A popular approach has been to use a transverse beam. The electrons are accelerated and focused parallel to the screen before being sharply deflected onto the phosphors (Figure 6.18). Lower voltage, high efficiency phosphors would help but focusing and deflection have their own problems with this arrangement, which has had only limited success.

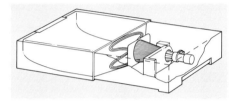

Figure 6.18 Flat CRT

A relatively flat cathodoluminescent display is much easier to conceive if we abandon the steered beam. We could consider a matrix of tubes, each containing a few pixels with their own filaments. Indeed this is how a very large screen, several square metres in area, can be made only tens of millimetres thick. Relatively 'flat' it may be, but not in my home — not yet. It's time to look at our alternative class of effects — those based on liquid crystals.

> SAQ 6.3 (Objective 6.3)
> (a) How does a black and white television tube differ from that required for colour images?
> (b) Refer back to 'The eye' (Section 6.3.2.) and Table 6.2 and suggest useful phosphors for the black and white tube.

> SAQ 6.4 (Objective 6.4)
> Are the following statements about CRT phosphors true or false?
>
> (a) Efficient phosphors convert a large fraction of incident beam energy into photons of visible light.
>
> (b) The emission spectrum of a phosphor is determined by the extra energy levels inserted by the activator atoms in the band gap of the host.
>
> (c) Pure, defect-free crystals are essential for long-persistence phosphors.

6.5 Liquid crystal displays (LCDs)

Liquid crystals are fluids consisting of stiff rod-like organic molecules which can form structures with some order. They therefore exhibit some of the properties of crystalline solids, hence their name. Ordering of the rod-like molecules means that their bulk properties are highly anisotropic. It is this anisotropy in such properties as dielectric permittivity and refractive index that is exploited in display devices. Of a variety of possibilities, Figure 6.19 represents a simple example — a seven-segment cell that could be used in a digital watch or pocket calculator.

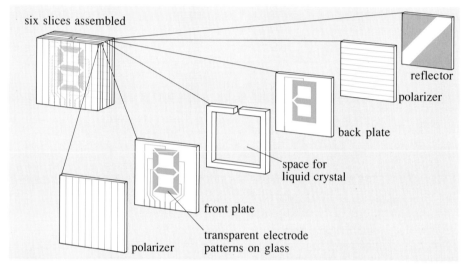

Figure 6.19 A simple type of LCD

The liquid crystal substance is sandwiched between transparent electrodes and two sheets of polarizer. Electrically, the cell looks like a parallel-plate capacitor with the liquid crystal as its dielectric. With no electric field across the capacitor, the molecular order is influenced by molecule–molecule interactions and by molecule–surface interactions, and light passes through the cell. Under a d.c. or an a.c. field (typically 3 V, 50 Hz) the molecular alignment is dominated by the external field and no light is transmitted. We explore what actually happens shortly. So by suitably shaping and addressing the electrodes, a display can be created. LCDs have developed a long way from this simple example, but before considering large displays we need to know more about the structure and properties of liquid crystals and how they can be engineered to meet the requirements of a display device.

6.5.1 The structure of thermotropic liquid crystals

The rod-like molecular structure of thermotropic liquid crystals is typified by the cyanobiphenyl compounds:

R—⬡—⬡—CN $\underset{\rightarrow}{n}$

where R is a short hydrocarbon chain, a common one being C_6H_{13}. The two rings are 'phenyl' and CN is the cyano group. The arrow shows the predominant direction of orientation (see below).

There are three major classes of liquid crystal, each manifesting a different degree of order. A simplified representation of each is shown in Figure 6.20. In distinguishing one class from another and in modelling their properties, it is useful to define a unit vector, called the director n, which describes the time-averaged orientation of the molecules in any small volume of the material.

Nematic phase (N)

Nematos is the Greek word for thread. Nematic liquid crystals are the least ordered and therefore most liquid-like. However, the rod-like shape

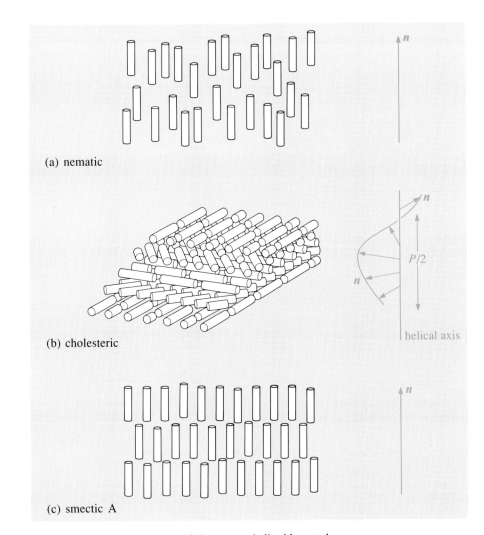

(a) nematic

(b) cholesteric

(c) smectic A

Figure 6.20 Three basic types of thermotropic liquid crystal

of the molecules induces them to line up in a thread-like fashion in which they are roughly parallel (Figure 6.20a). They can be likened to matches in a box, able to slide relative to one another and to rotate about their long axis, but tending to remain parallel. In reality, of course, thermal activation causes some deviation from the ideal of completely parallel alignment.

SAQ 6.5   (Revision)
(a) **Why do the molecules tend to lie parallel? Hint: Think of the intermolecular bonding between organic molecules like the cyanobiphenyl above.**
(b) **Why should such molecules align with an electric-field?**

Cholesteric phase (Ch)

This phase (Figure 6.20b) is really a special case of the nematic since on a local scale they look identical (it is sometimes called chiral nematic, N*). However, on a larger scale the director rotates from layer to layer around a helical axis, thus giving a spiral structure. The **pitch** (P) of the helix or screw is defined as the distance over which the director rotates through 360°. Figure 6.20(b) shows rotation of P/2 (half a turn). The pitch is an important quantity, particularly for the so-called 'supertwist' display device (see later). A pure material may have a pitch of a fraction of a micron. To complicate matters, if some cholesteric material is mixed into a nematic material, the result is a cholesteric with a pitch inversely proportional to the concentration of the dopant — but that sort of complication is worth having as it offers control. The pitch can be either right- or left-handed.

Smectic phase (S)

There are at least eight different types of smectic liquid crystal phase of which smectic A (S_A) shown in Figure 6.20(c) is the most common and the least ordered.

What is the difference between the smectic A structure illustrated in Figure 6.20(c) and the nematic structure of Figure 6.20(a)?

The smectic is layered.

Simple LCD displays, such as those in calculators, utilize the ▼Elastic and dielectric properties of nematics▲. The next section will show how.

SAQ 6.6   (Objective 6.5)
Compare the electrical (dielectric) and optical (refractive index) properties for two liquid-crystal phases, one having very short molecules which are permanent electric dipoles, the other having more acicular, polarizable molecules (but without a permanent dipole).

▼Elastic and dielectric properties of nematics▲

We expect a crystal to show elastic behaviour; that is, we expect an applied stress to produce a distortion, and we expect the structure to revert to its unstrained state when the stress is removed. Liquid crystals are no exception. We can envisage three modes of elastic distortion of a liquid crystal: splay, twist and bend, illustrated by Figure 6.21.

EXERCISE 6.16   Imagine that the switchback in parts (a) and (b) of Figure 6.22 is a scratched surface with which the liquid-crystal molecules are in contact. In (a) the molecular alignment follows the ridges in the substrate, whereas in (b) the alignment is parallel to the ridges and we see the molecules end-on. With splay, twist and bend in mind, which of the configurations of liquid crystal molecules, (a) or (b), is energetically favoured?

The possibility of imposing a particular molecular alignment on a liquid crystal by putting grooves on a substrate is crucial to the design of some display devices. The grooves can be made simply by rubbing a polymer coating on the substrate with a cloth, and the alignment can be maintained over many microns.

Suppose we have some liquid crystal which has been aligned by using a grooved substrate, as in Figure 6.22(b). The liquid crystal forms the dielectric of a capacitor. We find the capacitor has two capacitances, depending on whether the electric field runs parallel or perpendicular to the long axis of the molecules (Figure 6.23a and b). That is, the dielectric has relative permittivities $\varepsilon_\parallel$ for homeotropic alignment and $\varepsilon_\perp$ for heterotropic alignment. Why should structural anisotropy at the molecular level produce this dielectric anisotropy?

As you know from Chapter 2, permittivity is related to ease and orientation of polarization. In the molecules we are considering there are two possible contributions to polarization: molecular and electronic.

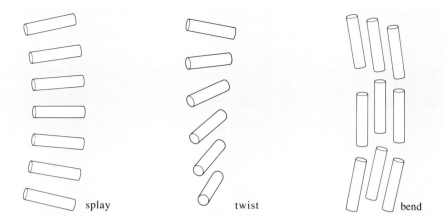

splay   twist   bend

Figure 6.21 Three elastic distortion modes of a liquid crystal

presence of an electric field. In rod-like organic materials, the electronic polarizability along the molecule α_l is greater than that across the molecule α_t (see Figure 6.23d). Hence the different contributions to dielectric anisotropy.

At optical frequencies, the refractive index (which is equal to the square root of the relative permittivity at optical frequencies) is dominated by electronic polarizability. Because of the dielectric anisotropy, the refractive indices differ parallel and perpendicular to the director n. Liquid crystals are thus optically anisotropic.

In Figure 6.23(b) the electric field and the molecular orientation are perpendicular. If the field is strong enough, it can make the polarized molecules turn through 90° and align with the field, as in Figure 6.23(a). So dielectric anisotropy allows us to align liquid crystals by applying electric fields, and hence to produce electro-optic effects.

Electronic polarizability

This is the other contribution to ε_r and it comes from the displacement of electrons along or across the molecule in the

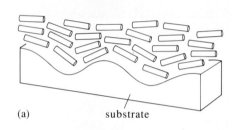

(a)   substrate

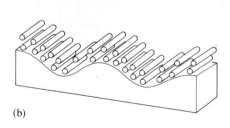

(b)

Figure 6.22 For Exercise 6.16

Molecular polarization

Fluorine, oxygen and nitrogen are very electronegative, and tend to hog the electrons when they are involved in covalent bonds. Consequently, organic molecules containing these elements can have unevenly distributed charge; that is, certain molecules are polar. The molecular polarizability arises from permanent molecular dipoles μ existing in the material. The angle that μ makes with the long axis of the molecule determines its contribution to $\varepsilon_{\parallel}$ and $\varepsilon_{\perp}$ (Figure 6.23c).

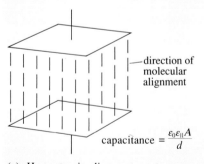

direction of molecular alignment

$$\text{capacitance} = \frac{\varepsilon_0 \varepsilon_{\parallel} A}{d}$$

(a) Homeotropic alignment

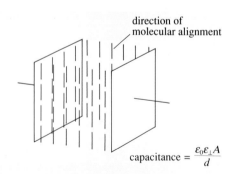

direction of molecular alignment

$$\text{capacitance} = \frac{\varepsilon_0 \varepsilon_{\perp} A}{d}$$

(b) Heterotropic alignment

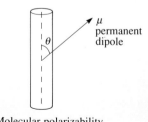

μ permanent dipole

θ

(c) Molecular polarizability

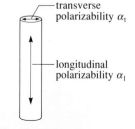

transverse polarizability α_t

longitudinal polarizability α_l

(d) Electronic polarizability

Figure 6.23 Liquid-crystal dielectric anisotropy

6.5.2 The twisted nematic (TN) device

The simple cell outlined in Figure 6.19 is based on a twisted nematic liquid-crystal display which is, by far, the most common LCD. Detail of one element is shown in Figure 6.24. A thin layer of rubbed polymeric material (usually polyimide) covers the inside surface of the electrodes; the rubbing directions are mutually perpendicular. The rubbing produces microscopic grooves which align the molecules, as described in 'Elastic and dielectric properties of nematics'. The glass plates are 10 μm apart.

Near each surface the nematic molecules align with the grooves and take up a smooth 90° twist in between. Figure 6.24(a) looks a bit like those domain walls in Chapter 3. Similar ideas of anisotropy and energy balance are at work. Because of the anisotropic refractive index, this twisted microstructure in the material causes the plane of polarization of the incident light to twist, rotating it through 90° as it is guided through. Just outside the top and bottom electrodes are polarizers, with crossed directions of polarization. So, the rotating effect on light polarization of the liquid-crystal phase enables light to pass through the whole device, and a greyish clear state results. This is the state of the device in Figure 6.24(a).

A relatively strong field across the device causes the molecules to re-orient along the field. Without the twisted crystal structure to rotate the polarization, light coming past the top electrode can't escape through the crossed polarizer at the bottom, where it is absorbed. The result is that no light passes through the cell. This gives the characteristic black characters on a grey background of digital watch and calculator displays.

Notice that the twist of a TN can be either right-handed or left-handed. If both occurred in a cell, the twists would interfere with one another and

▼ Response time ▲

What limits how quickly an electric field can cause the realignment of liquid crystals? Were it not for the fact that each molecule moves into space vacated by another, the process might be very rapid indeed. You can imagine the log-jam developing as the field is applied. But this delay through molecules interacting with each other is the stuff of viscosity. The rate of response to a field is indeed inversely proportional to the viscosity. Of course higher fields (larger switching voltages) will elicit a faster response. A cold LCD on old batteries will be somewhat slower than a warm one with fresh batteries.

The return to the twisted condition on removing the switching field is driven by the free-energy differences of the two states. Again lower viscosity aids this process.

Typical electrical response times are several milliseconds to rotate from the twisted condition and several tens of milliseconds to return to the twisted equilibrium when the switching voltage is removed. A 50 Hz (20 ms period) a.c. switching voltage is not instantaneously followed by such sluggish motions. The molecules respond only to the average effect of the switching voltage: this is given by the root-mean-square (r.m.s.) value of the voltage waveform. The result is molecular orientation perpendicular to the electrodes, with electrons rushing up and down the polarizable part of the molecule.

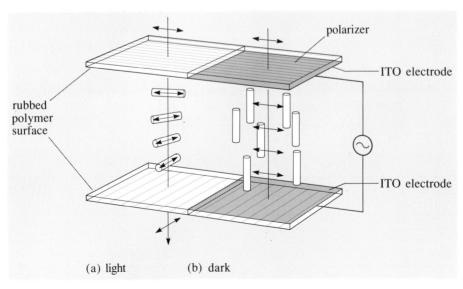

(a) light   (b) dark

Figure 6.24 Twisted nematic (TN) elements

produce variable modulation of light. Have you any ideas as to how we could persuade all the molecules to twist the same way? (Look back to Figure 6.20.)

The 'degeneracy' can be avoided by adding a small amount of a cholesteric material (typically 0.5–1.0%) to the nematic mixture. This adds a 'one-hand' twisted texture to the liquid.

In a typical device, a voltage of about 3 V needs to be applied to re-orient the crystal. If we use d.c., we risk degrading the material by electrolysis so in practice an equivalent a.c. voltage at about 50 Hz is used. The choice of frequency is influenced by the ▼Response time▲ of the device.

Because very little current flows through these devices, they consume very little power — a very attractive feature, especially for portable applications where battery size and life are important considerations.

If we were to use TN devices for a television display, we would need to arrange them into a matrix of pixels. In 'Forms of address' we saw that we need the switching voltage of each cell to be sharply defined to avoid having a whole row and column switched on when we all want switched on is the cell at their intersection. Figure 6.25 shows the transmission characteristic of a simple TN element. Clearly this falls into the 'not so useful' class of switching characteristics in Exercise 6.9. ▼Supertwist▲ looks at one way of sharpening the transmission characteristic; we will look at an alternative device, more suited to television screens, in Section 6.5.4.

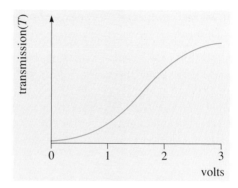

Figure 6.25 TN liquid crystal transmission characteristic

▼Supertwist▲

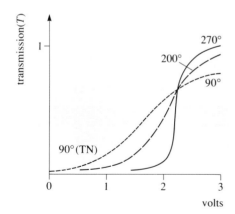

Figure 6.26 Schematic transmission characteristic for supertwisted nematic cell

TN devices with increased twist are generally known as **supertwist displays**. Figure 6.26 shows how the sharpness in the reorientation (switching) process increases as twist angle increases. At a twist angle of close to 270° the characteristic becomes almost vertical. At higher twist angles the characteristics are less steep again. Suitable mixtures of nematic and cholesteric materials can allow the switching to be very abrupt in supertwist displays, but there are still some problems to be overcome. Firstly the high twist devices no longer guide the plane of polarization of light through the twisted structure as in the 90° twisted nematic. The optics of the supertwisted layer are complex so we won't go further than to say that they can be used to make either a black and yellow display or else a blue and white display. In order to sustain a uniform appearance the cell thickness must be controlled very accurately (± 0.1 microns). Two other problems with supertwist devices are the control of the twist and the response time. In order to develop a high twist, cholesteric material must be added to a nematic, making a very accurate mixture with the correct pitch. This pitch is temperature dependent, which limits the operating range of the device, since the pitch has to be matched carefully to the thickness of the cell to ensure, for example, exactly 270° between the two walls of the cell, if it is to have the required shape of the characteristic. The response time of the device is also affected by the shape of the characteristic. The turn-off times can be as much as ten times slower for a supertwist compared with an ordinary twisted nematic display. However, in spite of these potential problems, large complex displays with up to 400 lines have been made. The good multiplexing performance and relatively low-voltage operation have allowed them to make a significant impact in the portable computer market, where relatively long response time and narrow temperature range are not significant disadvantages.

In the next section we look at the parameters involved in designing liquid crystals.

SAQ 6.7   (Objective 6.6)
Consider a simple twisted nematic display cell (Figure 6.19). Suppose that sodium ions diffuse from the glass at the boundaries into the liquid crystal. Explain why the display cell may now lose contrast and suggest how to guard against this mode of failure.

6.5.3 Materials engineering

The main elements in the specification of an LCD material are concerned with its operating temperature range, voltage and power requirements and the long lifetime necessary for a consumer product. These put particular emphasis on the chemical stability of the material under an applied voltage and also under external radiation (ultraviolet and visible).

Some of the early materials used commercially gave LCDs a bad name because they decomposed fairly rapidly in ultraviolet radiation and lost their nematic properties. Generally, the materials now used are made by mixing several pure compounds in order to optimize the physical properties required for specific applications. In particular, the operating temperature range can be increased by alloying. This was particularly important for the first stable family of liquid-crystal materials to be developed, the cyanobiphenyls, which became available in the mid-1970s. They exhibited liquid-crystal behaviour over a range of only about 10 K, but by judicious alloying this can be increased considerably. Materials have been developed with nematic properties over a 100 K range (typically $- 20°C$ to $+ 80°C$).

What about the relevant electro-optic properties? In essence, optical properties can be altered by, for example, changing the ring structures (that is, the phenyls in cyanobiphenyl), so manipulating the way electrons can move around the molecule. Also, the addition of side groups with permanent dipoles can be used to 'tune' the low-frequency dielectric anisotropy. Usually, a high dielectric anisotropy is needed for a low operating voltage, and such materials have been developed which can be added to a basic mixture. However, it is difficult to obtain all the right optical, elastic and viscous properties.

EXERCISE 6.17   Why is it difficult to achieve a material with a low viscosity (for ease of manufacture) and a high dielectric anisotropy (for good performance)?

A big problem with LCDs is how to generate colour, especially full colour, of controlled brightness and contrast. Two approaches show promise: incorporating dyes into the liquid crystals, and using independent (external) colour filters. We look at them in turn.

Dyes

It is important to understand here how a dye imparts colour. A red dye *absorbs* all colours except red. It appears red because only the red part of white light is reflected or transmitted to the eye, all other colours being absorbed.

In a liquid-crystal display, one idea is to have absorption (that is, colour) with the cell in the 'off' condition, and transmission (no absorption) with the cell 'on'. This is done by incorporating an anisotropic dye that dissolves homogeneously in the nematic liquid. What does that mean? Well, for success, the dye molecule must have two key characteristics. First, it must absorb light only when the plane of polarization of incident light is parallel to a specific direction in the dye. This direction is called the **transition moment direction** in the dye molecule. Second, this transition moment direction should be aligned with the nematic director *n*. The liquid-crystal molecules, in their nematic phase, act as hosts. The guest dye molecules politely join in the action.

A cell based on this guest-host principle could even be made without polarizers, just a pair of parallel rubbed inner surfaces and the usual transparent electrodes. Figure 6.27 shows how it would work. Since on average, half the light will be polarized parallel to the rubbing direction, without any cell voltage the dye will be able to absorb up to half the incident light (around its characteristic absorption wavelength) (Figure 6.27a). When the nematic phase is electrically reoriented (Figure 6.27b)

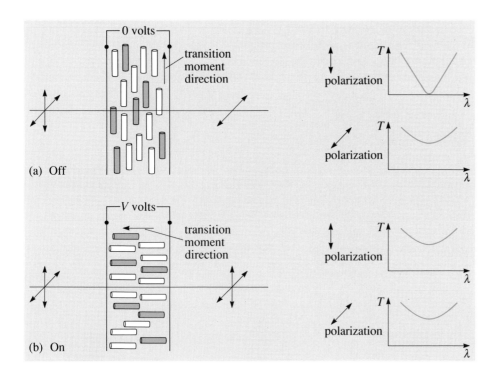

Figure 6.27 Principle of guest-host (dye) display. '*T*' stands for 'transmission'

the dye molecules get jostled round and are no longer effective. Their transition moment is at right angles to the incident light's planes of polarization. Virtually all light is now transmitted. So, with suitable monochromatic light we have a dull on bright contrast.

Can you see how the addition of one polarizer can be used to heighten contrast?

Use a single polarizer to cut out the constant horizontal component (that, is the component that is transmitted when the cell is 'off'). If we have an all-absorbing dye, rather than a colour dye, and a single polarizer, we can switch pixels between grey in the 'on' state (50% transmission of incident light) to black in the 'off' (absorbing all colours) state.

However, these levels of absorption and transmission are never achieved in practice because thermal activity leads to some disorder. This disordering effect can be reduced by incorporating dye molecules that are longer than those of the nematic liquid: greater thermal activity is necessary to move such molecules out of alignment than the shorter liquid crystal molecules.

Not surprisingly the dye molecules look a bit like the liquid-crystal molecules. The difference is they can absorb incident energy rather than simply passing it on. You can expect to find they offer lots of scope for distributed charge (phenyl groups, or alternating double and single bonds in a chain). Two examples are cited in Figure 6.28. The azo dye molecule in (a) is long and maintains a high degree of order (*azo* means the $- N = N -$ group). Unfortunately, the molecule is not very stable under ultraviolet light, so ultraviolet filters need to be incorporated in an assembly that uses these dyes. On the other hand, dyes based on the

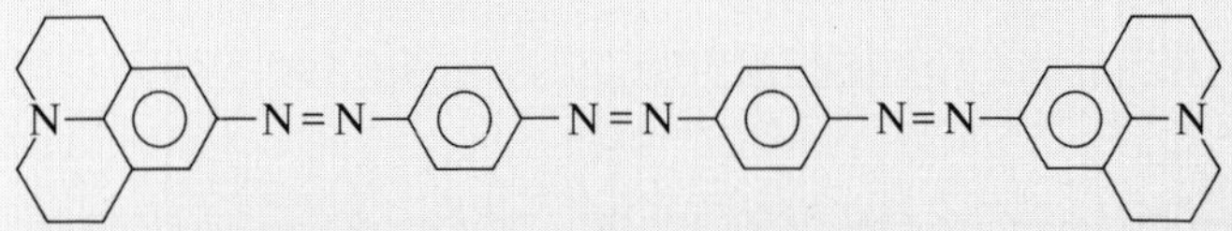

(a) Azo type - unstable but high order parameter

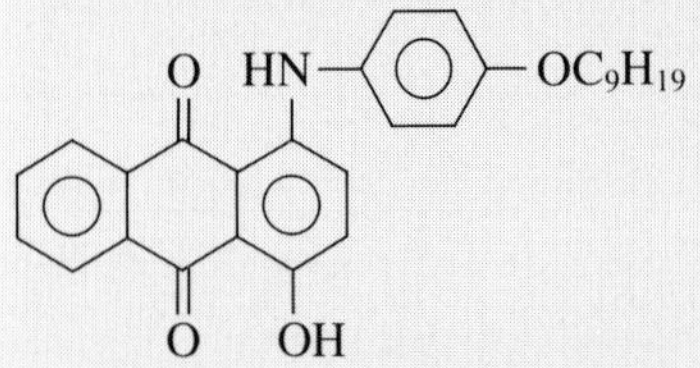

(b) Anthraquinone type - good stability, low order parameter

Figure 6.28 Dyes for use in liquid crystals

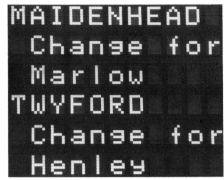

Figure 6.29 Paddington LCD signboard

anthraquinone molecule shown in Figure 6.28(b) are stable. But the basic molecule is obviously not rod-like and, therefore, does not have such a high degree of order. By substituting elongated chain groups at appropriate points in the molecule however, better materials have been produced.

Such dyes are available for producing red, yellow and blue, and as a mixture, black. They are used in large displays such as signboards. Figure 6.29 illustrates the one at Paddington Station.

EXERCISE 6.18   A guest-host display needs backlighting. Suggest a suitable source material for diffuse coloured light. (Hint: think of strip lights.)

Filters

An alternative to guest dyes in the liquid crystal is the inclusion of a dye in an external filter so that only a particular range of wavelengths (that is, a colour) enters each LCD cell. Liquid-crystal domestic television screens using this approach and based on TN cells have been developed. The structure of the display is illustrated in Figure 6.30(a). To give a better picture quality, the pixels are staggered by one-and-a-half dots (see Figure 6.30b). As with a group of three phosphors for each picture dot on a colour CRT, there needs to be a triad of filters. The size of filter and TN cell is dictated by the required resolution; so for conventional television we need $\lesssim$ 0.6 mm pixels as before. That's well within our photolithographic capabilities and it could even be printed so we'll leave the filter technology at that.

At the end of Section 6.5.2 we noted the poor suitability of the TN for matrix addressed systems, so why all this talk of television? We look next at how the transistor business can help.

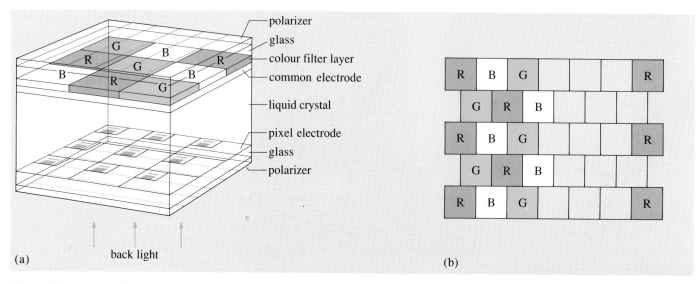

Figure 6.30 (a) An LCD cell in a television screen. (b) The colour pixel arrangement

SAQ 6.8 (Objective 6.7)

Consider a mixture containing a nematic phase; that is, the mixture is an alloy of liquid crystal with other, non-liquid-crystal substances. Link the following interrelated properties of the mixture with various features of its components.

(a) viscosity;

(b) switching speed and relaxation speed;

(c) switching voltage;

(d) stable temperature range;

(e) colour;

(f) refractive index and dielectric anisotropy;

(g) degree of order.

6.5.4 TFT displays

Let's begin by reviewing the position of liquid-crystal displays:

- There are several electro-optic effects which can be used to make displays (so far we have seen TN, supertwist, guest-host).
- For television we need to check the specification set out earlier. This is the business of Exercise 6.19.

EXERCISE 6.19 Complete Table 6.3, which assesses the potential of backlit TN cells for a television display.

Table 6.3

	Design requirements	Comment on suitability of TN cell
Viewing angle	as wide as possible (45° at least)	
Resolution	≲ 0.6 mm	
Speed	should be at least eye-response time	
Brightness	should be usable in moderate daylight	
Colour	RGB required (dye or filter?)	
Contrast	should be high	
Power	should be low (volts × amps = watts)	
Efficiency	should be high (anything get hot?)	

Exercise 6.19 highlights the major outstanding problem to be tackled: TN cells do not have a sharp enough threshold to suit simple matrix addressing. So here we will look at a way of sharpening the effective cell threshold without altering the liquid crystal at all.

Mention of thresholds may put you in mind of the MOSFET devices discussed in Chapter 5. These too have thresholds. The plan is to transfer the threshold problem to a tiny MOSFET switch at each pixel site and leave the liquid crystal to get on with the light modulation.

The ball stays firmly in the materials scientist's court because we are not suggesting glueing a small MOSFET onto each pixel (not for a display which may have over 500 000 pixels!). The MOSFET switches have to be built onto the glass substrate alongside the filters, electrodes, and so on. That means we cannot simply use the single-crystal microcircuitry of Chapter 5. We need a more general purpose MOSFET. Such a transistor is called a **thin film transistor** (TFT).

There have been three basic types of TFT employed in this application. The first uses the semiconductor cadmium selenide and the other two use silicon in amorphous and polycrystalline thin-film form. Table 6.4 compares their electron mobilities. These materials have been chosen because they can be deposited as accurate and reproducible thin films over the large areas required for displays. Deposition processes compatible with transparent substrates, ideally ordinary glass, have been developed for each of them and suitable transistor geometries optimized both to address the signals onto the final display and also to enhance the yield with which reproducible transistors can be fabricated over the display area.

Essentially they all work as gate-voltage operated MOSFET switches. I am going to concentrate on a TFT made from amorphous silicon, or more exactly amorphous silicon-hydrogen (αSi:H). The beauty of the system is that *all* the processes are within the well-tried practices of evaporation (and allied deposition methods), masking and etching, *and* all the materials are inherently cheap compared with single-crystal wafers.

Table 6.4

Material	Mobility/(m^2 V^{-1} s^{-1})
CdSe	0.02–0.05
Amorphous Si	0.00002–0.0003
Polycrystalline Si	0.001–0·01

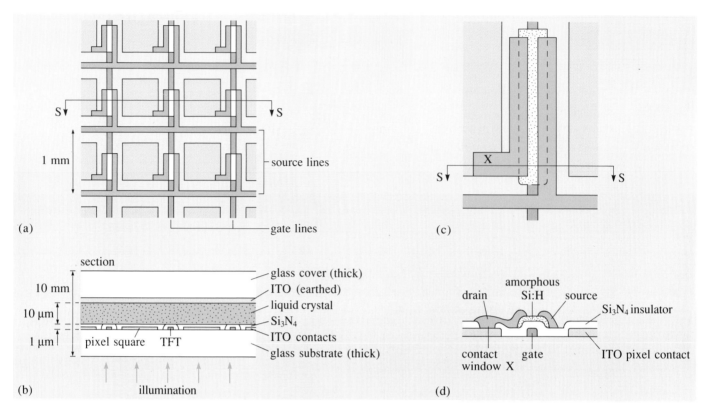

Figure 6.31 Construction of an LCD/TFT display panel. (a) Plan, showing several pixels. (b) Section through panel. (c) Plan of a TFT within a pixel. (d) Section through the line SS in (c)

Figure 6.31(a) shows a plan view of the panel, and Figure 6.31(b) a section. The panel is designed to be viewed in transmitted light. The substrate and top plate are of glass and are optical polarizers. The liquid crystal is continuous through the gap of about 10 μm between these plates and an electric field can be applied through the usual transparent conductor, indium-doped tin oxide (ITO). The substrate carries the TFTs and a pattern of ITO squares, while the top plate has a continuous film of ITO and is usually the earthed contact.

The details of construction can be followed on Figures 6.31(c) and (d). The first steps consist of depositing ITO and etching the patterns of the back contacts of the liquid-crystal cell. A chromium film is then evaporated and etched to provide the gate electrodes, which are typically 5–50 μm wide. The prepared substrate is then coated with silicon nitride insulator about 0.4 μm thick and about 0.3 μm of αSi:H, both deposited using a radio-frequency glow-discharge process. The unwanted αSi:H is etched away to leave small islands at each TFT site. Windows (shown in Figure 6.31d) are etched through the silicon nitride down to the ITO to allow contact to the drain. (Note that only the drain is electrically connected to the ITO electrode.) Finally, aluminium is evaporated to form the source and drain contacts. Amorphous silicon (αSiH) is an extremely efficient photoconductor and therefore stray light may affect the performance of these transistors. This is especially important in the off state (high-resistance state), where the off resistance may be reduced by an order of magnitude. But the problem can be eliminated by suitable masking.

To address information to the pixel a voltage is applied to the gate which activates the transistor so that the signal information applied to the source electrode is switched through to the drain and connected to the pixel pad. In a matrix display all the gates in one row are connected together to form a scanning electrode and the sources in each column are joined together to form the signal (or data) electrode.

The TFT cell successfully isolates each pixel in the matrix but we still need a range of contrasts (grey scale) for TV pictures. A grey scale is achieved by selection of particular voltage signals so that the r.m.s. voltage seen by the cell moves it to a particular point on the twisted-nematic transmission-voltage characteristic. Now, in contrast to the multiplexed twisted nematic, the optimum shape of the TN characteristic for TFT cells is that it should be as shallow as possible, so that it is easier to have a number of grey scale points on the characteristic at significantly different voltages.

The material used for the TFT and the geometry of the subsequent transistors are linked by critical material properties such as the electron mobility in the material and the need to address the display at video frame rates. Applying a voltage to a pixel is really just charging the capacitance of the liquid crystal elements to the necessary voltage to achieve the desired optical transmission level. If this is to be achieved in a short address time, the resistance of the source-to-drain channel of the transistor must be as low as possible. The resistance of the channel will be dependent on its area and length together with the mobility of electrons in the channel region. Low mobility materials need to have short but wide channels in order to have sufficiently low resistances to allow charging of the pixel.

That sounds easy, but don't forget that the material deposited to form the transistors must be uniform in properties over the whole display area so that the resultant transistors have uniform voltage characteristics and can be switched to develop the required voltages at each pixel. Steady improvements in material deposition, fabrication in very clean environments, together with improved geometries, redundancy and automatic repair techniques, have given sufficient yield to allow up to 3-inch diagonal colour displays based on this technology to be fabricated commercially. And devices of up to 18-inch are reported to be at the experimental stage at the beginning of the 1990s. Do you want to have a go at making one? ▼Manufacturing LCDs for TV▲ will get your business planning started.

▼Manufacturing LCDs for TV▲

Assembly of liquid-crystal displays must be undertaken in a dust-free, clean environment. The more critical the cell spacing specification, the cleaner the environment must be. Manufacture of low-complexity watch and calculator displays can be undertaken in cheap facilities with limited automation and still give good enough yields to be competitive; but high-technology television and computer displays require more investment in clean rooms and automation. These increase the volume of production and thereby improve yield sufficiently to enable the displays to be sold at an affordable price. In all cases the basic process is very similar and includes the following stages.

1 Cut substrates to correct size. This may be an individual display, but often small displays are processed in bulk on a single substrate and separated at a later stage. (The substrate glass may be precoated with silicon for TFT displays or ITO for other types.)

2 Clean substrates to remove contamination.

3 Photolithography, etching and depositions for electrode patterns TFTs, filters and so on.

4 Deposition and preparation of alignment layers — usually coating with a polymer such as polyimide and rubbing with a roller or cloth to define an alignment direction.

5 Assemble, and seal. This includes: aligning and registering electrode patterns on the top and bottom substrates; depositing material, such as small plastic beads or fibres of the correct size, to control the cell spacing; screen printing or some other technique for applying a continuous strip of adhesive around the outside of the cell active area, leaving a gap through which to fill; and sealing together, usually using heat to cure the adhesive. After this stage a batch of cells on one substrate will be separated prior to filling.

6 Fill with the liquid crystal, often using a vacuum technique to ensure no gas bubbles occur in the cells, and afterwards seal.

7 Apply required polarizers, inspect and test.

SAQ 6.9   (Objective 6.8)
In the end many electronic systems fail because of a connection failure.
One way to reduce the density of connections around a display screen,
and also to reduce production and assembly costs, is to build as many
signal processing circuits as possible using TFTs on the glass at the
edge of the screen. Specify two qualifications which the deposited thin
film and its circuitry must have for this purpose.

6.5.5 Ferroelectric and other displays

Thin film transistors enable the TN display to be compatible with the
matrix addressed demands of a TV display. There are problems though
with producing ever larger screens so it mustn't be supposed that the TV
market is sewn up.

Liquid crystals offer a medium as versatile almost as the mixed-oxide
ceramics of Chapter 4. It is not surprising that other effects can offer
alternative designs. The TFT cell acts as a light valve with one signal
selecting whether or not the valve *can* be on and another setting the
degree to which it is on. Now the idea of an on/off light valve was
discussed in Chapter 4 ('Flash goggles', which used PLZT). Also, we
know that anything shorter than several microseconds is averaged by the
eye. So the degree to which a cell 'appears' to be on might be controlled
through a duty factor. We need a material that can, within a few
microseconds, have its optical anisotropy rotated through 45° (as the
PLZT did), sandwiched between crossed polarizers.

The ferroelectric display uses long molecules, aligned by a grooved
substrate. These molecules have a polar side group, which can be
influenced by low frequency fields (hence 'ferroelectric'), and a pattern of
bonds, which allows electrons to move back and forth in response to
high-frequency (optical) fields. This optical axis is at an angle to the
length of the molecule. Now, when a switching field changes direction, it
grabs the side group and spins the molecule 180° about its long axis
without disturbing its alignment. The optical axis is then a mirror image
of the earlier setting.

An array of such shutters can be fabricated into a display suitable for
video-frame rates. They would not need the extra etching and deposition
steps involved in using TFTs, so yield and reliability should improve.
Furthermore, scaling up to larger screens should be easier.

6.6 Which display?

It is not possible to say clearly which of the CRT or an LCD is the
'better' display device — nor, even more generally, to decide between
light emitters and electromodulators. What is clear, though, is that the

wide variety offers us a plethora of potential solutions to the problems of display. In certain applications there will be distinct losers. For example, the CRT is set to lose the market for large-screen high-definition television displays because it is just too big. In other cases there may be clear favourites. For example, where volume or weight or power consumption are key factors the LCD has a head start. But otherwise it is a question of market forces and commerce. It will be interesting to watch what they make of these technologies.

Objectives for Chapter 6

You should now be able to do the following:

6.1 Distinguish between, and briefly account for, the different sorts of luminous solids in terms of excitation processes and energy bands. (SAQ 6.1)

6.2 Discuss the viewing requirements of flat-screen displays for television/video pictures in terms of angle, resolution, speed, brightness, colour, and contrast; and identify relevant aspects of power and efficiency. (SAQ 6.2)

6.3 Give a simple account of CRTs for television. (SAQ 6.3)

6.4 Explain the efficiency, colour and persistence properties of CRT phosphors in terms of photons, phonons and energy bands. (SAQ 6.4)

6.5 Relate the anisotropic properties of liquid-crystal phases to their molecular structure. (SAQ 6.6)

6.6 Give a simple account of a twisted nematic display element. (SAQ 6.7)

6.7 Describe the ideas underlying the engineering of liquid-crystal phases for various display purposes. (SAQ 6.8)

6.8 Discuss the electronic material requirements of a display cell incorporating a nematic liquid crystal phase and thin film transistors. (SAQ 6.9)

6.9 Expand the following shorthand:

CRT
LED
LCD
TN
TFT
ITO (also in Chapter 4)

Answer to Exercises

EXERCISE 6.1 We need to convert between wavelength and energy. Using the formula just before Exercise 6.1

$$E_p = \frac{hc}{\lambda}(\text{joules}) = \frac{hc}{e\lambda}(\text{eV})$$

The value of h/e is 4.1×10^{-15} J s eV^{-1}, so for violet light ($\lambda = 400$ nm)

$$E_p(\text{eV}) = \frac{4.1 \times 10^{-15} \times 3 \times 10^8}{4 \times 10^{-7}} \text{ eV}$$

$$= 3.1 \text{ eV}$$

For red light ($\lambda = 700$ nm)

$$E_p(\text{eV}) = \frac{3.1 \times 400}{700} \text{ eV}$$

$$= 1.8 \text{ eV}$$

EXERCISE 6.2

(a) The energy separations are as follows (i) 2 eV, (ii) 1.8 eV, (iii) 1.9 eV, (iv) 0.8 eV and (v) 1.2 eV. Only those in the range 1.8–3.1 correspond with visible photons so (i), (ii) and (iii) could be associated with visible light emission. Separations (iv) and (v) are too low in energy in this case.

(b) You may have suggested one of the following:

• There must be some way of getting electrons into the upper energy state (and, equivalently, there must be room in the lower energy state).
• The transition from upper to lower state with accompanying photon must be feasible (quantum mechanics does a lot of excluding).
• The emitted photon must be able to escape to the outside world without being absorbed on the way.

EXERCISE 6.3 The following comments refer to Figure 6.4.

Material (i). These conduction-to-valence band transitions are characteristic of intrinsic semiconductors.

Material (ii). This local level near the conduction band looks like a possible donor impurity level, so this could be an n-type semiconductor.

Material (iii). Likewise, a local level near the valence band can be an acceptor impurity level, so this material could be p-type semiconductor.

Material (iv), (v). A deep level in the middle of a conduction band, such as these have, usefully suppresses conductivity. This could be an insulator.

EXERCISE 6.4

(a) At the very high frequencies of visible light, only the electrons get disturbed. (You might also have suggested crystal boundaries as sources of refraction.)

(b) At lower frequencies ions and polar molecules can join in.

EXERCISE 6.5 Based on my experiences of good and bad television, I would expect a good quality picture to be:

• clear, that is not spotty or 'snowy';
• contrasty, with sharply distinct colours and a good range of shades;
• accurate in its rendering of colour;
• fast but steady, that is with no flicker.

EXERCISE 6.6 In Chapter 1 we discussed resolution and suggested that the dottiness of a picture might be 'seen' if dots were larger than 0.6 mm. By the same token, lines closer than this will be too close to be resolved. So, 2 lines per millimetre or more ought to do the trick. Comment: Smaller screens, with the full number of lines, will have smaller dots — but we view these screens more closely.

EXERCISE 6.7 A time of 0.04 second corresponds with 625 × 833 elements. Each element can therefore receive attention for

$$\frac{0.04}{625 \times 833} \text{ s} = 7.68 \times 10^{-8} \text{ second}$$

So, we have up to 77 nanoseconds to do things which will alter the state of the material of the element. Each element then waits

$$(0.04 - 0.000\,000\,077) \text{ s} \approx 0.04 \text{ s}.$$

EXERCISE 6.8 Displaying 625 lines in 0.04 second means 625/0.04 lines per second or, equivalently, 0.04/625 second per line. That works out at 64 μs for each line. The whole line is changed at once so that we have 64 μs to deal with the material at each element of the picture. The time before we can return to any line is (40 − 0.064) ms = 39.936 ms.

EXERCISE 6.9 Material A: Useless. At 2.5 volts the material is almost 'fully on', so all of row F and all of column 3 are as bright as the one intended element at the intersection.

Material B: Not so useful. At 2.5 volts the material is already 25% bright and at 5.0 volts it is only 75% bright so all of row F and column 3 are 25% bright (should be zero) except for the intersection which at 75% is not as bright as it could be.

Material C: Ideal. No response at 2.5 volts, full response at 5 volts.

EXERCISE 6.10 This is the way it is done. You should have picked up the key points.

The shadow mask is used as the photolithographic mask, with light sources located where each electron gun will sit. A positive photoresist is put onto the screen. For one set of dots, say the red ones, a light source at the red-gun position exposes the photoresist through the shadow mask. The resist is developed, giving holes in the resist at the places where red dots are needed. The red phosphor is settled on. After removal of the resist, the desired pattern of red dots remains. For the other colours, the process is repeated with fresh resist each time.

EXERCISE 6.11 Formula (c) is the only one which indicates the increase which is observed in practice with both current and voltage.

Comments: Formula (c) is the best of those given but in practice the light emission is not in simple proportion with either current or voltage. Also none of the formulae allows for saturation.

EXERCISE 6.12 You may have suggested one of the following, arising as a result of reduced penetration at low voltage:

(a) Crystal surfaces influence band structure (recall MOS devices) so energy gap may not be so clear.

(b) Impurities accumulate at crystal boundaries increasing risk of 'non-visible' recombination.

(c) Low-energy beams are partially reflected by the crystal surface.

EXERCISE 6.13 Three base colours are needed, matched to the red, green and blue sensitivity of the eye.

EXERCISE 6.14   You may have got one of these:

• A negative-ion vacancy could engage an electron for some time.
• Surface states, as encountered in MOS oxide interfaces.
• Impurity atoms interacting with the crystal structure.

EXERCISE 6.15

(a) There is less space for acceleration and focusing.

(b) Angle of scanning is wider.

EXERCISE 6.16   In (a) the molecular alignment is bent and splayed to follow the rise and fall of the substrate. Elastic distortion energy is stored in the crystal, so there is an energy cost associated with this configuration.

In (b) there is no distortion, so no energy cost. We assume that the liquid crystal will always adopt the lowest energy configuration, so (b) is favoured.

EXERCISE 6.17   A high dielectric anisotropy requires molecules with large electric dipoles. The resulting strong dipolar interactions between the molecules make flow difficult: the mixture has a high viscosity.

EXERCISE 6.18   A fluorescent lamp uses phosphors to give diffuse light. Colour CRTs use phosphors to give coloured dots. By combining the two, fluorescent lamps with coloured phosphors, we get diffuse coloured light. Alternatively we could use an external colour filter on an otherwise 'white' fluorescent source.

EXERCISE 6.19   See Table 6.5

Table 6.5

	Design requirements	Comment on suitability of TN cell
Viewing angle	as wide as possible	Just about acceptable (check with an LCD calculator display),
Resolution	≲ 0.6 mm	Easy by silicon technology standards. Need to have uniformity of cells over entire display area could be problematic for large displays.
Speed	should be at least eye-response time	A bit sluggish. Could blurr fast action.
Brightness	should be usable in moderate daylight	No problem — backlight available.
Colour	RGB required	Achievable with pixel filters
Contrast	should be high	Problem. Switching threshold not sharp enough, leading to poor contrast in matrix addressed system.
Power	should be low	TN cells draw almost no current. Power = voltage × current, so power consumption very low.
Efficiency	should be high	Almost no heat production, so efficiency should be high

Answer to self-assessment questions

SAQ 6.1   The reference to band structure implies that we are concerned with solids and luminescence. Electroluminescent phosphors and LEDs (excited by internal currents and fields) and cathodoluminescent phosphors (bombarded by externally accelerated electrons) all emit light in accordance with the band structure of the host material, influenced by dopants or activators. Thus the relevant entries are those for GaAs–GaP, ZnS–Mn, and ZnO–Zn.

SAQ 6.2   See Table 6.6

SAQ 6.3   (a) The CRT for monochrome pictures is simpler than that required for colour, since a single electron beam can scan the entire screen and stimulate a uniform phosphor coating. There is no need for the shadow mask therefore and

Table 6.6

	Same as domestic TV? yes/no	How different?
Viewing angle	no	wider angle
Resolution	no	relative to size of screen, same resolution as TV required; therefore, *absolute* resolution required of large screen is less than for TV
Speed	yes	
Brightness	no	must be brighter for outdoor use
Colour	yes	
Contrast	yes	
Power	no	much greater; power dissipation is at least proportional to area
Efficiency	at least	or higher

no need to coat the screen with the colour triads (RGB) of phosphor.

(b) For black-and-white displays the phosphor could be made from a homogeneous mixture of colour CRT materials; the first, third and last entries in Table 6.2 look useful, but notice the low efficiency of the red one. It is also possible to mix just two phosphors together: since a yellow (Y) phosphor will stimulate the eye just the same as a mixture of red and green phosphors, we can replace RG by Y and then blend with a blue (B) phosphor. The first two entries in Table 6.2 are used for this purpose.

SAQ 6.4

(a) True.

(b) True, although it does omit the special case of manganese activated phosphors.

(c) False. Long persistence requires the electron–hole recombination process to be held up, so low mobility is called for. Purity and structural perfection are requirements for high mobility.

SAQ 6.5

(a) Weak secondary (van der Waals) bonds can form between molecules. So they naturally lie parallel with one another. The short (stiff) rod shape also promotes parallel packing when dense enough — rather like matches in a box.

(b) In an electric field, some electrons will move along molecules: an electric dipole is induced with one end of a molecule being positive (electrons lacking) and the other negative (slight excess of electrons). The induced dipoles then tend to align with the electric field — the liquid structure allowing this to happen.

SAQ 6.6
Both phases are anisotropic dielectrics, since the molecules can be electrically oriented by an external electric field, through permanent or induced dipoles. The elongated molecule, with

freedom for electron displacement along it, is likely to confer on its phase the stronger optical anisotropy.

SAQ 6.7
Sodium ions dissolved in the liquid-crystal phase are likely to be fairly mobile. (In Chapter 5 we saw them able to ruin the characteristics of MOS structures by moving around in the oxide.) If they oscillate back and forth with the a.c. voltage which is switching a cell, they are likely to knock the more sluggish liquid crystal molecules out of their alignment. So the liquid-crystal molecules don't stay neatly out of the path of the incident light, as they should. Also, the sodium-ion current wastes power. In sufficient quantities, the sodium ions will also affect the threshold voltage of the cell.

The expensive precaution is an ion-free glass (quartz) or one in which ions are relatively immobile. A cheaper solution is to coat the inner glass surfaces with a diffusion barrier.

Comment: In practice the rubbed polymeric layer illustrated in Figure 6.19 serves both to isolate the glass and to align liquid crystal molecules with the surface.

SAQ 6.8

(a) Viscosity is a consequence of inter-molecular forces. It will depend on the nature and quantity of each type of molecule present.

(b) Switching and relaxation are both viscosity-dependent properties. Switching is also influenced by the polarizability (and permanent dipolar) properties of the constituents. See (f).

(c) Switching voltages arise as electrical reorientation competes with the molecule/surface interaction. Relatively strong intermolecular forces and relatively weak dielectric anisotropy (non-polar and low polarizability) will both increase threshold voltages.

(d) The nematic phase arises when it is energetically favourable to adopt an ordered if fluid structure. Just as with solid crystals, stronger intermolecular attraction will require more thermal energy to disrupt it. (And of course there's entropy to consider.)

(e) The colour of a 'transparent' liquid is that which it does not absorb from the incident light. Any dye molecules in the mixture may impart colour to it.

(f) Rings and chains with the right sort of bonds between will give rise to molecules along which electrons can 'skate'. At optical frequencies such components give the material an anisotropic refractive index. At lower frequencies, together with any molecular polarization components, there is an associated dielectric anisotropy: the relative permittivity is different in different directions in a liquid crystal.

(g) The ordered crystalline tendency is perturbed by thermal energy. Long, thin, sticky molecules are harder to jostle out of alignment than short, fat, slippery ones (but if they are too long they are hard to align).

SAQ 6.9
You may have suggested from among the following:

(a) The thin film must be uniform in thickness and electronic quality across the area of the screen and across that area required for the additional circuits.

(b) The film must have sufficiently good electronic properties (mobility) to allow the extra devices to function. Single MOS switches are one thing, complex arrays fast enough to be useful are another.

(c) The thin-film circuits should be compatible with the processing used for the liquid-crystal devices (without detriment to the performance of the display elements).

(d) The thin-film circuitry must be designed with sufficient redundancy to sustain yields in spite of the more complex processing.

Chapter 7
Warm superconductors — so what?

7.1 Super stuff

So far we have considered electrical conductivity, magnetism, and some optical properties in connection with our goals which are the electronic functions of transduction, memory and display. Our discussions have been about adjusting certain material properties to the benefit of function.

In this closing chapter we are going to look at one more electronic phenomenon, superconductivity, but considering markets, as well as functions and products. Superconductivity allows the continued passage of a steady current without a potential difference to drive it. (True superconductors also have a specific interaction with magnetic fields which we will come to in due course.)

> EXERCISE 7.1   If a current *I* flows in a wire (length *l*, cross-sectional area *A*) without there being any potential difference whatsoever, what can you say about
>
> (a) the resistance of the wire (assuming Ohm's law to hold)?
> (b) the resistivity of the material from which the wire is fabricated?
> (c) the power dissipated in the wire?

Until 1986, the better known superconductors were very low temperature metallic intermediate compounds and metal alloys (such as niobium–titanium below 8 K) but since then a new category of ceramic oxide alloy materials has been found to superconduct at much higher temperatures (for example yttrium–barium–copper oxide below about 100 K). See ▼Causes and effects▲. The extremely low operating temperature of the traditional materials has restricted them to very specialist applications where the costs of refrigeration plant and space are of secondary importance. The ceramic oxides are less inconvenient and hold out hope that perhaps one day we will see superconductivity at room temperature. But there is a gulf between materials curiosity and commercial engineering reality, which is why I ask — so what? We are going to look from the 'electronics for function' angle to see how this new material, optimistically dubbed 'warm superconductor', might make an impact.

We will focus on just one application of superconductors to see what suite of properties is involved (it's not just zero resistivity) and to what extent there is scope for new designs with new materials. The application is the generation of extremely stable magnetic fields for medical imaging scanners based on so-called nuclear magnetic resonance (NMR). (The function is essentially one of transduction turning the presence and position of certain atomic nuclei into electronic signals.) Commercial NMR systems have long exploited superconducting magnets operating at 4.2 K, so a market and rival products already exist.

▼ Causes and effects ▲

It is not unusual in materials technology to find that a variety of actions can be used to promote the same effect. For example the strengthening of some materials can be achieved either by removing impurities, or by the controlled addition of impurities and also by purely physical refinements of the microstructure. At the scientific level there is a similar diversity of causes for one effect (or property).

In Chapter 3 for instance we saw that ferromagnetism is associated with alloys of a few special transition metals. Atoms are coupled by the so-called exchange interaction. Comparable bulk magnetic properties are achieved in ferrimagnetic crystalline oxides in which various transition metal ions are coupled by a super-exchange interaction. (As we will see there are also other circumstances in which atoms can exhibit magnetic effects.) We have found an even broader range of contributors to electrical conduction.

EXERCISE 7.2   List the three contributions to electrical conduction in oxide ceramics discussed in Chapter 4.

The attributing of an effect or a property to more than one source really shows that we are not yet dealing with the component parts of matter, but we'll leave that to elementary particle physicists. Electrical conductivity is not only associated with metals (remember graphite, indium-doped tin oxide and doped silicon). Magnetism is not only a property of rocks. Our models of matter have linked both of these phenomena to the behaviour of electrons within the constituent atomic and molecular structures. But we find the results of different electronic behaviour can lead to similar bulk properties (such as conductivity or magnetization). It is therefore vital that we keep an open mind and that we are not surprised when some

phenomenon is not exclusively associated with one category of materials. As you should know, trying to class real materials as metallic or polymeric or ceramic is largely a subjective activity. To take an agricultural analogy, you don't have to grow sugar cane to produce sugar — you could grow sugar beet, and even cabbages contain some sugar!

We should therefore be relieved (rather than surprised) to hear that superconductivity is found to be a property of some ceramic oxides as well as a few metals and alloys. In fact in 1964 strontium titanate (perovskite structure) was found to superconduct below 0.3 K and in 1973 lithium titanate (spinel structure) was shown to superconduct below 14.7 K. In 1986–1987, ceramic alloys overtook metallic alloys in terms of the upper temperature limit when some perovskite-structured materials were found to superconduct up to around 100 K.

First though let's just see what superconductivity is. After that we will investigate the background to the proposed medical imaging system. The superconducting options will then be considered. Finally I'll point out the nature of other products which may incorporate material with this special property.

7.1.1 Superconductivity

The Dutch physicist Heike Kamerlingh-Onnes (1853–1936) stumbled across superconductivity in 1911 while investigating the effects of low temperatures on electrical resistivity. Three years earlier he had mastered practical thermodynamics and been able to liquefy helium at a mere 4.2 K (helium atoms are so 'content' with their own electronic structure that its only at very low temperatures that they can be persuaded to associate with each other!).

EXERCISE 7.3   What would you expect to happen to the resistivity of a metal like platinum at low temperatures? (Remember, resistance arises from electrons interacting with imperfections and from the thermal energy of the lattice.)

Kamerlingh-Onnes found that for platinum the resistivity levelled out at a residual value owing to impurities in the metal. Turning to mercury which was available in purer form, he found that below 4.2 K, the resistivity had suddenly become imperceptibly small. Yet above this temperature he could easily measure a value (see Figure 7.1). Clearly the impurities don't suddenly vanish below a critical temperature, so something else must happen which makes them irrelevant. That thing is called superconductivity. It is a low temperature property of at least a quarter of the elements, many metallic alloys, various intermediate compounds, and also some mixed oxides and sulphides.

The onset of superconductivity is similar to that of ferromagnetism and ferroelectricity. There is a critical temperature above which thermal energy prevents associations but below which co-operations between certain electrons spontaneously give rise to a new regime of behaviour.

There are in fact three quantities which have to be below a critical value for the material in order for the superconducting state to exist. They are

(a) The temperature of the material ($T < T_c$)
(b) The current density in the material ($j < j_c$)
(c) The magnetic field adjacent to the material ($B < B_c$).

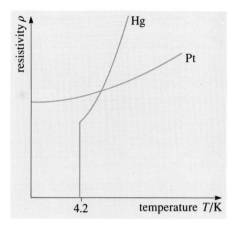

Figure 7.1 Platinum and mercury resistivity with temperature

If any one of these criteria is not met, then the material won't be superconducting. The critical temperature, current density and flux density all depend on each other so you can imagine a set of axes of T, j and B with critical contours forming a critical surface (see Figure 7.2).

There is more to superconductivity than just zero resistivity. Suppose I drop a small, short, fat permanent magnet onto a piece of superconductor. As it falls, currents will be induced in the surface of the superconductor, just like eddy currents, by the action of the moving magnetic field (see Figure 7.3). Faraday's law tells us that an e.m.f. will be developed and currents will immediately follow. In the superconductor, provided the critical values are not exceeded, these currents will rise almost instantly to a level which exactly cancels the effect of the field they are in, so that there is no net field in the bulk of the superconductor. The supercurrents make the material appear as if it were itself a permanent magnet of equal magnitude but of opposite direction to the local field. At some point the repulsive magnetic force will be strong enough for it to balance the gravitational force so that the magnet will be held suspended above the material.

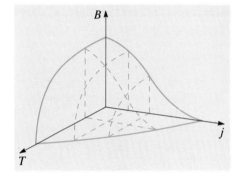

Figure 7.2 T–j–B critical surface

All this would happen because the material has zero resistivity and therefore zero power dissipation. The special bit, which is the hallmark of superconductivity, is that if we start with the material above its critical temperature, with the magnet resting on top, currents will be induced in the material when it is cooled below T_c. The magnet will levitate to the same position as before, so that the magnetic field inside the superconductor is again zero. This exclusion of magnetic field is called the **Meissner effect** and you need more than just zero resistivity for it to occur. You also need cooling through T_c to generate currents. That is why it is called superconductivity rather than infinite or perfect conductivity. Remember though that beyond a critical magnetic flux density, the material reverts to normal behaviour.

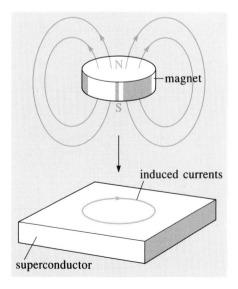

Figure 7.3 Superconducting levitation

Some superconductors, instead of simply giving up at a certain flux density, tolerate pockets of magnetic flux (not unlike domains) up to a considerably higher second critical flux density where they eventually submit and revert to the normal state. This type of extended tolerance of magnetic flux is a useful attribute for superconductors intended for magnetic applications.

▼Superconduction▲ goes just a little further into what is going on. You should remember the following about superconductors:

Owing to the special behaviour of a few paired electrons

(a) they have zero resistivity
(b) they expel magnetic flux (Meissner effect)
(c) there are critical levels of temperature, current density and magnetic flux above which superconductivity vanishes.

SAQ 7.1   (Objective 7.1)
Write a brief description of superconductivity using about 50 words.

▼Superconduction▲

Superconductivity is not quite the same as infinite conductivity so I can't just say

$$\sigma = n_e\, e\, \mu_e = \infty$$

and leave it at that. Some charges are somehow being transported. But since lattice imperfections and impurities don't suddenly vanish, any free charges ought to go on having collisions even below the critical temperature. A useful picture of superconductivity deals with this by supposing that below the critical conditions, the current is carried by charge carriers (electrons or maybe holes) working in pairs, weakly bound to each other through a complicated interaction with the crystal lattice. The pairs are usually dubbed **Cooper pairs** in recognition of one of the subject's theoreticians. The normal result of a collision is the transfer of energy and momentum from a charge carrier to the lattice. A superconducting pair of charges has dramatically different properties from a single charge. Unlike an electron, a superconducting electron pair has no net spin and no momentum either. It's also a good bit more nebulous. Most important of all, it doesn't get scattered. A Cooper pair just doesn't 'see' the lattice the way a single electron does. We have just got to accept it as a new form of charge carrier. (To get much further you need quantum mechanics or a lot of faith.)

So what do we need to do to encourage these special pairings?

First, we don't want too many electrons around because unpaired electrons tend to get in the way. We don't need many pairs because their infinite mobility make any pair a potent contributor to current!

Secondly, as I indicated, we need the atomic structure (the lattice) to co-operate, so you can guess that certain crystal structures may be more favourable than others. A few electrons strongly coupled to the lattice sounds at first like a recipe for high resistivity — it is (unless we have superconducting pairs). Superconductors tend to be formed by materials with high resistivity above the superconducting transition.

Lastly, we could do without too much thermal energy around as we don't want the pairs shaken apart.

In practice, predicting the metallic and ceramic alloys which are superconductors is a tricky business. However we ought not to be surprised to find transition metals with their rich 'd shell' chemistry in many a successful blend.

7.2 Magnetic resonance imaging

We turn now to the background of the application to medical scanners.
So far our studies of electronic materials have been centred around
properties associated with electrons on or missing from atoms and ions.
Insulators, conductors, magnets, transducers, semiconductors, and liquid
crystals have all been viewed in this way. The rest of the atom (the
nucleus) has so far been largely ignored when considering the electronic
properties of materials.

EXERCISE 7.4 Chemical properties are generally reckoned to arise
through the interactions of outer shell electrons in making and
breaking bonds. What physical effects can you think of which depend
upon the character of the nucleus? (Try to list three.)

The nucleus is more than just dead weight — as the answers to Exercise
7.4 will have reminded you, the nucleus has properties sufficient for its
own book on 'nuclear materials'. One of the properties in my answers to
Exercise 7.4 that may have taken you by surprise is nuclear magnetism.
We need that here, so I will explain it a little further.

7.2.1 Nuclear magnetism

The nucleus contains protons and neutrons, which like electrons have
some of that mysterious quantity called spin. Protons also have positive
charge. Do you remember what we had to say in Chapter 3 about the
negatively charged electron 'spinning' on its axis? It behaves like a tiny
magnet and in solids it is the chief source of magnetism, giving rise in
particular to ferromagnetism and ferrimagnetism. Protons are almost
2000 times heavier than electrons and are confined to the nucleus, but as
spinning charges they too can give rise to magnetic effects.

Even neutrons, though neutral overall, comprise spinning charges.
Nuclear magnetism is the net result of protons, neutrons and quantum
mechanics. It turns out that only nuclei with an odd number of nucleons
(neutrons and protons) have a magnetic effect.

The nuclear magnets are much weaker than those associated with an
atom's electrons and also there is no equivalent of the exchange or superexchange interaction to couple one nuclear magnet with its neighbours.
However, nuclear magnets are not without their uses, especially since
they allow some of the key biological elements like hydrogen and
phosphorus to be detected by means of Nuclear Magnetic Resonance
(▼NMR▲).

In a steady magnetic field, a nuclear magnet has a natural (or resonant)
frequency dependent only upon its mass and the local field strength. As
with any resonant system, we can detect resonances by exciting the
system with a range of frequencies and seeing which ones strike a chord.
So by mapping the occurrence of certain resonances, images can be built
up, based on say the density of hydrogen nuclei.

▼NMR▲

In a steady magnetic field, a nuclear magnet behaves rather like an atomic magnet, in that it settles either parallel or antiparallel to the field. Quantum mechanics accordingly allocates it two energy levels which have the nuclear magnet set at a fixed angle to the external field as illustrated in Figure 7.4. The energy difference between the two allowed states is small compared with kT at room temperature.

The 'spin' of a proton in a magnetic field behaves like a tiny gyroscope. If its axis is

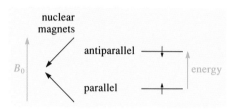

Figure 7.4. Parallel and antiparallel nuclear magnets

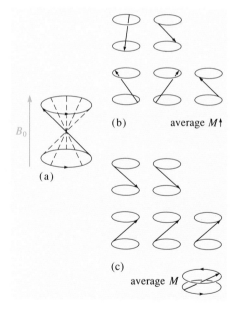

Figure 7.5 (a) Precession of a nuclear magnet. (b) An incoherent group with steady average bulk magnetization M. (c) A coherent group precessing with average magnetization M

tipped away from the vertical, as it is in the quantum mechanical energy state, a precession takes place so that the axis of the spinning motion describes a cone (see Figure 7.5a).

The simplest case to consider is that of hydrogen because the nucleus is a single proton. When in a uniform magnetic field of 1 tesla, the nuclear magnet of any hydrogen atom has a natural precession frequency f of 42.58 MHz. At 2 tesla it is twice as much, since the precession frequency is proportional to the flux density. Consider for example the hydrogen atoms in a common hydrocarbon such as polyethylene. At room temperature, they are in thermal equilibrium when there is a fractional excess (one in a few million) of the nuclear magnets in the lower energy state, that is parallel to the external field. The polyethylene will therefore have a very small net magnetization. The energy difference between parallel and antiparallel states also turns out to be simply related to the precession frequency ($\Delta E = hf$).

There are two ways in which an assembly of nuclear magnets can resonate with some external frequency. First the precession of the assembly could lock in to produce a coherent precession of the magnetization (Figure 7.5b and c). Secondly, the magnetization can be changed by exciting nuclear magnets from the lower (parallel) to the higher (antiparallel) energy state. Both of these happen simultaneously and can be excited and detected with the arrangement shown in Figure 7.6. The coil around the sample creates a weak magnetic field which points one way then the other at frequency f_0 and at right angles to the steady field around which the nuclear magnets precess (also with frequency f_0). The coil is excited by a burst of signal long and intense enough to equalize and lock together the parallel and antiparallel spins. Then the coil becomes the detector, being sensitive to the coherently precessing nuclear magnets: A voltage at frequency f_0 is induced in the coil. The real precession frequency of any magnet depends upon its local field and each nucleus might see a slightly different field, so the magnets don't stay exactly in phase. Coherence is lost and

the induced signal decays. Also, there will be a return to the thermal equilibrium condition with a fractional excess at lower energy orientations and this too will reduce the induced signal.

NMR signals appear at frequencies characteristic of the resonant nuclei. The exact frequency depends upon the local environment of a nucleus, so for example, hydrogen in $\cdots$C—H bonds might be distinguished from hydrogen in $\cdots$O—H bonds. The initial strength of the induced signal is proportional to the number of nuclei present and its decay associated with differences in the environments for the resonant nuclei and the return to thermal equilibrium.

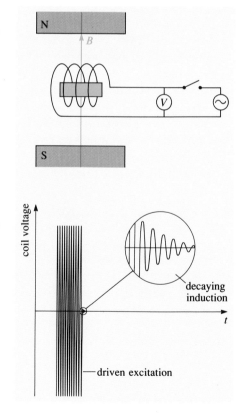

Figure 7.6 NMR excitation and detection

7.2.2 The basis of imaging

Suppose we have a sample of water (H_2O) in a PTFE (no hydrogen) box, between the poles of a strong magnet. At 1.000 tesla, protons are resonant at 42.58 MHz. Now there may be in our sample protons with an NMR resonant frequency of 42.50 MHz, where the local field is just less than 1.000 tesla. Suppose also that the sample sits inside a coil which excites the nuclear magnets with a signal at 42.58 MHz. That is, they are encouraged to precess at this frequency. This is so close to their natural frequency that they won't object, but when the excitation is removed, they will quickly slip back into their natural 42.50 MHz mode. Other protons may have slightly different resonances because they are in slightly different fields.

> EXERCISE 7.5   A proton nuclear magnet is resonant at 42.58 MHz in a field of 1.000 tesla. What can you say about the field around a proton for which the natural NMR frequency is 42.62 MHz?

Figure 7.7 shows the band of NMR frequencies present, just after excitation ceases, from a sample of rectangular section in a uniform field with a superimposed gradient. The protons sit in fields, say, between 0.98 and 1.02 tesla (position (a) in the figure), so we see frequencies between 41.74 and 43.42 MHz. When rotated through 90° (to position (c)), the sample now intercepts fields between 0.99 and 1.01 tesla, so the induced signal frequencies are between 42.16 and 43.00 MHz. In effect the frequency axis represents distance along the field gradient. At other angles the amount of material in a given field varies so the 'frequency spectrum' will have different amplitudes present, since signal strength is proportional to the number of resonant nuclei. So we can in principle build up an NMR image of the object by looking at the strength and range of frequencies present in the NMR spectrum when the object is viewed from different angles.

That's fine for a sample with uniform density, but what if the proton density also varies? There is obviously much more to real imaging techniques and systems. You should remember that we can use frequency to find position, and strength to indicate amount. We will stop here because we now have enough to begin specifying a magnet for the imaging system and it is in the magnet that we find an application for superconducting materials.

> SAQ 7.2 (Objective 7.2)
> Link the following phrases in a short account of magnetic resonance imaging. The order of appearance in the list is not significant:
>
> excitation and detection   nuclear magnetism
> hydrogen nuclei (protons)   proton density
> magnetic field gradient   uniform magnetic field
> nuclear magnetic resonance

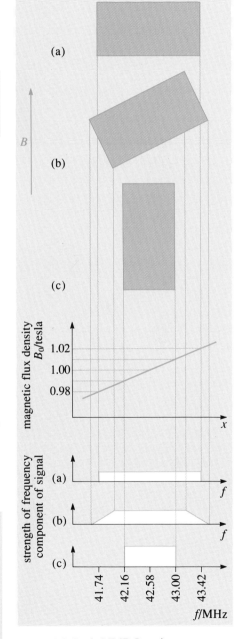

Figure 7.7 Basic NMR Imaging

7.3 Magnets for clinical magnetic resonance imaging

Living matter contains a lot of protons, mostly as part of water molecules. In biological tissue the amount of water take up varies according to type and condition, so a proton map of a biological sample may discriminate features on this basis. Furthermore, phosphorus nuclei also show a magnetic resonance and although much more thinly distributed than protons, it is possible in some circumstances to locate phosphorus nuclei and discriminate according to the type of molecule of which they are a part. (If you've done a little biology you will guess that I'm hinting at being able to differentiate between ATP and ADP.)

Magnetic resonance images are a valuable complement to clinical X-ray pictures because one maps soft tissue while the other depicts denser material such as bone.

In this section the goal is to check the specification for a magnet suitable for magnetic resonance imaging of live human specimens. (The medics speak of *in vivo* analysis.) Figure 7.8 shows an NMR proton 'picture' of a human head.

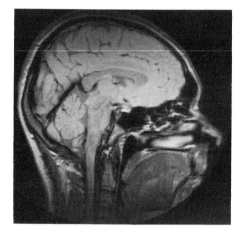

7.3.1 The magnet specification

To record NMR images of human subjects requires a magnet big enough, strong enough and steady enough to give unambiguous data.

The first item to address is the value of the steady and uniform magnetic flux density which the magnet must provide. The higher the flux density, the higher the NMR frequency of any nucleus detected (we'll presume hydrogen).

The detection mechanism relies on there being magnetization changes which induce a voltage in a detector coil. As we know from Chapter 3, larger rates of change of magnetization (that is higher frequencies) will induce larger voltages. So higher sensitivity (more signal) will be obtained by working with higher frequencies which will need higher magnetic fields. Therefore the higher the better. The upper limit is part economics and part feasibility. The larger the field, the more expensive it will be; the cost rises roughly in proportion to the flux density cubed! Higher flux density calls for more ampere-turns (recall 'Ampere's law' from Chapter 3). One of the problems with high magnetic fields generated by large currents is that currents and fields at right angles produce forces (recall how a motor works) and the need to resist these forces becomes a major element in the structural design and cost of large electromagnets. We must also be careful to limit the field's potential to cause damage. Stray

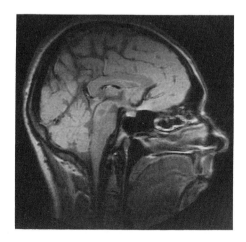

Figure 7.8 A human head *in vivo*, sections approximately 2 mm apart. (Courtesy of Picker International Ltd)

fields of a millitesla or so can disturb the normal operation of CRTs, X-ray tubes, pacemakers etc. A very important upper limit to the field is set by 'clinical benefit'. If the operator discerns no extra information from a stronger field, then there's no reason to use one. Better still, if the same information can be gained in a weaker (cheaper) field then we should reconsider the specification. Around a tesla is reckoned to be satisfactory, so we'll say 1.0 tesla to be definite.

The second feature to consider is that of access — if a patient has to be put into the high field region, the magnet design must allow this. Let's specify a clear 1.0 m diameter, 2.0 m long cylindrical open core for this purpose with currents in surrounding coils to generate the field.

Within the 'operational region' we require the main field to be uniform (imaging gradients can be superimposed later). The degree of uniformity is the subject of the next specification as this affects the spatial resolution. We can always say the flux density falls in a band centred on the nominal value $B = B_0 \pm \Delta B$. So we can say that the corresponding resonances will lie in the frequency range $f_0 \pm \Delta f$. The 42.58 MHz frequency for protons in one tesla is quoted to four figures and therefore indicates that we should be concerned about the odd ± 0.005 MHz — that is 5 parts in 50 000 (or 100 parts per million). The specification ought therefore to call for something like a central spherical volume of about 0.5 m diameter within which field variations are much less than this, say 10 parts per million.

The fourth specification is for the imaging gradient which is used to modify the uniform magnetic field. It must be steep enough to allow us to spread the resonance frequencies out so that we can resolve reasonably small distances, but not so steep that it becomes almost as strong (and expensive) as the main field itself. If the main field is 1.0 tesla then a gradient of 10^{-2} tesla per metre is about right for millimetre resolution.

EXERCISE 7.6

(a) If the imaging gradient is $10 \, \text{mT m}^{-1}$ how much does the flux density change in 1 mm?

(b) How much would the NMR frequency of protons change in one millimetre if the main field (B_0) were one tesla?

Finally we must specify the field stability. Where signals are weak it may take many seconds to acquire data. A reasonable criterion would be for no more than one millimetre's loss of resolution arising from field drift in one hundred seconds. The one millimetre resolution we have just associated with $10 \, \mu\text{T}$, so the drift rate should be less than $10 \, \mu\text{T}$ in 100 seconds ($10^{-7} \, \text{T s}^{-1}$). That's an amazingly steady field, and one of the reasons for using a superconducting magnet is that they can be used to generate extremely stable fields.

SAQ 7.3   (Objective 7.3)
Collect values from the foregoing text and complete the specification
and comment columns in Table 7.1.

Table 7.1

Quantity	Specification	Comment
Main field	1 T	
Access		whole body access
Uniformity		
Imaging gradient		with given uniformity, resolution is about 1 mm
Stability		

The sorts of calculations we did about currents and magnetic fields in
Chapter 3 can be used here to estimate the size of the currents and coil
needed for the imaging magnet. You should be able to follow the
reasoning in the following example but don't worry if you can't
reproduce it.

Figure 7.9 (a) A real magnetic resonance imaging magnet being delivered. (Courtesy
of Oxford Instruments Ltd)

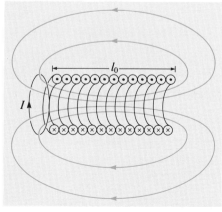

(b) Schematic approximation to the
magnet's structure

Worked example

Estimate the total current which must circulate round a cylindrical core (1 m diameter, 2 m length) to obtain an axial flux density of 1 T (see Figure 7.9).

Worked answer

Figure 7.9(b) shows the sort of structure (it's usually called a solenoid). The field won't be as uniform as we require but it is a good model for an estimate. (The actual magnet windings will differ from those of a simple solenoid in order to achieve the specified base uniformity and gradient — ▼Designer fields▲.)

Let's follow an Ampere's law path up the axis and back somewhere outside the solenoid. The main contribution to the $\Sigma\, H_i l_i$ of Ampere's law is just that part along the axis where the flux density B_0 and field strength $H_0 = B_0/\mu_0$ are high. If the current in each of the N turns of wire is I we therefore have

$$\frac{B_0}{\mu_0} \times l_0 \approx NI$$

For

$$B_0 = 1.0\,\text{T},$$
$$l_0 \approx 2.0\,\text{m}$$
$$\mu_0 = 4\pi \times 10^{-7}$$

we therefore find

$$NI \approx \frac{1.0 \times 2.0}{4\pi} \times 10^7$$
$$\approx 1.6 \times 10^6 \text{ ampere turns.}$$

We could have, for example, 100 turns with 16 kA in each one.

Now it's your turn:

> EXERCISE 7.7   A steady current of 16 kA flows in 2×10^{-2} m diameter copper wire, 300 m long (sufficient for about 100 turns at 1 m diameter). How much power is dissipated? ($\rho = 1.7 \times 10^{-8}\,\Omega\text{m}$ for copper.)

It is not so easy to produce such high fields in small spaces. Ordinary conductors will have to be cooled and this takes space and may represent a significant running cost.

▼ Designer fields ▲

If you want a certain magnetic field distribution in a volume of space you start by deciding upon your sources. Iron–cobalt alloys, as you know, will provide permanent magnets with remanent flux densities just over 2 tesla. For stronger fields or for extra flexibility, currents in wires are the only option. With care, the effects of one source of field can be added to those of others. So by successive accumulations of effects due to currents here and maybe lumps of iron there, the chosen pattern can be built up. Computer-aided design is in its element here.

To design an NMR magnet with its specified high uniformity takes a little more than keyboard skills and the right software, but these are a good start. The art comes in restricting the size of the overall design. In general, the more remote the source (currents in coils or lumps of iron), the more smoothly its influence is introduced. But it will be hard to sell a clinical magnet which is bigger than a standard hospital radiology room.

7.3.2 Superconducting metals

Superconductors are an attractive proposition for high-field magnets because not only do we avoid ohmic heating, which is expensive, but also we can avoid the continuous demand for ripple-free power (look at the stability specification). The idea is that we set currents flowing in superconducting windings and then leave them alone. So what do we need?

EXERCISE 7.8   What three critical quantities will enter the specification for metal-based superconducting coils to carry up to 16 kA in generating fields of up to 1.0 T?

As we saw in Section 7.1.1, it is necessary to work within the critical T–j–B surface, but just where is a matter of choice. In fact because helium liquefies at 4.2 K, liquid helium is a useful coolant for a wide range of superconductors. In general we want materials with as high a nominal critical temperature (T_c at $j = 0$ and $B = 0$) as possible since this reflects superior values of j_c and B_c at 4.2 K. Table 7.2 lists some candidate metal-based materials.

Table 7.2   Some metal-based superconducting materials

Material	T_c/K ($j = 0, B = 0$)	j_c/A m^{-2} ($T = 4.2$ K, $B = 0$)	B_c/tesla ($T = 4.2$ K, $j = 0$)
Elements			
V	5	—	<1
Nb	9	—	<1
Alloys			
50%Nb–50%Ti	8	3×10^9	10
60%Nb–40%Zr	11	3×10^9	8
Metallic intermediate compounds			
V$_3$–Si	17	—	22
Nb$_3$–Sn	18	3×10^{10}	26

EXERCISE 7.9   Niobium–titanium alloy is suitable for our purpose and allowing a good safety margin should be usable at 3×10^8 A m^{-2}. What is the cross-sectional area of superconductor if we stick with 100 turns at 16 kA?

In addition to being designed to operate within the critical surface, some safety margin is needed to prevent a catastrophic failure. For example, when initially 'charging' the superconducting magnet (Figure 7.10) as currents, fields and forces build up, there is a risk of mechanical settling when the structure takes the strain. If any settling motion is sudden the mechanical energy introduced may momentarily knock a region of superconductor back to its normal resistive state — ▼Superwires▲ explains the construction of special cables to avoid such catastrophes.

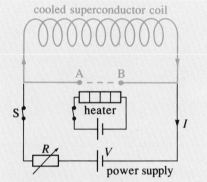

Switch S is closed and the superconductor between A and B is heated to take it into its normal resistive state. Current I flows into the zero resistance coil.

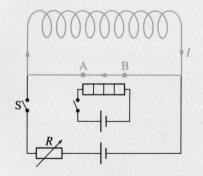

Once the full current has been reached, switch S is opened while AB cools to its superconducting state, diverting current through AB.

Figure 7.10 Charging a superconducting magnet

▼Superwires▲

If superconducting coils are to be made, a useful starting material is a superconducting wire, but the wire has to incorporate some protective features. Figure 7.11(a) shows a cross-section through a niobium–titanium (Nb–Ti) based wire. The main thing to notice is that the Nb–Ti alloy is deployed as 42 µm diameter filaments in a very pure (low resistivity) copper matrix. The filamentary structure prevents a runaway failure by providing an escape path for heat and current in the event of local loss of superconductivity.

Multifilament Nb–Ti was developed in the late 1960s for high-field magnet windings. Nb–Ti alloy rods are inserted into drilled out channels in a high purity (so-called oxygen-free high conductivity) copper billet. Extrusion and cold drawing are used to reduce the diameter to that suitable for winding. An important property of the finished wire is its suitability for joining — copper is etched away and the cleaned filaments are twisted and squeezed or crimped together. After all, a complete circuit has to be made and of course external connections to the charging supply are necessary.

A 'superior' superconductor for specialist applications is the metallic intermediate compound Nb_3Sn. The method of manufacture into wires is ingenious and needs to be because metallic intermediate compounds more usually appear as

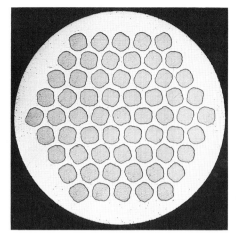
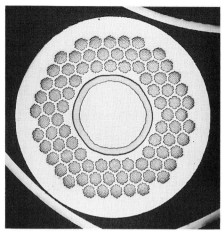

Figure 7.11 (a) A section through a composite Nb–Ti/Cu wire of about 0.5 mm diameter. (b) A section through a Nb_3Sn/Cu wire of about 0.8 mm diameter. (Courtesy of IMI Titanium Ltd)

localized phases in metal alloy microstructures rather than as ingots or powders ready for use. Furthermore, as you should know they are very brittle and unlikely candidates for wire drawing. The method is as follows. Pure niobium rods are inserted into holes drilled in an ingot of copper–tin (bronze) alloy, and then extruded and drawn into wires (as with the Nb–Ti alloy wires). This composite wire is quite flexible and can be formed as

necessary. The final step is a heat treatment (several days at around 1000 K) during which tin from the bronze diffuses into and reacts with the niobium to form filaments of the desired intermediate phase. The only drawback is that the remaining copper alloy is more resistive than we would have liked, so we have to incorporate some veins of pure copper as well — the central circular section in Figure 7.11(b).

Superconducting 4.2 K magnets are a firmly established technology. The penalties associated with such a low operating temperature are chiefly concerned with capital costs arising from the need to install a liquid helium cooling system. The running costs are not excessive and are not much compared with, say, operator salaries.

SAQ 7.4 (Objective 7.4)
(a) If a filament of Nb–Ti superconductor in a copper matrix wire inadvertently enters the normal (resistive) state, what will happen to the flows of current and heat?

(b) What are the properties of copper which are crucial to its use as the host matrix in Nb–Ti alloy wires?

▼ Warm superconductors ▲

YBa₂Cu₃O₇₋ₓ materials

These materials are sometimes known as '1–2–3' compounds after the ratio of their positive (metallic) ions. The x in the formula indicates oxygen vacancies, but notice that just mixing $\frac{1}{2}$ (Y_2O_3) $2(BaO)$ and $3(CuO)$ would give $Y\,Ba_2Cu_3O_{6.5}$ so one of the ions (copper) must be changing its mind about how much oxygen it can handle. We've seen this before in oxide ceramics. Figure 7.12 shows the crystal structure. There are three perovskite-like unit cells stacked one above the other. The upper and lower cells have barium ions at their centres and copper ions at each corner. There are oxygens half-way along the cell edges, but with several of them missing from the end plane in each cell and from the yttrium plane. The net effect again is that of planar arrays of copper and oxygen atoms with the addition, in this particular structure, of copper–oxygen chains perpendicular to the planes.

Y can be readily replaced by most of the rare earths (La, Eu, Dy, Gd etc.) without any major changes in properties (T_c is still in the range 80–95 K), but with some modification in susceptibility to attack by atmospheric moisture. Replacement of the Cu in $YBa_2Cu_3O_{7-x}$ by Ag and various transition metals (Ni, Fe, Cr) leads to a decrease in T_c in every case.

Bi–Ca–Sr–Cu–O materials

The mixed bismuth–copper oxides are flaky, having layered crystal structures with a tendency to cleave, similar to that of mica. The structures of the following compositions form a sequence with increasing numbers of juxtaposed copper–oxygen layers: $Bi_2CaSrCuO_x$ (denoted 2111 by metal ion ratio), $Bi_2CaSr_2Cu_2O_y$ (2122) and $Bi_2Ca_2Sr_2Cu_3O_z$ (2223). The copper–oxygen layers are similar to those in the $YBa_2Cu_3O_{7-x}$ structure, where copper atoms are surrounded by 4 oxygen atoms in the plane. As shown in Figure 7.13, the 2111 composition has one copper–oxygen (CuO₂) layer between the Bi₂O₂ layers, the

2122 has two, and the 2223 has three copper–oxygen layers before the Bi₂O₂ layers repeat. The measured T_c value increases as we pass along this sequence, from about 80 K to 105 K, and finally 110 K as the highest value.

Tl–Ca–Ba–Cu–O materials

Once again layered structures are observed, with thallium (Tl) and barium (Ba) being equivalent to Bi and Sr in the previous family. The analogy is completed by a similar sequence of structures up to the $Tl_2Ca_2Ba_2Cu_3O_z$ structure with three copper–oxygen layers shown at the right in Figure 7.14. The critical temperature for this material is 125 K. Thallium is a toxic element, having been used in the past as rat poison, so care is required in handling the powders during preparation.

EXERCISE 7.10 What critical quantities will enter the specification for oxide ceramic superconducting coils to carry up to 16 kA in generating fields up to one tesla?

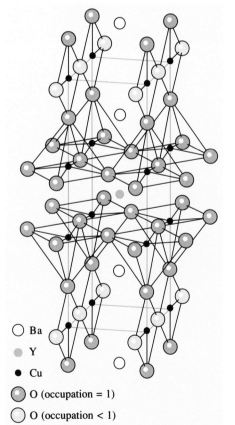

○ Ba
● Y (grey)
● Cu
◎ O (occupation = 1)
○ O (occupation < 1)

Figure 7.12 The YBa₂Cu₃O₇₋ₓ structure

Figure 7.13 The Bi–Ca–Sr–Cu–O structure

2111	2122	2223
⋮	⋮	⋮
	Ca	Ca
	CuO₂	CuO₂
	SrO	SrO
⋮	Bi₂O₂	Bi₂O₂
CaO	SrO	SrO
CuO₂	Bi₂O₂	CuO₂
SrO	SrO	Ca
Bi₂O₂	CuO₂	CuO₂
CaO	Ca	Ca
CuO₂	CuO₂	CuO₂
SrO	SrO	SrO
Bi₂O₂	Bi₂O₂	Bi₂O₂
CaO	SrO	SrO
CuO₂	CuO₂	CuO₂
⋮	⋮	⋮

Figure 7.14 The Tl–Ca–Ba–Cu–O structure

2021	2122	2223
⋮	⋮	⋮
	Ca	Ca
	CuO₂	CuO₂
	BaO	BaO
⋮	Tl₂O₂	Tl₂O₂
BaO	BaO	BaO
CuO₂	Tl₂O₂	CuO₂
BaO	BaO	Ca
Tl₂O₂	CuO₂	CuO₂
BaO	Ca	Ca
CuO₂	CuO₂	CuO₂
BaO	BaO	BaO
Tl₂O₂	Tl₂O₂	Tl₂O₂
BaO	BaO	BaO
CuO₂	CuO₂	CuO₂
⋮	⋮	⋮

A materials issue

In 1986 a new class of superconductors based on a mixed lanthanum–copper oxide with critical temperatures around 35 K were discovered. (This discovery by Bednorz and Müller won them a Nobel prize in 1987.) Further research produced materials with T_c around 100 K and the prospects for switching to a much more convenient liquid nitrogen coolant (77 K) seemed good. There is more to the electronic materials industry, however, than the odd favourable prospect. For one thing there is a need for proven feasibility — could there be a 77 K superconducting magnet? Then there is corporate strategy — are we special metals fabricators or superconductor suppliers? Or should we also be magnet makers or refrigerated system designers? Well, let's see if we can help with the feasibility issue. I prefer to leave corporate strategy to the corporations.

7.3.3 Superconducting ceramics

We are concerned here with families of mixed oxide ceramics. Prepared in the right way, blends of copper oxide with certain other metal oxides form ceramics with superconducting properties. Critical temperatures are around 100 K, substantially above those of the 'best' metal-based material.

By the end of 1989, three main families of mixed oxides had shown interesting superconducting properties above 77 K. They all have anisotropic crystal structures, with certain planes rich in oxygen ions; ▲Warm superconductors▲ looks at their structures. These compounds are related to the early lower T_c perovskites: strontium titanate (0.3 K, 1964); barium lead bismuth oxide (14 K, 1975); and lanthanum barium copper oxide (35 K, 1986). Fortunately we don't need to understand why they superconduct to consider further their use in imaging magnets.

With such high transition temperatures there is a great temptation to seek a warmer coolant than liquid helium. Here are three good reasons:

1   Figure 7.15 shows how the specific heat capacity c of a solid declines as temperature falls. At very low temperatures, the low specific heat capacity means that a small amount of energy will produce a large temperature rise. So a higher temperature system will be less susceptible to loss of superconductivity because it will need more energy per kelvin to heat it up past the transition temperature of the superconductor.

2   The specific latent heat of vaporization of liquid helium is $21\,\mathrm{kJ\,kg^{-1}}$. For comparison, liquid nitrogen absorbs more energy on vaporizing: $199\,\mathrm{kJ\,kg^{-1}}$. Also, liquid helium costs about as much as good brandy but liquid nitrogen costs about as much as good milk. So there is nothing particularly to recommend liquid helium as a coolant except its relatively low boiling temperature.

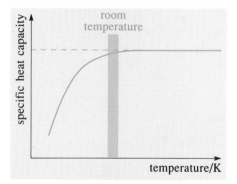

Figure 7.15 Specific heat capacity and temperature for a solid

3   With operating temperatures significantly above 4.2 K, it would be feasible to integrate a refrigerator into the assembly of a superconducting magnet. A user need only keep it plugged in to a suitably robust electrical supply to have a 'permanent' magnet.

There is an equal number of snags:

1   The critical superconducting properties of some ceramic superconductors at 77 K (liquid nitrogen — a 'cheap' coolant) are substantially inferior to those of the metal-based material at 4.2 K. Figure 7.16 shows the form of the T–j–B surfaces. We've seen before that thermal energy is a mixed blessing. It is desirable to operate well below T_c.

Thallium-based material ($T_c \approx 125$ K), which offers a reasonably high transition temperature to operating temperature ratio, may however exhibit adequate properties for magnet applications at 77 K.

2   The need to make electrical joints onto ceramic superconductors, and the brittle nature of ceramics in general, are likely to force radical design changes. So we cannot simply substitute oxide alloys for metallic alloys.

3   The ceramic crystals are highly anisotropic in their superconducting properties (remember those oxygen-rich planes). Polycrystalline material therefore is unlikely to achieve the performance of single crystal specimens on which the optimistic data are based.

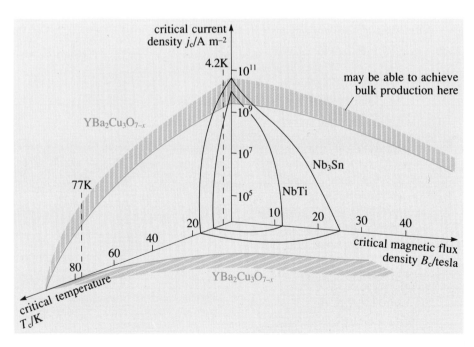

Figure 7.16 The T–j–B critical surfaces for some 'low' and 'high' T_c superconductors

▼Superfibres and supertapes▲

There are three categories of fabrication process which overlap somewhat; namely sintering, melting, and the deposition of layers.

Sintering

Sintering, which is the basis of most ceramics fabrication, is the conversion of a fine compacted powder to a relatively dense solid. The objective is normally pore-free, fine-grained polycrystalline solid. The powder is heated to a temperature where the diffusivities of the atoms in the material are high; there may also be partial melting. In either case, the driving force for densification is the reduction of surface energy associated with the elimination of the surfaces of the particles. The choice of the best sintering time and temperature is largely empirical.

The need to control the oxygen stoichiometry of the '1–2–3' superconductors in particular means that it is often desirable to have a few percent of open porosity in the ceramic so that oxygen can rapidly diffuse into the structure. A consequence of the residual porosity is a lowering of the mechanical strength and toughness, which are correspondingly poor compared with niobium alloy superconductors. One way of compensating for this lack of strength is to pack the oxide powder inside a metallic silver tube which is then formed to the final shape before it is heated to the sintering temperature of the oxide. This is then annealed in oxygen or air. The oxygen atoms readily diffuse through silver to give the correct stoichiometry. Such a wire has good mechanical properties, but the critical current density is low due to the random isotropic grain structure of the ceramic. Most other fibre or wire production routes show the same limitation.

Melting

Most of the superconducting oxide compositions can be melted by heating to temperatures in the range 1100–1420 K. In theory therefore, they could be cast into moulds like common metallic alloys; in practice, such processing is hindered by two factors. First as usual the primary phase on cooling through the liquidus is not of the same composition as the starting material, nor necessarily of the phase which is aimed for. However, prolonged annealing at a lower temperature can compensate for this. Secondly, the oxygen stoichiometry is in practice difficult to compensate when the material is in a dense form in large cross-sections.

Semicontinuous fibres of the bismuth based superconductors may be produced from a sintered rod precursor of the appropriate composition and about 1 mm in diameter. The rod is mounted vertically and a molten pool formed on the top surface by heating with focused laser beams. A seed crystal is lowered to touch the top surface of the melt pool and withdrawn slowly. Oriented fibres with the copper–oxygen planes parallel to the growth direction have been produced in this way. Single crystals of the oxide superconductors can also be grown from the melt rather like other electronic materials such as silicon and garnets.

Deposition

As we have seen, many electronic components such as ICs and displays employ the deposition of layers or coatings during the fabrication cycle. The industry 'feels comfortable' with such technology, and in many cases has extensive capital investment in a background knowledge of the techniques.

Thick film superconducting layers can be laid down as rather coarse circuitry using appropriately blended inks.

Sputtering is also an attractive process for these materals. Energetic gaseous ions, such as argon, are made to impact a solid, mixed-oxide target where they dislodge atoms giving them energies of typically ten to a hundred times those of thermally vaporized atoms. Substrates positioned in front of the target or targets of suitable composition are coated with the atomic debris. One major advantage is that the substrates can be maintained at a relatively low temperature, so we could contemplate coated tapes for magnet windings.

If a superconducting winding is constructed from ceramic material we are going to have to think of 'fibre' or 'tape', rather than 'wire' which has a curiously metallic ring to it: see ▼Superfibres and supertapes▲.

SAQ 7.5   (Objective 7.5)
Starting from Figure 7.12, write down the yttrium barium copper oxide ($YBa_2Cu_3O_{7-x}$) layer structure in the manner of Figures 7.13 and 7.14 and identify the feature common to three classes of warm superconductor. (Hint: count the atoms in the planes within the thin green frame in Figure 7.12 which contains one formula unit of atoms. For example, the upper and lower end planes each contribute $4 \times \frac{1}{8}$ Cu atoms and $2 \times \frac{1}{4}$ O atoms with occupations < 1 so write $2 \times \frac{1}{4}O_{1-x}$.)

7.3.4 Innovation or status quo?

It's time now to try to resolve the issue raised in Section 7.3.2. Given the advent of an alternative winding material for a superconducting magnet for NMR imaging, is there a case for deploying it? (I'm not going to answer that!)

There may be a case but it is unlikely to be a simple matter of substitution as we have seen before. Electronic materials are incorporated in systems which provide a function. The same function can be had from different systems involving different materials.

SAQ 7.6   (Objective 7.6)
Which of the following examples, encountered earlier, most closely resemble the current dilemma? Briefly justify your answer making reference to products, processes principles and properties.

(a) LCD or CRT for displays?
(b) Polymer or ceramic substrates for circuit assemblies?
(c) NdFeB or Alnico for motor magnets?

The commercial position is further complicated by other factors. For example in imaging, advances in software can enhance image analysis enabling the recovery of the same information in weaker fields and less time. Indeed, there may be no need for the superconducting magnet in some forms of the product.

Can we ignore this new technology? I advise against too much status quo — we might get left behind. But the skeletons by the roadside are also those of the innovators. Perhaps we should see what comes and pray we can get round the patents — it seems to have worked in the past, for most leading semiconductor manufacturers have had a share in most of the electronics revolution without necessarily having been innovators. However, there has been a quite unprecedented frenzy of research associated with these novel materials and the web of patents is already tight. You can see why I want to leave the issue unresolved. We conclude therefore with a glimpse of what else we can do with superconductors.

7.4 The potential of superconductors

Superconducting material is more widely useful than for lossless, high field, high stability magnet coils. We can identify three other categories in which superconducting properties are essential or at least desirable. These are:

- Zero-resistance components
- Novel electronic devices
- Photon sensors.

EXERCISE 7.11   Which of the following could benefit from the use of superconducting material for the component in parentheses?

(a) Electric motors (rotor and stator windings)
(b) Incandescent light bulbs (filament)
(c) National grid (overhead/underground cables)
(d) Semiconductor circuits (interconnection between chips and devices)
(e) Automobile heated windows (the conducting track embedded in the glass).

In Exercise 7.11, I have referred to semiconductor circuits working with superconducting connections. Don't forget, extrinsic semiconductors have a working temperature range: too hot and they revert to intrinsic, temperature-dependent behaviour; too cold and the dopant is inactive. (Guess where 77 K puts silicon!)

The sort of novel devices I have in mind in the second category are those which exploit ▼The Josephson effects▲ and also a class of device which is not quite 'in my mind' nor anyone else's yet because we must allow for new inventions. Nevertheless it is here that warm superconductors may make their biggest impact.

▼The Josephson effects▲

In the 1960s, Brian Josephson made predictions about how superconducting electrons would behave if two superconductors are separated by a thin insulating barrier (about 1 nm wide). If an insulating barrier between two ordinary metals is thin enough, electrons, are said to 'tunnel' through as they pass the barrier even though they have insufficient energy to get over it. It is one of those quantum mechanical phenomena. Were it not for this, many metal–metal and metal–semiconductor junctions would fail to establish good (ohmic) electrical contact owing to residual surface oxides and other impurities.

For the superconducting state, similar tunnelling can allow a supercurrent of Cooper pairs to cross such a thin barrier at zero potential difference. But if the supercurrent exceeds a critical value I_0, a voltage ΔV suddenly appears across the junction. Figure 7.17 shows the current–voltage characteristic for a Josephson junction. The two branches of the I–V curve arise as follows: Normally currents will only flow across the non-superconducting, insulating barrier if

ordinary electrons tunnel through. There is a sort of energy gap in superconductors rather like that in semiconductors and it turns out that for the Josephson junction, tunnelling only really gets under way when a voltage ΔV equivalent to the gap energy $e\Delta V$ is applied. But at sufficiently low currents (well below critical levels for the superconductors) with sufficiently thin barriers, pairs of electrons can tunnel together — it is as if they persuade the insulator to be a superconductor and allow pairs to pass unhindered. So below a critical junction current I_0, no voltage is needed to sustain supercurrent across the junction. Above I_0, the insulator gives up its superconducting pretensions and the normal curve is resumed.

Another thing which can arise for two closely spaced superconductors is that magnetic fields arising from current in one superconductor can cause the other to move outside its critical surface. The change between superconducting and normal states is an extremely fast change of state — just the job for high speed logic circuits.

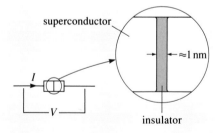

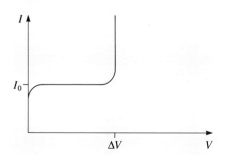

Figure 7.17 Josephson junction and I–V characteristic

The third category (photon sensors) reminds us that electromagnetic radiation (with the exception of gamma rays) interacts with matter through a material's electrons. Chapter 6 discussed examples in connection with the visible spectrum and 'ordinary' matter. Superconductors are no exception and are sensitive to a wide range of photon energies from as low as 10^{-2} eV, reflecting the delicate and weak nature of the coupling between pairs of electrons in the superconducting state.

Specialist applications already exploit examples in each class. There can be no doubt that products with superconducting components will find increasingly common applications in all the categories as the ingenuity of materials scientists develop the processing and that of physicists develops the principles of this particular electronic property. But all that will have to be another story.

SAQ 7.7   (Objective 7.7)
Give three general properties (including aspects of processability) which would be desirable for the building of microcircuits that include superconducting material.

Objectives for Chapter 7

After studying this you should be able to:

7.1 Give a simple account of superconductivity. (SAQ 7.1)

7.2 Describe the principles of magnetic resonance imaging. (SAQ 7.2)

7.3 Justify the specifications (flux density, uniformity, imaging gradient, stability) for an NMR imaging magnet. (SAQ 7.3)

7.4 Explain the use of a composite filament/matrix structure in wires made of NbTi and Cu, for use at 4.2 K. (SAQ 7.4)

7.5 Discuss the structure of some mixed oxide superconductors in terms of constituent layers. (SAQ 7.5)

7.6 Examine the contrasting options of ceramic and metallic superconductors (and other electronic materials) in terms of products principles properties and processes. (SAQ 7.6)

7.7 Briefly discuss the wider application of superconductors. (SAQ 7.7)

7.8 Define or explain the following terms or concepts:
Meissner effect
Cooper pairs
nuclear magnetism
NMR

In perspective

This book has stressed the need to appreciate the electronic functions expected of devices which use electronic materials. To this end, there has been a particular emphasis on ways of achieving transduction, memory and display. It is now worth stepping back from this device-centred view to look for patterns in the materials science on which all the devices and technologies depend.

Particular materials properties, such as resistivity, polarization (dielectric constant) and magnetization have been identified as being exploitable in electronic devices. Each of these properties is the consequence of the behaviour of electrons at a microscopic level. For example, resistivity depends on the quantity of charge carriers (n), their nature (q), and their local interaction with the environment (μ). Similarly, polarization and magnetization are quantified in terms of 'how many?', 'how strong?' and 'how coupled?'

This trinity of 'number', 'strength' and 'coupling' is fundamental to the understanding of the electronic properties of matter. Materials science is often seen as being concerned with atomic microstructure; but for electronic properties it is also important to understand a material's electronic microstructure — that is, where and how the charge carriers and dipoles are arranged. ('Dipoles' here refers to both the magnetic and the electric sort.)

The basic electronic properties are manipulated using the same principles, more of less, as in any other branch of materials: through composition, temperature, and stress (both mechanical and electromagnetic). An awareness of these fundamental principles assists in developing both the processing and the performance specifications of all electronic devices, whether for transduction, memory, display or some other purpose.

Answers to exercises

EXERCISE 7.1

(a) The resistance R must be zero since $V = IR = 0$ and I is not zero.

(b) The only way in which $R = \rho l/A$ can be zero (if l is not zero and A is not infinity) is if the resistivity ρ is zero.

(c) A wire with absolutely no resistance dissipates no power ($I^2R = 0$). Or, a wire with no voltage across it dissipates no power ($VI = 0$).

EXERCISE 7.2

(a) Oxygen ion diffusion.

(b) Extrinsic n-type conduction in oxygen deficient materials. Such materials accumulate spare electrons on cations of variable valency.

(c) Extrinsic p-type conduction where cation deficiency leads to some oxygen ions being short of electrons (i.e. they have holes).

Intrinsic conduction by thermally excited electron/hole pairs is specifically discounted for these materials with wide band gaps.

EXERCISE 7.3 Since the thermal energy of the lattice decreases with temperature, we can expect the resistivity to fall with temperature. But lattice imperfections arising from impurity atoms and defects in the structure won't get any less, so at very low temperature impurity effects will dominate and the resistivity will 'bottom out' at a constant level.

EXERCISE 7.4

• Effects involving atomic mass (which is dominated by that of the nucleus) — diffusion, ion acceleration, electrical resistance through collisions with massive atoms.

• Effects involving the nucleus directly — radioactive decay, nuclear magnetism.

• Other effects involving nuclear reactions — fusion (in which light nuclei combine), fission (in which heavy nuclei break up).

EXERCISE 7.5 The NMR frequency of a proton is directly proportional to the field in which it sits. If the frequency rises by 0.04 parts (42.62–42.58) in 42.58, then the field must be locally stronger by the same proportion. That is, it must be 1.001 tesla.

EXERCISE 7.6

(a) $10\,\mathrm{mT\,m^{-1}}$ is $10^{-2}\,\mathrm{T\,m^{-1}}$. So in $10^{-3}\,\mathrm{m}$ (1 mm) the field change is $10^{-3} \times 10^{-2} = 10^{-5}\,\mathrm{T} = 10\,\mu\mathrm{T}$. Comment: This is a good choice of gradient given that the specified spatial uniformity for a one tesla main field is also $\pm 10\,\mu\mathrm{T}$ (ten parts per million).

(b) The proton NMR frequency at $1.0000\,\mathrm{T}$ is $42.58\,\mathrm{MHz}$. Since the frequency is proportional to flux density, a change of 10^{-5} will shift the frequency by $10^{-5} \times 42.58 \times 10^6\,\mathrm{Hz} = 425.8\,\mathrm{Hz}$.

EXERCISE 7.7 The resistance $\rho l/A$ of $300\,\mathrm{m}$ of $0.02\,\mathrm{m}$ diameter copper wire is

$$\frac{(1.7 \times 10^{-8})\,\Omega\,\mathrm{m} \times (3 \times 10^2)\,\mathrm{m}}{\pi \times 1 \times 10^{-4}\,\mathrm{m}^2}$$

$$\approx 1.6 \times 10^{-2}\,\Omega$$

The power dissipation I^2R is

$$(1.6 \times 10^4\,\mathrm{A})^2 \times 1.6 \times 10^{-2}\,\Omega$$

$$\approx 4.1 \times 10^6\,\mathrm{W}$$

Comment: Unless we cool it very well, any patient will be cooked as well as scanned. Try thicker wire?

EXERCISE 7.8 The three are operating temperature ($T < T_c$), current density ($j < j_c$) and critical (adjacent) flux density ($B < B_c$).

EXERCISE 7.9 The area required to carry $1.6 \times 10^4\,\mathrm{A}$ at a density of $3 \times 10^8\,\mathrm{A\,m^{-2}}$ is

$$\frac{1.6 \times 10^4\,\mathrm{A}}{3 \times 10^8\,\mathrm{A\,m^{-2}}}$$

$$\approx 5 \times 10^{-5}\,\mathrm{m}^2$$

Comment: This is equivalent to a single circular section of about $4\,\mathrm{mm}$ radius, but as we shall see the material is not deployed as a single strand.

EXERCISE 7.10 The three are again operating temperature ($T < T_c$), current density ($j < j_c$) and critical (adjacent) flux density ($B < B_c$).

EXERCISE 7.11 (a), (c) and (d) do waste power in ohmic heating (I^2R) and (d) also involves delay through RC time constants. All three would therefore benefit from a reduction or elimination of resistance in the specified components.

(b) and (e) require ohmic heating (I^2R) to achieve the elevated temperature necessary for their operation and would not work with superconducting components!

Answers to self-assessment questions

SAQ 7.1   Here is my attempt in 47 words. You should have made reference to the words or phrases in italics.

Superconductivity is the ability to carry *current* without *resistive loss*. Steady currents do not need to be sustained by steady voltages. The superconducting state involves *electrons being specially paired* and it is destroyed by excessive thermal energy (*temperature*), *current density* and *magnetic field*. Superconductors *expel magnetic flux*.

SAQ 7.2   Here is my attempt.

Some clinical imaging techniques are based on *nuclear magnetic resonance*. The principle is based upon the *excitation and detection* of natural resonances of *nuclear magnetism*, particularly those of *hydrogen nuclei (protons)*. The object to be imaged is placed in a uniform *magnetic field*. By superimposing a weak *magnetic field gradient* the resonances can be spread out in frequency in accordance with their position in space. The signal strength is dependent on local *proton density*, so a density map can be built up.

SAQ 7.3   See Table 7.3.

SAQ 7.4

(a) If a filament goes resistive, current will be diverted to surrounding lower resistivity copper. Heat will be conducted away from the region by the copper matrix. As current density and temperature in the filament return to subcritical values, superconducting conditions are established once more.

(b) Copper is used as the matrix for Nb–Ti alloy wires as a precaution against catastrophic failure. It is chosen for its relatively high electrical and thermal conductivities.

Comment: High electrical conductivity requires high purity.

SAQ 7.5   Referring to Figure 7.12 we have:

$$\left.\begin{array}{l} BaO \\ CuO_{1-x} \\ BaO \\ CuO_2 \\ Y \\ CuO_2 \\ BaO \\ CuO_{1-x} \end{array}\right\} \text{formula unit}$$

Comparison with Figures 7.13 and 7.14 indicate that the CuO_2 planes are common to all structures.

SAQ 7.6   This is my answer. Yours ought to have some of these (or similar) points.

(a) LCD or CRT — both provide the display function but are entirely different products incorporating different materials.

(b) Polymer or ceramic substrates — these are components of electronic products. The associated processing differs because of material differences, but they both require to be insulators with some degree of temperature stability to accommodate soldering. Ceramics are part of hybrid circuit technology, whereas polymeric boards use prepackaged components.

(c) NdFeB or Alnico — these permanent magnets were considered as options for superior product performance. The change of material allowed there to be weight (and volume) savings in a modified design.

The current dilemma is 'metallic or ceramic' for superconducting imaging magnets. Thus, unlike case (a), it is the same product. Cases (b) and (c) are similar in that they are apparently examples of competing materials offering the same dominant properties. Either can be selected as most like the present case — (b) is a class change (ceramic for polymer) similar to the ceramic for metallic superconductor case, with attendant processing variations and (c) is a case where substitution forces design change.

SAQ 7.7   In microcircuits we can expect both novel device structures and interconnecting tracks to make use of superconductor; other areas will use semiconductor, insulator and maybe ordinary conductors. You may have noted some points from among the following.

i Thermal compatibility with the substrate — we may process above room temperature but expect it to remain serviceable at much lower temperatures

ii Environmental compatibility (critical temperatures) — semiconductors won't work if they are too cold and superconductors won't work if they are too warm. We must also pay attention to heat capacities and thermal conductivities in connection with cooling and power dissipation and the general electrical environment (critical fields and current densities).

iii Processing compatibility — suitability for thick or thin film deposition and patterning techniques.

iv Electrical compatibility ('contactability') — we have to be able to get electrical signals into and out of some of the material.

Table 7.3

Quantity	Specification	Comment
Main field	1 T	restricted by price and clinical benefit
Access	1 m dia × 2 m length open core	whole body access
Uniformity	< 10 parts per million (central sphere 0.5 m dia.)	affects spatial resolution
Imaging gradient	$10^{-2}\,T\,m^{-1}$	with given uniformity, resolution is about 1 mm
Stability	$< 10^{-7}\,T\,s^{-1}$	allows time to acquire data

Acknowledgements

Grateful acknowledgement is made to the following sources for material reproduced in this book.

Figure 1.2, courtesy of the Trustees of the Science Museum. Figure 1.21, courtesy of Plessey plc. Figure 2.1, courtesy of R. Hawes, British Vintage Wireless Society. Figure 2.2, M. Levers, Open University. Figure 2.3, R. Black, Open University. Figure 2.4, courtesy of D. K. Hamilton, Department of Engineering Science, Oxford University. Figure 2.5, courtesy of Menvier Hybrids, Banbury. Figure 2.9(a), L. M. Levinson *Electronic Ceramics—Properties, Devices and Applications*, © 1988 Marcel Dekker Inc., N.Y. Reprinted courtesy of Marcel Dekker Inc. Figure 2.26, courtesy of Plessey plc. Figure 2.28, N. Williams, Open University. Figure 2.34, C. Gagg, Open University. Figures 3.2 and 3.4, N. Williams, Open University. Figure 3.25(a), R. W. De Blois and C. D. Graham Jr *Journal of Applied Physics*, **29**, 528, © 1958 American Institute of Physics. Figure 3.25(b), C. A. Fowler, E. M. Fryer and D. Treves 'Domain structures in iron whiskers as observed by the Kerr method', *Journal of Applied Physics*, **32**, 296S, © 1961 American Institute of Physics. Figure 3.32, B. Elschner and W. Andra, 'Magnetische Elementarbezirke', *Fortschritte der physik*, **3**, 163, © 1955 Forschritte der Physik. Figure 3.39, courtesy of Mullard Ltd. Figure 3.41, courtesy of British Steel. Figures 3.55(b), 3.56(b) and 3.57, T. H. O'Dell *Magnetic Bubbles*, Macmillan, London and Basingstoke, © 1974 T. H. O'Dell. Figure 4.5, R. M. Dell and A. Hooper in P. Hagenmuller and W. Van Good (eds.) *Solid Electrolytes—General Principles, Characterization, Materials, Applications*, figure 42, p. 291, © 1978 Academic Press. Figure 4.12, E. M. Logothetis *Ceramic Engineering Science Proceedings*, **1**, 281, © 1980 American Ceramic Society, reprinted by permission. Figure 4.23, E. Sawaguchi 'Ferroelectricity versus antiferroelectricity in the solid solutions of $PbZrO_3$ and $PbTiO_3$', *Journal of the Physical Society of Japan*, vol. 8, pp. 615–629, © 1953 Physical Society of Japan. Figure 4.24, J. M. Herbert *Ferroelectric Transducers and Sensors*, p. 282, © 1982 Gordon and Breach Science Publishers Inc. Figure 4.25, J. T. Cutcher, J. O. Harris and G. R. Laguna 'Electro-optic devices using quadratic PLZT devices', *Sandia Laboratories Technical Report*, SLA–73–0777, © 1973 Sandia Laboratories. Figure 4.26 and 4.28, R. C. Buchanan (ed.) *Ceramic Materials for Electronics—Processing, Properties and Applications*, © 1986 Marcel Dekker, Inc., N.Y. Reprinted by courtesy of Marcel Dekker Inc. Figure 5.1, courtesy of Plessey plc. Figures 6.7 and 6.29, M. Levers, Open University. Figure 7.8, courtesy of Picker International Ltd. Figure 7.9(a), courtesy of Oxford Instruments Ltd. Figure 7.11, courtesy of IMI Titanium Ltd.

Index of materials

Subject index

Page numbers prefixed by ▼ refer to green texts